The MySQL Workshop

A practical guide to working with data and managing databases with MySQL

Thomas Pettit

Scott Cosentino

BIRMINGHAM—MUMBAI

The MySQL Workshop

Publishing Product Manager: Heramb Bhavsar
Senior Editor: David Sugarman
Content Development Editor: Joseph Sunil
Technical Editor: Rahul Limbachiya
Copy Editor: Safis Editing
Project Coordinator: Aparna Nair
Proofreader: Safis Editing
Indexer: Sejal Dsilva
Production Designer: Aparna Bhagat
Marketing Coordinator: Nivedita Singh

First published: April 2022

Production reference: 1130422

Published by Packt Publishing Ltd.
Livery Place
35 Livery Street
Birmingham
B3 2PB, UK.

ISBN 978-1-83921-490-5

www.packt.com

To my girlfriend, Emma, and my cats, Jazz and Noodle, who have helped me stay motivated. And to my family and friends who have helped me to achieve my goals.

– Scott Cosentino

Contributors

About the authors

Thomas Pettit began developing software as a hobby. He changed tracks from being a truck driver to being a software developer by earning a graduate degree in software development at the age of 35. He taught basic computing skills at a community adult education center in Melbourne for 2 years before commencing his software development career. Tom has worked for several government agencies, including defense, law enforcement, and transport, as well as large and small private businesses. Tom has mentored several up-and-coming software developers during his career and takes great joy in assisting others in improving their skills and furthering their career prospects.

Scott Cosentino is a software developer and teacher currently working in computer security. Scott has worked extensively with both low- and high-level languages, working on operating system- and enterprise-level applications. Scott has a passion for teaching and currently writes and creates videos on computer security and other programming topics. He has developed an extensive library of courses and has taught over 45,000 students through courses with Udemy, Packt, and CodeRed. He maintains a blog on Medium, and is active on YouTube and LinkedIn, where he enjoys creating content and interacting with students.

About the reviewer

Vlad Sebastian Ionescu is a university lecturer with a Ph.D. in machine learning as well as being a freelance software engineer. He has over 10 years of computer science teaching experience in a variety of roles: schoolteacher, private tutor, internship mentor, university TA, and lecturer. Over the years, Vlad has worked with many cutting-edge technologies in areas such as frontend development, database design and administration, backend programming, and machine learning.

Table of Contents

3

Using SQL to Work with a Database

4

Selecting, Aggregating, and Applying Functions

Section 2: Managing Your Database

5

Correlating Data across Tables

6

Stored Procedures and Other Objects

7

Creating Database Clients in Node.js

8

Working with Data Using Node.js

Section 3: Querying Your Database

9

Microsoft Access – Part 1

10

Microsoft Access – Part 2

11

MS Excel VBA and MySQL – Part 1

12

Working With Microsoft Excel VBA – Part 2

Section 4: Protecting Your Database

13

Getting Data into MySQL

14

Manipulating User Permissions

15

Logical Backups

Appendix

Index

Other Books You May Enjoy

Preface

Do you want to learn how to create and maintain databases effectively? Are you looking for simple answers to basic MySQL questions, as well as straightforward examples that you can use at work? If so, this workshop is the right choice for you.

Designed to build your confidence through hands-on practice, this book uses a simple approach that focuses on the practical, so you can get straight down to business without having to wade through pages and pages of dull, dry theory.

As you work through bite-sized exercises and activities, you'll learn how to use different MySQL tools to create a database and manage the data within it. You'll see how to transfer data between a MySQL database and other sources and use real-world datasets to gain valuable experience in manipulating and gaining insights from data. As you progress, you'll discover how to protect your database by managing user permissions and performing logical backups and restores.

If you've already tried to teach yourself SQL but haven't been able to make the leap from understanding simple queries to working on live projects with a real database management system, *The MySQL Workshop* will get you on the right track.

By the end of this book, you'll have the knowledge, skills, and confidence to advance your career and tackle your own ambitious projects with MySQL.

Who this book is for

This book is for anyone who wants to learn how to use MySQL in a productive, efficient way. If you are totally new to MySQL, it'll help you get started, while if you've used MySQL before, it'll fill in any gaps, consolidate key concepts, and offer valuable hands-on practice. Prior knowledge of simple SQL or basic programming techniques would be beneficial to help you quickly grasp the concepts covered, but are not strictly necessary.

What this book covers

Chapter 1, Background Concepts, introduces the concepts of databases, database management systems, relational databases, and the general structure of MySQL.

Chapter 2, Creating a Database, discusses how a database is created in MySQL. We will look at how to create a database and a table, how to set up indices and keys, and how to model database systems using ER and EER diagrams.

Chapter 3, Using SQL to Work with Databases, shows how SQL can be used to work with MySQL databases. We will look at ways to back up and restore databases. We will also look at ways to create databases and tables, as well as inserinserting data, updating data, altering table structures, truncating tables, deleting data, and dropping tables.

Chapter 4, Selecting, Aggregating, and Applying Functions, discusses methods of selecting and analyzing data from databases. We will look at selecting and filtering data, as well as methods to apply functions and aggregations to data.

Chapter 5, Correlating Data across Tables, discusses methods of joining tables together. We will also look at subqueries and common table expressions.

Chapter 6, Stored Procedures and other Objects, discusses the various types of database objects that exist in MySQL. This includes views, functions, store procedures, triggers, and transactions.

Chapter 7, Creating Database Clients with Node.js, discusses the methods of using Node.js with a MySQL database. We will look at setting up development MySQL servers, the basics of Node.js, and the methods of connecting to MySQL to create databases and tables.

Chapter 8, Working with Data Using Node.js, expands our knowledge of using Node.js to interface with MySQL. We will see how we can insert, update, and display data through Node.js. We will also learn how to set up and use ODBC connections.

Chapter 9, MS Access Part 1, shows how to interface with MySQL through MS Access. We will look at methods for configuring MS Access, adjusting field properties, and migrating data to link with MySQL.

Chapter 10, MS Access Part 2, looks at advanced topics of MS Access interactions with MySQL. This will include working with passthrough queries, calling MySQL objects, and working with MS Access forms.

Chapter 11, MS Excel VBA and MySQL Part 1, works with MS Excel, using VBA to connect with MySQL databases to retrieve and alter data.

Chapter 12, MS Excel VBA and MySQL Part 2, expands our knowledge of MS Excel to discuss methods of reading, inserting, updating, and pushing data from Excel to MySQL.

Chapter 13, Further Applications of MySQL, looks at various applications we can use to further our MySQL skills and abilities. We will learn how to use X DevAPI and examine concepts such as inserting documents, loading data from CSVs, and exporting/importing various file formats.

Chapter 14, User Permissions, shows how user permissions are used to provide secure access to MySQL databases. We will look at how users are created, how permissions are granted, and how users can be used with a MySQL database.

Chapter 15, Logical Backups, shows how to create logical backups in MySQL. We will learn about different types of restores and methods for scheduling backups on a MySQL server.

To get the most out of this book

Software/hardware covered in the book	Operating system requirements
MySQL Community Server 8.0.28	Windows, macOS, or Linux
MySQL Workbench 8.0.28	
Node.js 16.14.2	
MS Office 2016	

If you are using the digital version of this book, we advise you to type the code yourself or access the code from the book's GitHub repository (a link is available in the next section). Doing so will help you avoid any potential errors related to the copying and pasting of code.

Download the example code files

You can download the example code files for this book from GitHub at `https://github.com/PacktWorkshops/The-MySQL-Workshop/`. If there's an update to the code, it will be updated in the GitHub repository.

We also have other code bundles from our rich catalog of books and videos available at `https://github.com/PacktPublishing/`. Check them out!

Download the color images

We also provide a PDF file that has color images of the screenshots and diagrams used in this book. You can download it here: `https://static.packt-cdn.com/downloads/9781839214905_ColorImages.pdf`.

Conventions used

There are a number of text conventions used throughout this book.

`Code in text`: Indicates code words in text, database table names, folder names, filenames, file extensions, pathnames, dummy URLs, user input, and Twitter handles. Here is an example: "We used WHERE to filter out the rows we were interested in."

A block of code is set as follows:

```
SQL = "SELECT Count(capacityindicatorsstats.ID) AS RecCount
FROM capacityindicatorsstats;"
Call CreatePassThrough(SQL, "CISCount", True, False)
Set RS = CurrentDb.OpenRecordset("CISCount", dbOpenDynaset)
```

When we wish to draw your attention to a particular part of a code block, the relevant lines or items are set in bold:

```
RS.MoveFirst
Me.cntSeries = RS.Fields("SeriesCount")
RS.Close
```

Any command-line input or output is written as follows:

```
$ mkdir css
$ cd css
```

Bold: Indicates a new term, an important word, or words that you see onscreen. For instance, words in menus or dialog boxes appear in **bold**. Here is an example: "If not, right-click on it in the **Navigation** Panel and select **Design View**."

> **Tips or important notes**
> Appear like this.

Get in touch

Feedback from our readers is always welcome.

General feedback: If you have questions about any aspect of this book, email us at customercare@packtpub.com and mention the book title in the subject of your message.

Errata: Although we have taken every care to ensure the accuracy of our content, mistakes do happen. If you have found a mistake in this book, we would be grateful if you would report this to us. Please visit www.packtpub.com/support/errata and fill in the form.

Piracy: If you come across any illegal copies of our works in any form on the internet, we would be grateful if you would provide us with the location address or website name. Please contact us at copyright@packt.com with a link to the material.

If you are interested in becoming an author: If there is a topic that you have expertise in and you are interested in either writing or contributing to a book, please visit authors.packtpub.com.

Share Your Thoughts

Once you've read *The MySQL Workshop*, we'd love to hear your thoughts! Scan the QR code below to go straight to the Amazon review page for this book and share your feedback.

https://packt.link/r/1-839-21490-2

Your review is important to us and the tech community and will help us make sure we're delivering excellent quality content.

Section 1: Creating Your Database

This section covers the basics of MySQL, relational databases, and database management systems. We will discuss the ways you can create databases and insert, modify, query, and delete data contained within them.

This section consists of the following chapters:

1
Background Concepts

In this chapter, you will gain an understanding of the basic types of databases and how people tend to use them. You will learn how MySQL implements specific concepts such as database structures, layers, organization, and what its architecture looks like. You will explore what a **relational database management system** such as MySQL is, and how it differs from a standard **database management system**. You will also learn about data normalization and data modeling.

By the end of this chapter, you will have a good overview of what a database is and its different components. You will also learn what makes MySQL special and how it fits into this ecosystem.

This chapter covers the following topics:

- Introducing databases
- Exploring MySQL
- Exercise 1.01: Organizing data in a relational format
- Exploring MySQL architecture
- Storage engines (InnoDB and MyRocks)

- Data modeling
- Normalization
- Activity 1.01: Creating an optimized table for an employee project

Introducing databases

Information is abundant, an ever-growing pile of little bits of data that drives every aspect of your life, and the bigger that pile of data grows, the more valuable it becomes to yourself or others. For example, consider a situation where you need to search the internet for a specific piece of information, such as how to create a MySQL database. To do this, you would send a query to a search engine, which then parses large sets of data to find the relevant results. Putting all that data into some form of useful context manually, such as inputting it into spreadsheet software, is time-consuming.

Using databases, it is easier to automate the input and processing of data. Now you can store all that data into ever-growing databases and push, pull, squeeze, and tug on the data to get information from it that you could never dream of getting before, and in the blink of an eye. A database is an organized collection of structured data. The data becomes information once it is processed. For example, you have a database to store servers and their information, such as processor count, memory, storage, and location. Alone, this data is not immediately useable for business decisions and analysis. However, detailed reports about the utilization of servers at specific locations contain the information that can be fetched from the database.

To ensure fast and accurate access and to protect all the valuable data, the database is usually housed in an external application specifically designed to efficiently store and manage large volumes of data. MySQL is one such application. In almost all cases, the database management system or database server is installed on a dedicated computer. This way, many users can connect to a centralized database server at the same time. Irrespective of the number of users, both the data and the database are important—as sensitive data and useful insights are stored in it—and must be suitably protected and efficiently used. For example, a database can be used to store log information or the revenue of a company.

In this book, you will build up your knowledge to manage your database. You will also learn how to deploy, manage, and query the database as you progress in the book.

The following section will describe databases in greater depth.

Database architecture

A database is a collection of related data that has been gathered and stored for easy access and management. Each discrete item of data in a database is, in itself, not very useful or valuable, but the entire collection of data as a whole (when coupled with ease of use and fast access) provides an exceptionally powerful tool for business and personal use. For example, if you have a set of data that shows how much time a user spends on a specific page, you can track user experience on your application. As the volume of data grows and its historical content stretches further back in time, the data becomes more useful in identifying and predicting past and future trends, and the value of the data to its owner increases. Databases allow the user to logically separate data and store it in a well-structured format that allows them to create reports and identify trends.

To understand the advantage of databases, consider a telephone book that is used to store people's names, addresses, and phone numbers. A phone book is a good example of a manual data store, in which data is organized alphabetically to find the information easily (albeit, manually). With a phone book, storing large sets of data creates a bulky physical object, which must be manually searched to find the data we want. The process of searching the data is time-consuming, and we can only search the data by name since this is how it is organized.

To help improve this process, you can utilize computer-based information systems to store the data either in tables or flat files. Flat files store data in a plain text format. Files with the extensions `.csv` or `.txt` are usually flat files.

```
test,password1
test2,password2
test3,password3
```

Figure 1.1 – An example of a flat file

Tables store data in rows and columns, allowing you to logically separate data and store them.

username	password
test	password1
test2	password2
test3	password3

Figure 1.2 – An example of a table

You use databases in almost everything you do in your life. Whenever you connect to a website, the screen layout and the information displayed in front of the screen are fetched from the database. The cell phone you use in your day-to-day life stores the contact numbers in a database. When you watch a show on a streaming service, your login details, the information about the show, and the show itself are stored in a database.

There are many different types of database systems out there. Most are quite similar in some ways, though quite different in others. Some are geared toward a specific type of activity, and others are more general in their application. You will look at the two most common database management systems used by businesses today, DBMS and RDBMS, in the upcoming sections.

A centralized database is one that is located, stored, and maintained at a single site. The simplest example of a centralized database is an MS Access file stored on SharePoint that is used by multiple people. A distributed database is more complex as the data is not stored in a single place, but rather at multiple locations. A distributed database helps users to fetch the information quickly as the data is stored closer to the end users.

For example, if you have a database that is distributed across America, Europe, and Asia, American users will access the database stored in America, European users will access the one stored in Europe, and so on. However, this does not mean that Americans cannot access data in Europe or Asia. It's just that accessing data closer to them is faster.

Relational and object-based databases are ideas as to how the data is stored behind the scenes. Relational databases include databases such as MySQL and MSSQL, whereas object databases include databases such as PostgreSQL. Relational databases use the concept of the relational database model explained in this chapter, while object-based databases use the concept of intelligent objects and object-oriented programming, where the elements know what their purpose is and what they are intended to be used for.

In the next section, you will look at a few examples of common database management solutions used by developers.

MS Access as a database

MS Access is a database application from Microsoft. It is one of the simplest examples of a database. It allows users to manipulate data with macro functions, queries, and reports, to be able to share it via different visualization techniques, such as graphs and Venn diagrams. It is a number cruncher and is excellent for analyzing numbers, forecasting, and performing what-if scenarios.

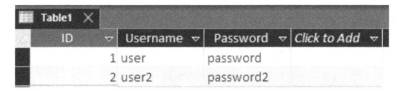

Figure 1.3 – MS Access file

However, MS Access is not the best database available, due to certain limitations in terms of functionality. For example, if offices of your company are present at multiple locations, it is possible to share an Access database. However, there is a limit to the number of users who can connect at a single time. In addition, there are limitations on the size of Access database files, making it only possible to store limited datasets. Access works best in situations where the groups accessing the database are small, and also the dataset is small, within the range of 1 million records or less.

Take, for example, a situation where an insurance company is creating a database for customer service to access customer data for insurance policies. If the team starts small, with 3 customer service agents and 300 records, MS Access works well, since the scope of usage is limited. However, as the company grows, more customer service agents may be added, and more records may be created. As the database grows, MS Access becomes less practical and eventually, Access will no longer work for the application.

Because of these limitations, alternative database management systems are preferred.

Database management system

A **database management systems (DBMSs)** aim to provide its end users with fine-tuned access to data based in a controlled environment. These systems allow you to define and manage data in a more structured manner. There are many different types of DBMSs used in applications, each with distinct pros and cons. When selecting a DBMS, it is important to determine the best choice for a given problem.

Take the previous example of an insurance company creating a database for customer service agents. If the developers wanted to transition away from MS Access, they could store data within a generic DBMS. These systems can help to organize data in a similar fashion to the Access database, while removing the size and connection caps created by Access. This solves the problem of the database system being limited; however, there are still limitations in terms of the data's structure based on the generic DBMS solution. Some DBMS solutions will simply organize data in tabular formats without any structural advantages. These situations are less ideal for large sets of data. These issues can be eliminated by **relational database management systems (RDBMSs)**.

Examples of DBMS include your computer's filesystem, FoxPro, and the Windows Registry.

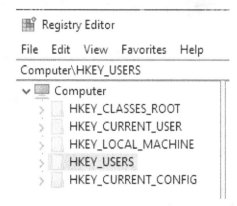

Figure 1.4 – Windows Registry is an example of a basic DBMS

RDBMS

A relational database stores data in a well-structured format of rows, columns, and tables. A row contains a set of data related to a single entity. A column contains data about a single field or descriptor of the data point. Take, for example, a table that contains user data. Each row will contain data about a single user. Each column will describe the user, storing data points such as their username, password, and similar information. Different types of relationships can be defined between tables, and specific rules enforced on columns. This is an improved version of the DBMS concept and was introduced in 1970. It was designed to support client-server hierarchy, multiple concurrent users or application access, security features, transactions, and other facilities that make working with data from these systems not just safe but efficient as well.

An RDBMS is more robust than a general DBMS or MS Access database. With the insurance database example, you can now create a structure around the data being stored for the customer service representatives. This structure represents the relationships between different datasets, making it easier to draw conclusions from related data. Additionally, you still get all the advantages of a DBMS, giving you the best system to fit your needs.

The following figure is an example of a database in MySQL. As you can see, the database has multiple tables (countrylanguage, country, and city), and these tables are linked to each other. You will learn how to link different tables later in *Chapter 10, MS Access, Part 2.*

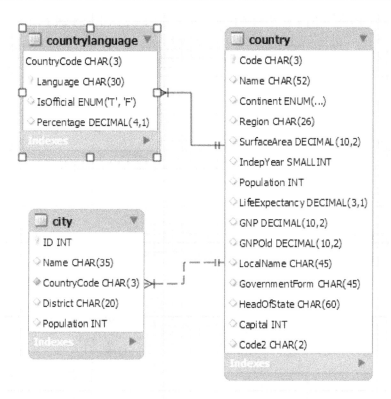

Figure 1.5 – RDBMS entity relationship diagram

Some popular RDBMS systems are MySQL, Microsoft SQL Server, and MariaDB. You will learn about MySQL in the following section.

Exploring MySQL

MySQL is an open source RDBMS that uses intuitive keywords such as SELECT, INSERT INTO, and DELETE to communicate with the database. These keywords are used in queries that instruct the server on how to handle data, how to read and write the data, or to perform operations on the database objects or the server, such as creating or modifying tables, stored procedures, functions, and views. The database objects are defined and manipulated using SQL commands and all communication and instructions issued to the database by the client applications are done using SQL code.

MySQL has a wide range of applications in business. This includes data warehousing, inventory management, logging user sessions on web pages, and storing employee records.

MySQL is based on the client-server model. The client-server model makes it possible for MySQL to handle concurrent connections from multiple users and host a great number of databases, each with their own tables and fine-tuned security permissions to ensure the data is only accessed by the appropriate users.

In the next section, you will explore some of the data types that are used in MySQL for storing data.

Data types

Each column in a database table requires a data type to identify the type of data that will be stored in it. MySQL uses the assigned data type to determine how it will work with the data.

In MySQL version 8.0, there are three main data types. These data types are known as string, numeric, and date and time. The following table describes these types in more detail.

- string: Strings are text-based representations of data. There are various types of string data types, including CHAR, VARCHAR, BINARY, VARBINARY, BLOB, TEXT, ENUM, and SET. These data types can represent data from single text characters in CHAR types to full strings of text in VARCHAR types. The size of string variables can vary from 1 byte to 4 GB, depending on the type and size of the data being stored. To learn more about these data types, you can visit https://dev.mysql.com/doc/refman/8.0/en/string-types.html.

- numeric: Numeric data types store numeric values only. There are various types of numeric data, including INTEGER, INT, SMALLINT, TINYINT, MEDIUMINT, BIGINT, DECIMAL, NUMERIC, FLOAT, DOUBLE, and BIT. These data types can represent numbers of various formats. Types such as DECIMAL and FLOAT represent decimal values, whereas INTEGER types can only represent integer values. The size range stored is dependent on the numeric data type assigned to the field and can range from 1 to 8 bytes, depending on whether the data is signed, and whether the type supports decimal values. To learn more about these data types, you can visit https://dev.mysql.com/doc/refman/8.0/en/numeric-types.html.

- date and time: There are five date and time data types: Date, Time, Year, DateTime, and TimeStamp. Date, Time, and Year store different components of date in separate columns, DateTime will record a combined date and time, and Timestamp will indicate how many seconds have passed from a fixed point in time. Date-based data types typically take up around 8 bytes in size, depending on whether they store the time as well as the date. Visit the following link for further details: https://dev.mysql.com/doc/refman/8.0/en/date-and-time-types.html.

As the developer, it is your responsibility to select the appropriate data type and size for the information you will be storing in the column. If you know a field is only going to use 5 characters, define its size as 5.

In the next exercise, you will learn how to organize a set of data in a relational format, with proper data types for each field.

Exercise 1.01: Organizing data in a relational format

Suppose you are working for a company, ABC Corp. Your manager would like to develop a database that stores clients' contact information, as well as the orders a client has made. You have been asked to determine how to organize the data in a relational format. In addition, the company would like you to define the data types that are appropriate for each field. The following is a list of properties that are to be stored in the relational model:

- **Customer Data:**

  - Customer ID

  - Customer Name

  - Customer Address

  - Customer Phone Number

- **Order Data:**

  - Customer ID

  - Order ID

  - Order Price

Perform the following steps to create a relational database structure:

1. First, determine the data types that are appropriate for the data. The ID fields should be `int` data type, since IDs are typically numeric. For fields containing names, addresses, and phone numbers, a `varchar` data type is appropriate since it can store general text. Finally, a price can be defined as `double`, since it needs to be able to store decimal values.

2. Determine how many tables you should have. In this case, you have two sets of data, which means you should have two tables – `CustomerData` and `OrderData`.

3. Consider how tables are related to each other. Since a customer can have an order in the order data, you can conclude that customers and orders are related to one another.

4. Next, look at what columns are the same between the two sets of data. In this case, both tables contain the `CustomerID` column.

Finally, combine all the information. You have two tables, `CustomerData` and `OrderData`. You can relate them by using the column they share, which is `CustomerID`. The relational model would look like the following:

CustomerData			OrderData		
CustomerID	int		CustomerID	int
CustomerName	varchar		OrderID	int
CustomerAddress	varchar		OrderPrice	double
CustomerPhoneNumber	varchar			

Figure 1.6 – The data for customers and orders organized in a relational format

With this, you now have a fully defined relational structure for your data. This structure with data types can be used to construct a proper relational database.

Now, you will delve into the architecture of MySQL in the following section.

Exploring MySQL architecture

Under the hood, all computer systems consist of several layers. Each layer has a specific role to play within the system's overall design. A layer is responsible for one or more tasks. The tasks are broken down into smaller modules dedicated to one aspect of the layer's role. An operation needs to get through all the layers to succeed. If it fails at one, it cannot proceed to the next and an error occurs.

MySQL server also has several layers. The physical layer is responsible for storing the actual data in an optimized format. The physical layer is then accessed through the logical layer. The logical layer is responsible for structuring data in a sensible format, with all required permissions and structures applied. The highest layer is the application layer, which provides an interface for web applications, scripts, or any kind of applications that have the API to talk to the database.

As discussed before, an RDBMS system typically has a client-server architecture. You and your application are the client, and MySQL is the server.

The MySQL layers

There are three layers in the MySQL server:

- Application layer
- Storage layer
- Physical layer

These layers are essential for understanding which part is responsible for how your data is treated. The following is a graphical representation of the basic architecture of a MySQL server. It shows how the different components within the MySQL system relate to each other.

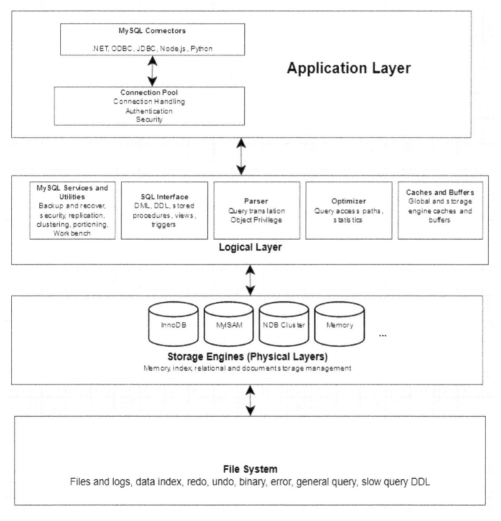

Figure 1.7 – MySQL architecture

Application layer – Client connection

The application layer accepts a connection using any one of the client technologies (JDBC, ODBC, .NET, PHP). It has a connection pool that represents the API for the application layer that handles communication with different consumers of the data, including applications and web servers. It performs the following tasks:

- **Connection handling**: The client is allocated a thread while creating a connection; think of it as a pipeline into the server. Everything the client does will be over this thread. The thread is cached so the client does not need to log in each time they send a request. The thread is destroyed when the client breaks the connection. All clients have their own threads. When a client wants to connect to a database, they will start by sending a request to the database server using their credentials. Typically, the requests will also include details about which database they specifically wish to connect to on the server. The server will then validate their request, establish a session with the server, and return a connection to the user.

- **Authentication**: When the connection is established, the server will then authenticate the client using the username and password details sent with the request. If the login details are incorrect, the client will not be allowed to proceed any further. If the login details are correct, the client will move to the security checks.

- **Security**: When the client has successfully connected, MySQL will check what the user account is permitted to do in it. It will check their read/write/update/delete status, and the security level for the thread will be set for all requests performed on this connection and thread.

When a client connects to the server, several services activate in the connection pool of the server layer.

MySQL server layer (logical layer)

This layer has all the logic and functionality of the MySQL RDBMS. Its first layer is the connection pool, which accepts and authenticates client connections. If the client connects successfully, the rest of the MySQL server layers will be available to them within the constraints. It has the following components:

- **MySQL services and utilities**: This layer provides services and utilities to administer and maintain the MySQL system. Additional services and utilities can be added as required; this is one of the main reasons why MySQL is so popular. Some of the services and utilities include backup and recovery, security, replication, clustering, portioning, and MySQL Workbench.

- **SQL interface**: SQL is a tool to provide interaction between the MySQL client and the MySQL server. The SQL tools provided by the SQL interface layer include, but are not limited to, **Data Manipulation Language** (**DML**), **Data Definition Language** (**DDL**), stored procedures, views, and triggers. These concepts will be taught thoroughly throughout the course of this book.

- **Parser**: MySQL has its own internal language to process data requests. When a SQL statement is passed into the MySQL server, it will first check the cache. If it finds that an identical statement has previously been run by any client, it will simply return the cached results. If it does not find the query that has been previously run, MySQL parses the statement and compiles it into the MySQL internal language.

  The parser has three main operations it will perform on the SQL statement:

  - A lexical analysis takes the stream of characters (SQL statement) and builds a word list making up the statement.

  - A syntactic analysis takes the words and creates a structured representation of the syntax, verifying that the syntax is correctly defined.

  - Code generation converts the syntax generated in *Step 2* into the internal language of MySQL, which is a translation from syntactically correct queries to the internal language of MySQL.

- **Optimizer**: The internal code from the parser is then passed into the optimizer layer, which will work out to be the best and most efficient way to execute the code. It may rewrite the query, determine the order of scanning the tables, and select the correct indexes that should be used.

- **Caches**: MySQL will then cache the complete result set for the SELECT statements. The cached results are kept in case any client, including yourself, runs the same query. If they do so, the parsing is skipped, and the cached results are returned. You will notice this in action if you run a query twice. The first time will take longer for the results to be returned; subsequent runs will be faster.

Storage engine layer (physical layer)

The storage engine layer handles all the insert, read, and update operations with the data. MySQL uses pluggable storage engines technology. This means that you can add storage engines to better suit your needs. Storage engines are often optimized for certain tasks or types of storage and will perform better than others at their "specialty."

Now, you will look into different types of storage engines in the following section.

Storage engines (InnoDB and MyRocks)

MySQL storage engines are software modules that MySQL server uses to write, read, and update data in the database. There are two types of storage engines – **transactional** and **non-transactional**:

- **Transactional storage engines** permit write operations to be rolled back if it fails; thus, the original data remains unchanged. A transaction may encompass several write operations. Imagine the transfer of funds from one account to another in the company accounting system; debiting funds from one account and crediting them to another is a single transaction. If the failure happens near the end of the transaction, all preceding operations will be rolled back, and nothing in the transaction will be committed. If all write tasks were successful, the transaction would be committed, and all changes will be made permanent. Most storage engines are transactional, like InnoDB.

- **Non-Transactional storage engines** commit the data immediately on execution. If a write operation fails toward the end of a series of write operations, the preceding operations will need to be rolled back manually by code. To do so, the user will likely need to have recorded the old values elsewhere to know what they were. With the accounting example, imagine that the funds were debited from the first account but failed to be credited to the second, and the initial debit was not reversed. In this case, the funds would simply disappear. An example of this type of engine is MyISAM.

Another consideration when selecting a storage engine is if it is ACID-compliant.

ACID compliance

ACID compliance ensures data integrity in case of intermittent failures on different layers, such as broken connectivity, storage failure, and server process crash:

- **Atomicity** ensures all distinct operations within a transaction are treated as a single unit, meaning that if one fails, they all fail. This ensures no transaction is left partially done. If the transaction is successful, the changes are committed to the storage layer, and data is guaranteed to be correct.

- **Consistency** ensures a transaction cannot bring the database to an invalid state. Any data written must comply with all defined rules in the database, including constraints, cascades, triggers, and the referential integrity of the primary and foreign keys. This will prevent the corruption of data caused by an illegal transaction.

- **Isolation** ensures that no part of the transaction is visible to other users or processes until the entire transaction is completed.

- **Durability** ensures that once the transaction is committed, it will remain committed even in the event of a system failure, or power failure. The transaction is recorded in a logon store that is non-volatile.

The default storage engine of MySQL is InnoDB, and it is ACID-compliant. There are other types of storage engines as well that store and manipulate the data differently. If you are interested in learning more about what type of storage engines are available for MySQL, you can refer to the following link: `https://dev.mysql.com/doc/refman/8.0/en/storage-engines.html`.

In the next section, you will take a look at how different applications can connect to your database through the application layer

Data modeling

Data modeling is the conceptual and logical representation of the proposed physical database provided in a visual format using **entity relationship** (**ER**) diagrams. An ER diagram represents all the database entities in a way that defines their relationships and properties. The goal of the ER diagram is to lay out the structure of the entities such that they are easy to understand and are implemented later in the database system.

To understand data modeling, there are two crucial concepts you need to be aware of. The first is the primary key. Primary keys are used to uniquely identify a specific record or row in your database. For now, you should know that it enforces the table to have no duplicate rows with the same key. The other concept is the foreign key. The foreign key allows you to link tables together with a field or collection of fields that refer to a primary key of another table.

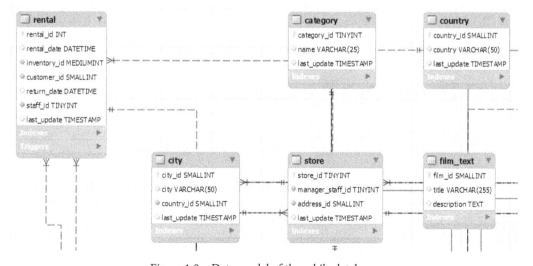

Figure 1.8 – Data model of the sakila database

The preceding screenshot shows you parts of the data model for the sakila database. It shows how different tables are connected and what their relationships are. You can read the relationships through the fields shared between the connected tables. For example, the rental table and category table are connected by the last_update field. The category table is then connected to the country table through the same last_update field. This demonstrates the general structure of the table relationships.

The data model ensures that all the required data objects (including tables, primary keys, foreign keys, and stored procedures) are represented and that the relationships between them are correctly defined. The data model also helps to identify missing or redundant data.

MySQL offers an **Enhanced Entity Relationship Diagram** for data modeling with which you can interact directly to add, modify, and remove the database objects and set the relationships and indexes. This can be accessed through the Workbench (this is explained in detail in the next chapter). When the model is completed, it can then be used to create the physical database if it does not exist or update an existing physical database.

The following steps describe the process by which a database comes into existence:

1. Someone gets an idea for a database and application creation.
2. A database analyst or developer is hired to create the database.
3. An analysis is performed to determine what data must be stored. This source information could come from another system, documents, or verbal requirements.
4. The analyst then normalizes the data to define the tables.
5. The database is modeled using the normalized tables.
6. The database is created.
7. Applications that use the database for reporting, processing, and computation are developed.
8. The database goes live.

For example, suppose that you are working on a system that stores videos for users. First, you need to determine how the database will be structured. This includes determining what data needs to be stored, what fields are relevant, what data types the fields should have, and the relationships between the data. For your video database example, you may want to store the video's location on the server, the name of the video, and a description of the video. This might link into a database table that contains ratings and comments for the video. Once this is produced, you can create a database that matches the proposed structure. Finally, you can place the database on a server so that it is live and accessible for users.

In the next section, you will learn about database normalization, which is the act of creating an optimized database schema with as few redundancies as possible with the help of constraints and removing functional dependency by breaking up the database into smaller tables.

Normalization

Normalization is one of the most crucial skills for anyone planning to design and maintain databases. It's a design technique that helps eliminate undesirable characteristics such as insert, update, and delete anomalies and reduces data redundancy. Insert anomalies can come from the lack of primary keys, or the presence of functional dependency. Simply put, you will have duplicate records when there should be none.

If you have a big table with millions of records, the lookup, update, and deletion operations are very time-consuming. The first thing you can do is to give more resources to the server, but that does not scale well. The next thing to do is to normalize the table. This means you try to break up the big table you have into smaller ones and link the smaller tables by relationships using the primary and foreign keys.

This technique was first invented by Edgar Codd, and it has seven distinct forms called **normal forms**. The list goes from **First Normal Form (1NF)** to **Sixth Normal Form (6NF)**, and one extra one, which is **Boyce-Codd Normal Form (BCNF)**.

The first normal form states that each cell should contain a single value and each record should be unique. For example, suppose you have a database that stores information about employees. The first normal form implies that each column in your table contains a single piece of information, as shown here.

EmployeeName	Designation	EmployeeLocation
Jeff	Database Administrator	Canada
Sarah	Programmer	United States
Bob	Accounting	Europe
Jane	Operations	Canada

Figure 1.9 – Example of a table in 1NF

The second normal form means the database is in first normal, and it must also have a single-column primary key. With the previous example, you don't currently have a single unique column, since the employee name could duplicate, as well as the title and location. To convert it into a second normal form, you can add an ID as a unique identifier.

EmployeeID	EmployeeName	Designation	EmployeeLocation
1	Jeff	Database Administrator	Canada
2	Sarah	Programmer	United States
3	Bob	Accounting	Europe
4	Jane	Operations	Canada

Figure 1.10 – Example of a table in 2NF

The third normal form requires the database to be in the second normal form and it is forbidden to have transitive functional dependencies. A transitive functional dependency is when a column in one table is dependent on a different column that is not a primary key. This means that every relationship in the database is between primary keys only. A database is considered normalized if it reaches the third normal form. The table here is in the third normal form, as it has a primary key that can be used to relate to any other tables, without the need for a non-key field:

EmployeeID	EmployeeName	Designation	EmployeeLocation
1	Jeff	Database Administrator	Canada
2	Sarah	Programmer	United States
3	Bob	Accounting	Europe
4	Jane	Operations	Canada

Figure 1.11 – Example of a table in 3NF

For further details, you can visit the following site: `https://docs.microsoft.com/en-us/office/troubleshoot/access/database-normalization-description`.

Now that you have learned all about working with datasets, let's perform an activity to recap everything we have learned so far in this chapter.

Activity 1.01: Creating an optimized table for an employee project

Your manager asked you to create a database that holds information about network devices in your corporate network site. You may have multiple devices with the same name in the same location. You are required to make the tables conform to the 3NF to make them as efficient as possible. In addition to this, you need to determine the proper data types for each column in the table. Finally, you are required to determine which columns should be primary keys, such that 3NF is satisfied. You have decided to perform the following steps.

1. Analyze the following table:

Hostname	Location	OperatingSystem	Layerlevel
PINKY	Ground Floor A	IOS	L2
PINKY	Ground Floor A	NXOS	L2
HERETIC	First Floor A	JUNOS	L3
HERETIC	First Floor A	NXOS	L3
HERETIC	Ground Floor B	IOS	L3

Figure 1.12 – A table of devices on the network

2. Identify patterns to determine the data types and possible primary keys. You may need to add a column to the table if an appropriate primary key does not already exist. Next, bring the table to 1NF

3. Bring it to 2NF, break down the table, and try to bring it to the 2NF form according to the rule.

4. Bring it to 3NF, break it down even further, and bring it to 3NF so the table is in 2NF with the appropriate constraints.

> **Note**
> The solution to this activity can be found in the Appendix.

Now you have an optimized table set up, you will be able to use this technique to efficiently optimize your database before you start filling it up with data and deploying it in production.

Summary

In this chapter, you have learned what a relational database is and what the differences are between a DBMS database and an RDBMS database. You learned about the client-server model used by MySQL and had a brief introduction to the MySQL architecture to see how MySQL works.

You then explored what layers make up MySQL, how to define different data models, and added tables to those data models. You also went through the basic concepts of ACID and how to initialize your database.

In the next chapter, you will further improve your knowledge of data modeling, entity relationships, and how to use the MySQL Workbench to set up/configure databases.

2
Creating a Database

In this chapter, you will learn about data modeling and the differences between a physical database and a conceptual database. Additionally, you will learn about **Entity-Relationship (ER)** diagrams and **Enhanced Entity-Relationship (EER)** diagrams and how they assist with data modeling. Following this, you will create indexes and foreign keys and learn how to generate them using the Workbench **Graphical User Interface (GUI)**.

By the end of the chapter, you will have gained an understanding of how you can interact with a MySQL server. This will include building new databases, creating tables, structuring tables, building, and visualizing relationships, and creating indices.

In this chapter, we will cover the following main topics:

- Developing databases
- The MySQL Workbench GUI
- Creating a database
- Using Workbench to add a table
- MySQL table indexes and foreign keys
- Reverse engineering a database

Developing databases

Chapter 1, Background Concepts, defined databases and their types. You learned how to create data models and add tables to those models.

During database development, you will be expected to work with or upgrade existing databases. There are chances that these databases would have been developed without modeling or proper planning. To understand how you can cope in such situations, in this chapter, you will create the physical database and use reverse engineering to generate an EER diagram and model. Often, reverse engineering is used when you are required to work on an existing database that is not documented.

However, when you develop a new system, you should take a bottom-up approach, beginning with the analysis and modeling. Then, you should design EERD and the database after which you should develop your applications.

In this chapter, you will begin using the MySQL Workbench GUI to create a new database. Additionally, you will create and populate a database using a `.sql` script. Then, you will focus on creating an EER diagram and a database model, which you will use to update the live database.

The MySQL Workbench GUI

MySQL Workbench is a GUI from Oracle. It allows users to interact with multiple instances of the MySQL databases. This interaction can be user management, database management, and essentially, anything that a person wants to interact against the MySQL instance. This makes it easier than having to perform the same steps via the command line.

There are two versions of the MySQL Workbench GUI: **8.0** (32-bit) and **8.0** (64-bit). The version you have will depend on the bit version of MySQL that you have installed on your system. Both versions look and work the same across all Windows, Mac, and Linux OSs:

Figure 2.1: MySQL Workbench 8.0

> **Note**
>
> You can download the relevant version of MySQL Workbench at `https://dev.mysql.com/downloads/workbench/`.

When you open the MySQL Workbench GUI, a screen similar to the following screenshot should be visible:

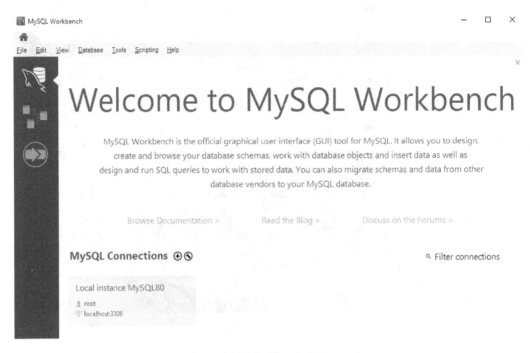

Figure 2.2: The MySQL Workbench GUI opening screen

You can find the connections to any databases you have set up in the central area of the screen. In the preceding screenshot, there is only one connection, which is the local instance of the MySQL server. This local instance runs on port 3306 and is authenticated to the root user on the database server. You can also connect to multiple servers on your LAN, or even remotely, at the same time.

To the left-hand side of the screen, in the gray bar, there is a series of commonly used utilities and websites that you might find useful. Any database models that have been created will be displayed at the bottom of the screen.

As you work through the chapters of this book, you will be introduced to several options and tools that are available to you in Workbench. For now, let's look at how to connect the Workbench GUI to the MySQL server.

Connecting the Workbench GUI to MySQL

Before you work with the MySQL server and the databases stored on it, you need to connect to it. Whenever you connect to the MySQL server, whether it be with Workbench, MS Access, Excel, or a third-party GUI, all of the applications will require some necessary information to initiate the connection. Please ensure you have the following details available before attempting the next exercise:

- **The IP address of the MySQL Server instance**: If MySQL is installed on your system, you can use `localhost` or `127.0.0.1`.

- **Port number**: The port number should be `3306` unless you changed it during the installation.

- The account name and password of the SQL server account connecting to the database.

> **Note**
> You will need these details as you progress through the book, so keep them handy.

In the following exercise, you will set up a connection to the MySQL server. This process will allow you to access your SQL server to be able to start creating databases for your projects. With this connection, we will start to learn about the fundamental ways of interacting with MySQL databases through the MySQL Workbench GUI.

Exercise 2.01 – creating a connection with the MySQL Workbench GUI

So, you have installed a copy of the MySQL server, and in order to interact with it, you will need to connect with it. In this exercise, you will set up a connection to the MySQL server through MySQL Workbench. Once you have set up a connection in Workbench, you can save the profile so that you don't have to enter all the details again the next time. Note that you need to change the IP address and use the username and password of the MySQL account you are connecting with.

To create a connection to the MySQL server, perform the following steps:

1. Open **MySQL Workbench**. Click on the + button to the right-hand side of the **MySQL Connections**, as shown in the following screenshot:

Figure 2.3: Plus (+) for creating a new connection

This will open the **Setup New Connection** screen.

2. Fill in the connection details, as shown in *Figure 2.4*. Fill in **Connection Name** as My First Connection, **Connection Method** as **Standard (TCP/IP)**, **Hostname** as <your server IP>, **Port** as <your port number>, and **Username** as <your username>:

Figure 2.4: The Setup New Connection screen

> **Note**
>
> The information for **Hostname**, **Port**, and **Username**, as shown in the preceding screenshot, are for demonstration purposes only. Your specific details are likely to be different, so be sure to use your own details. Additionally, your **Port** value should be 3306 unless you changed it during your server installation.

Please note that you can also set up multiple connections to the same MySQL server using different account names or set up connections to several MySQL servers, as shown in *Figure 2.3*.

> **Note**
>
> While you could now test and save the connection, you have not yet entered a password. This is because every time you connect, you will be prompted to enter the password. You might want this for security reasons if other people use your computer.

3. To save the password for the account being used with this connection, click on **Store in Vault…**, as shown in *Figure 2.4*. This will open the **Store Password For Connection** screen:

Figure 2.5: Entering the password for the displayed account

4. Enter the password, click on **OK**, and the screen will simply close.

5. To test the connection, click on the **Test Connection** button at the bottom of the setup screen. If all the details are correct, a screen similar to the following should appear:

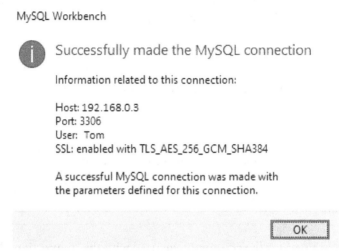

Figure 2.6: The connection was successful

If the test fails, check your details and try again, as you might have mistyped your credentials or the IP address.

6. Leave the **Default Schema** option in *Figure 2.4* blank. Click on the **OK** button to save the connection, and a new connection button will appear on the Workbench screen, as you can see in the following screenshot:

Figure 2.7: The new connection will appear on the Workbench screen

7. Click on **My First Connection**. A new tab will open displaying the interface and displaying information about the database, as shown here:

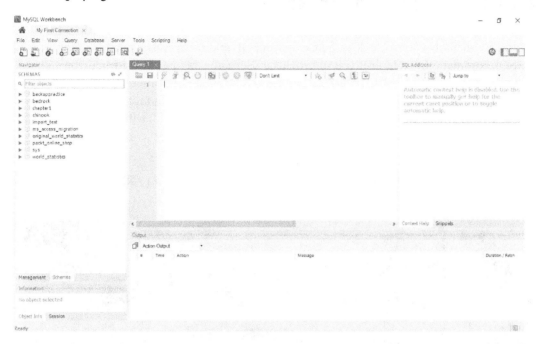

Figure 2.8: The interface to the MySQL server

All databases on the server will display in the **SCHEMAS** panel on the left-hand side. Initially, you might only have the **sys** database. Don't modify anything in the **sys** database; it is maintained by the server and messing around in there could damage the MySQL server.

Creating a connection to the database will enable you to open the database with a simple mouse click. Once you have gained access to the database, you can then add, modify, or remove databases at will.

In the next section, you will learn how to connect to a database using the command line.

Accessing MySQL through the command-line interface

MySQL is also accessible through your computer's **command-line interface (CLI)**. This interface will allow you to quickly and easily run SQL queries against a database. The MySQL command line requires you to provide a username and password when you launch it. The `-u` argument specifies the username, and the `-p` argument specifies the password. So, for example, `mysql -u root -p 123456` will sign into MySQL using the username, `root`, and the password, `123456`. By default, MySQL will have an account with a username, `root`, and no password. So, the `mysql -u root` command will allow you to enter the default installation of MySQL.

Once you have successfully launched the MySQL command line, you will see an interface that is similar to the following screenshot:

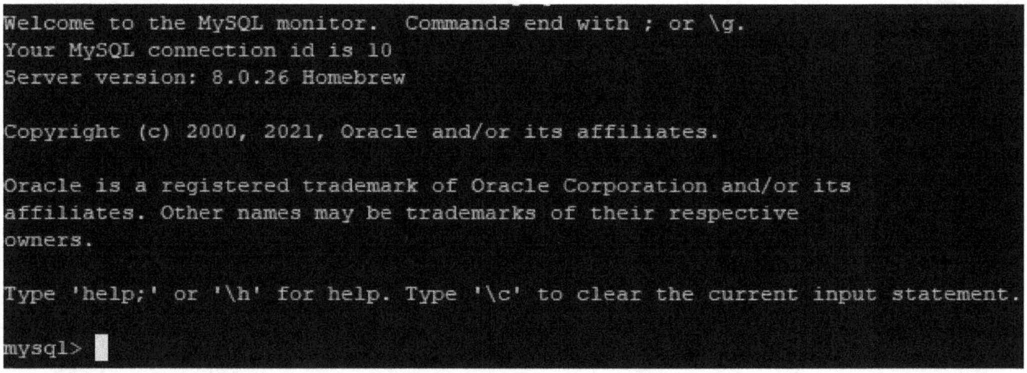

```
Welcome to the MySQL monitor.  Commands end with ; or \g.
Your MySQL connection id is 10
Server version: 8.0.26 Homebrew

Copyright (c) 2000, 2021, Oracle and/or its affiliates.

Oracle is a registered trademark of Oracle Corporation and/or its
affiliates. Other names may be trademarks of their respective
owners.

Type 'help;' or '\h' for help. Type '\c' to clear the current input statement.

mysql>
```

Figure 2.9: The command line for MySQL

When you will start learning the SQL syntax, you will understand how the code works in both MySQL Workbench and the CLI. For now, you will use the Workbench GUI, starting with learning how to create a database to work with.

Creating a database

A MySQL server requires a database to store and organize data. Your MySQL server can hold many databases and will efficiently work with all of the databases and multiple client connections simultaneously. In fact, each client will seemingly have exclusive access to the server and database; however, often, they will be sharing it with many or perhaps even hundreds of other people.

Databases are logical containers that group tables together to achieve a goal by providing special access rights, user management, and many other useful features.

A database schema is the collection of data tables, views, stored procedures, and functions that make up the database. To make the MySQL server useful, you need to have created at least one database.

In the following exercise, you will create the `autoclub` database.

Exercise 2.02 – creating the autoclub database

In this scenario, you are the database administrator of an automobile club. Every database system starts with the simple task of creating a new and empty database. You are asked to create a membership database to store the details of the members of the club.

To achieve the aim of this exercise, perform the following steps:

1. Open MySQL Workbench, and click on **My First Connection**.

2. Click on the **Create new schema in the connected server** option to open a new database window.

   You could also right-click on the white space of the **Schema** panel and select **Create Schema**. This brings you to the following screen:

Figure 2.10: The new_schema tab

3. Name the database `autoclub`, as shown in *Figure 2.10*.

4. Select the **utf8mb4** collation, as shown in *Figure 2.10*.

5. Click on **Apply** at the bottom of the screen.

6. Click on **Apply** to run the script to create the database. The **Apply** button will execute the SQL commands that are required to create the database.

7. Click on **Finish**.

    Once the database has been created, you can see it in the schema panel, as shown in the following screenshot:

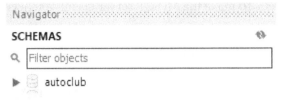

Figure 2.11: The autoclub database in the schema list

In this exercise, you learned how to create a database in a MySQL Server using MySQL Workbench.

Now, you will use MySQL Workbench to add tables to the database.

Using Workbench to add a table

Now, you will use MySQL Workbench to create a table in the `autoclub` database. You will learn how to add different columns with different datatypes. Additionally, in this section, you will learn the screen layout of the table creation screen.

To create a table, perform the following steps:

1. If you do not already have them opened and connected, open **MySQL Workbench** and click on **My First Connection**.

2. Open the **autoclub** database. Right-click on **Tables** and then select **Create Table**:

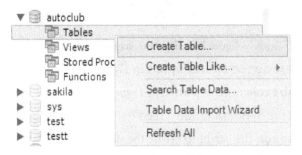

Figure 2.12: Insert a new table in the autoclub database

A new tab will open, displaying the table design screen, as shown in the following screenshot:

Figure 2.13: The table design screen

This screen is where you will design your tables. It consists of four main areas, which you will look at before you continue with creating a table:

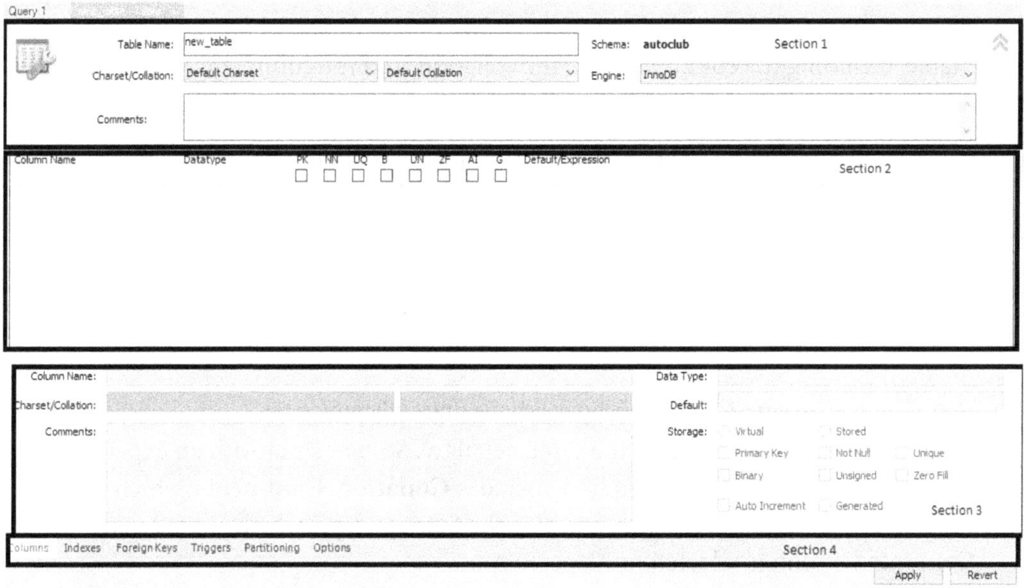

Figure 2.14: The creating a table window section

The following list showcases the various sections of the window:

- **Section 1**: This section of the screen allows you to set the name of the table, the collation, and add comments about the table. Additionally, if required, you can change the engine that the table uses:

Figure 2.15: Section 1 showing the table details

- **Section 2**: This section contains all of the columns in the table, along with their datatypes, default values, and properties:

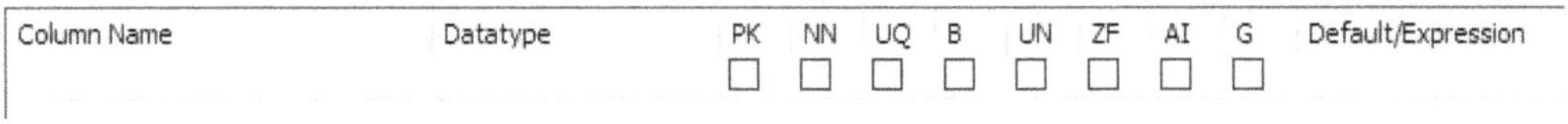

Column Name	Datatype	PK	NN	UQ	B	UN	ZF	AI	G	Default/Expression
		☐	☐	☐	☐	☐	☐	☐	☐	

Figure 2.16: Section 2 showing the grouped column details

This section is where you define the columns. The columns will be listed vertically, and in the order that they will appear in the table. You can set the name and datatype along with several options for the column. Additionally, you can provide a default value to the column.

- **Section 3**: This section contains the column or column details for the selected table, including the column names, the collation, and the comments, along with an expanded view of the datatype and the properties of the columns:

Figure 2.17: Section 3 showing the single column details

Note that this section contains the same details as *Section 2* but with an expanded set of details. Other information here includes **Collation**, **Comments**, which you can edit for the selected column, and also the **Storage** type, which is used when the **Generated** option is selected for the column.

- **Section 4**: This section contains the tabs to view the columns, indexes, foreign keys, and more for the table:

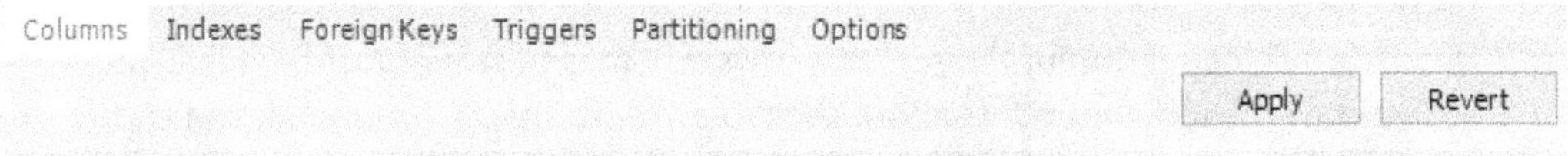

Columns   Indexes   Foreign Keys   Triggers   Partitioning   Options

Apply   Revert

Figure 2.18: Section 4 showing access to the Columns, Indexes,
Foreign Keys, Triggers, Partitioning, and Options tabs

> **Note**
> We will explain the indexes and foreign keys in more detail later.

The **Columns** tab is the default screen, as shown in *Figure 2.18*. You can use the other tabs to display the screens to allow you to set the other features of the table. We will be using some of these tabs in later exercises and activities.

Now that you have had a quick run-through of the screen layout, let's continue with creating a table.

3. The table needs a name in **Table Name**, which is located in *Section 1*. Type in Members as the table name.

4. Now you need to enter the columns. Click on the first row of **Column Name** in *Section 2*. As this is the first column of the table, MySQL will expect you to enter an **ID** column, so it will insert the name idMembers and set the datatype to **INT**, by default. If you choose to keep this name, just press *Enter*; otherwise, rename it. When you press *Enter* in the first column, MySQL will also set the **PK** (**Primary Key**) and **NN** (**Not Null**) options for you. Additionally, you will need to set the **AI** (**Auto-Increment**) option, as shown in the following screenshot:

Column Name	Datatype	PK	NN	UQ	B	UN	ZF	AI	G
idMembers	INT	☑	☑	☐	☐	☐	☐	☑	☐
		☐	☐	☐	☐	☐	☐	☐	☐

Figure 2.19: The first column has the Primary Key, Not Null, and Auto-Increment options

5. To add another column, click inside the next row of the **Column Name** column. This time, MySQL will name the column Memberscol. Change it to FirstName. The default datatype that is set by MySQL is VARCHAR(45). VARCHAR is correct, but 45 characters might be a little long. You can shorten it by clicking on the column's datatype and adjusting the number. You will probably not want the column to be empty. So, to ensure data is entered, tick **NN**, as shown in the following screenshot:

Column Name	Datatype	PK	NN
idMembers	INT	☑	☑
FirstName	VARCHAR(20)	☐	☑

Figure 2.20: The FirstName column with a shortened size limit, and Not Null set to ensure it contains data

> **Note**
>
> The value indicated in the VARCHAR () datatype is the only maximum size of the data. It does not indicate how much storage space the column will take. Had it been left at 45, a five-character name would only take up five characters of storage space. However, the column could have accepted a name with 45 characters.

6.   Now, add a **Bit** column. Click on the next row and name the column Active. Select **BIT()** from the datatype drop-down menu, click on the column, and add 1 inside the brackets. The datatype should now be Bit(1). The Bit datatype will hold one of the three values: Null, 0 (false), or 1 (true). Since you do not want the field to be blank, tick the **NN** option. Add 1 (true) as the default value to indicate that the member is active. Your new column should look similar to the following screenshot:

Column Name	Datatype	PK	NN	UQ	B	UN	ZF	AI	G	Default/Expression
idMembers	INT	☑	☑	☐	☐	☐	☐	☑	☐	
FirstName	VARCHAR(20)	☐	☑	☐	☐	☐	☐	☐	☐	
Active	BIT(1)	☐	☐	☐	☐	☐	☐	☐	☐	1

Figure 2.21: The Active column with the BIT(##1) datatype, Not Null set, and a default value of 1 (true)

7.   Add the TIMESTAMP columns. The WhenAdded column will record the date and time the record was added to the database. Its default value is CURRENT_TIMESTAMP. The LastModified column will record the date and time when the record was last modified in any way. Its default value is CURRENT_TIMESTAMP ON UPDATE CURRENT_TIMESTAMP:

Column Name	Datatype	PK	NN	UQ	B	UN	ZF	AI	G	Default/Expression
idMembers	INT	☑	☑	☐	☐	☐	☐	☑	☐	
FirstName	VARCHAR(20)	☐	☑	☐	☐	☐	☐	☐	☐	
Active	BIT(1)	☐	☐	☐	☐	☐	☐	☐	☐	b'1'
WhenAdded	TIMESTAMP	☐	☐	☐	☐	☐	☐	☐	☐	CURRENT_TIMESTAMP
LastModified	TIMESTAMP	☐	☐	☐	☐	☐	☐	☐	☐	CURRENT_TIMESTAMP ON UPDATE CURRENT_TIMESTAMP

Figure 2.22: Adding the WhenAdded and LastModified columns

8. Click on **Apply** to save the table. MySQL will open the **Apply SQL Script to Database** window, which looks similar to the following screenshot:

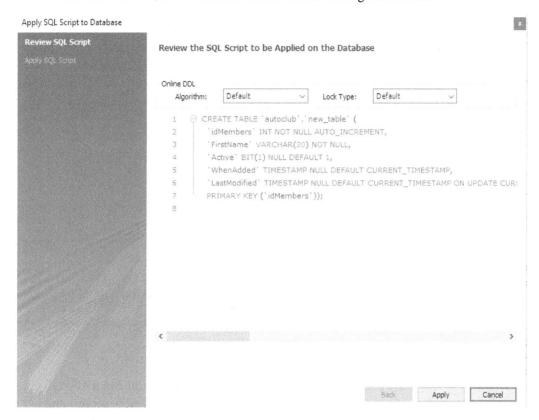

Figure 2.23: Reviewing the SQL script generated by MySQL to create a table based on your selections

9. Click on **Apply** to execute the script and apply the changes. When the screen returns a success message, click on **Finish**.

10. If the table is not displayed in the **Schema** panel, refresh the panel. The created table will be displayed, as shown in the following screenshot:

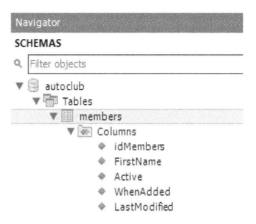

Figure 2.24: The members table in the autoclub database

> **Note**
>
> If you need to modify a table after it has been created, right-click on the table in the schema panel and select **Alter Table** to open the design tab.

Importing objects from a SQL script file

SQL script files (.sql) are a series of SQL statements that can be executed in a MySQL server. The file can be generated from different sources (including a MySQL server) and can be used to create a database that includes all of its objects and data. They are used to back up the database, copy a database into another server, add objects to an existing database, or modify the design of database objects.

> **Note**
>
> You will learn about SQL script files in greater detail in *Chapter 3*. Here, we have introduced it to ensure the autoclub database is in a complete state for the remainder of this chapter.

In the next exercise, you will import tables from an SQL script file.

Exercise 2.03 – importing tables from an SQL script file

In this exercise, you will run a .sql script file to add the ancillary tables to the autoclub database. The tables were created in another database, populated with data, and then exported to a single .sql file. Now, you need to bring those tables and their data to the autoclub database.

To run the SQL script file, perform the following steps:

1. Download the SQL script file, Chapter2 Ancilliary Tables.sql, from https://github.com/PacktWorkshops/The-MySQL-Workshop/ blob/master/Chapter02/Excercise%202.03/Chapter2%20 Ancilliary%20Tables.sql.

2. Select the **autoclub** database using **Workbench Schema Panel**, and open the **Tables** list. Note the current tables in the list, as shown in the following screenshot:

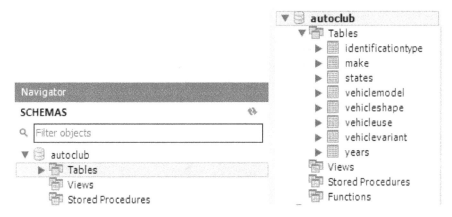

Figure 2.25: The current table in the autoclub database

3.   From the top menu, select **Server** and then select the **Data Import** option, as shown in the following screenshot:

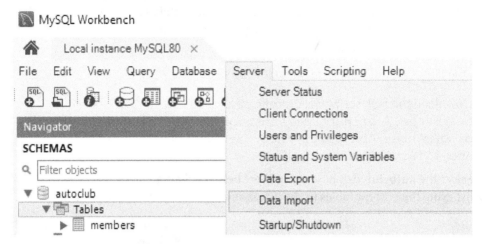

Figure 2.26: Select Server and then Data Import

This will open the **Data Import** tab, which looks similar to the following:

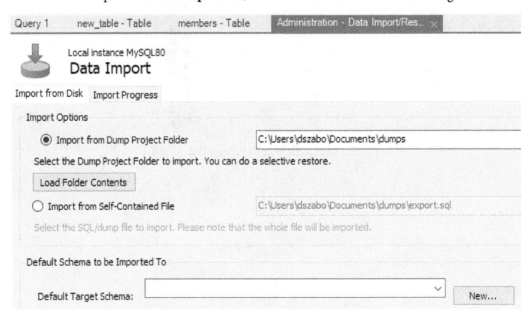

Figure 2.26: The Data Import tab

> **Note**
>
> This book is written using the Windows OS. So, all local file paths use the Windows filesystem naming convention. If you are using Mac or Linux, you will need to replace any file paths using the naming convention of your OS.
>
> The file path references can appear in the images, system resource files, or the text of the chapters.

4. Click on the **Import from Self-Contained File** option, as shown in *Figure 2.29*, and select the ellipses (**…**) at the end of the file path box.

5. Set the path of the `.sql` file.

6. Select `autoclub` for the `Default Target Schema` option. Then, click on `Start Import`. The file will be imported. You can see its progress in the following screenshot:

Local instance MySQL80
Data Import

Import from Disk   Import Progress

Import Completed

Status:
1 of 1 imported.

Log:

```
C:\Users\dszabo\Documents\GitRepos\The-MySQL-Workshop\Chapter02\Exercise 3\Exercise 3 – Create a new database.sql does not contain schema/table information
14:11:31 Restoring C:\Users\dszabo\Documents\GitRepos\The-MySQL-Workshop\Chapter02\Exercise 3\Exercise 3 – Create a new database.sql
Running: mysql.exe --defaults-file="c:\users\dszabo\appdata\local\temp\tmp4rhb2f.cnf"  --protocol=tcp --host=localhost --user=root --port=3306 --default-character-set=utf8 --
comments  < "C:\\Users\\dszabo\\Documents\\GitRepos\\The-MySQL-Workshop\\Chapter02\\Exercise 3\\Exercise 3 – Create a new database.sql"
14:11:32 Import of C:\Users\dszabo\Documents\GitRepos\The-MySQL-Workshop\Chapter02\Exercise 3\Exercise 3 – Create a new database.sql has finished
```

Figure 2.27: The progress screen showing Import Completed

7. Now, select the `autoclub` database, and then right-click on the **Tables** tab. Select the **Refresh All** option, which will refresh the list of tables. The updated list of tables can be seen here:

Figure 2.28: The new tables are displayed when Tables is refreshed

As you can see in the preceding screenshot, new tables such as `make`, `states`, `vehicle`, `vehiclemodel`, `vehicleshape`, `vehicleuse`, `vehiclevariant`, and `years` were added to our `autoclub` database.

8. Close the **Data Import** tab.

9. Finally, check each of the imported tables and ensure that the **Primary Key** option has been set and that the **ID** columns are set to **Auto Increment**. If they are not set, be sure to set them and save each table.

You have now successfully imported the ancillary tables, along with their data, to the `autoclub` database.

> **Note**
>
> SQL files for complete databases and sample databases are available on the internet. Sometimes, if you find one that is close to your final requirements, you can download that script to generate the database and modify the tables and objects to suit your needs. This can save you quite a bit of development time.

In this section, you learned how to save a lot of time by importing a prepared SQL script file. This topic will be discussed in greater detail when you progress to *Chapter 3*.

Now that you have a complete set of tables, in the next section, you will learn how to set table indexes and foreign keys in Workbench.

MySQL table indexes and foreign keys

In this section, you will learn about table indexes and foreign keys and how to set them up in your database using MySQL Workbench. Most importantly, you will learn why you should use them. The following section begins with indexes.

Indexes

Do you remember the last time you looked up information in a large book (the old style one that is made of paper)? Perhaps it was a directory and you needed to look up a single person. Note that you didn't start on page one and read through every entry until you finally found them. Instead, you went to the index and scanned an alphabetized list until you found their name and the relevant page number. Then, you went directly to the page and found the person's details there.

An index of a database table is the same. The MySQL server can maintain an index on a column, and when you look for a record in that column, MySQL will find it for you more quickly. You can set up multiple indexes on most columns in your tables. However, don't set up too many as that could even slow things down. You can set indexes on the columns that you are most likely going to search or filter on. Additionally, an index can control what data is stored inside the column. For example, you could set a unique index in a `Drivers' License Number` column. If an attempt was made to enter a record with a duplicate license number, MySQL would reject it, thus maintaining the integrity of your data.

Here is a list of some index types:

- `Index`: This is also known as a simple, regular, or normal index. The values do not need to be unique, and they can be `NULL`.

- `Unique`: The values in the column *MUST* be unique.

- `Fulltext`: This is used for columns containing text. It will index individual words within the text and aid in searching large volumes of text for words, groups of words, or phrases.

- Spatial: This is used for geometric and geographical data.

- Primary: All columns involved in a primary index must contain data and cannot be null.

Now, in the upcoming exercise, you'll create an index to get more hands-on experience.

Exercise 2.04 – creating an index

In this exercise, you will future-proof your autoclub database to allow rapid customer lookups based on their first name. You will create an index inside the members table using MySQL Workbench. Additionally, you will learn to understand the **Workbench** tab as you go through the process.

To create an index, perform the following steps:

1. Open **MySQL Workbench** and log in if you need to.

2. Open the **Tables** list in the **SCHEMAS** panel for the autoclub database:

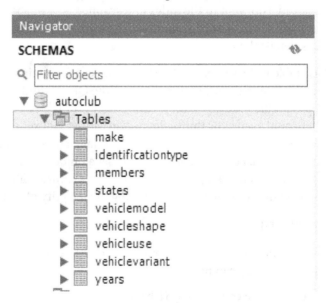

Figure 2.29: The Tables list for autoclub

3. Right-click on the **members** table and select **Alter Table**. The **Columns Design** tab will be displayed:

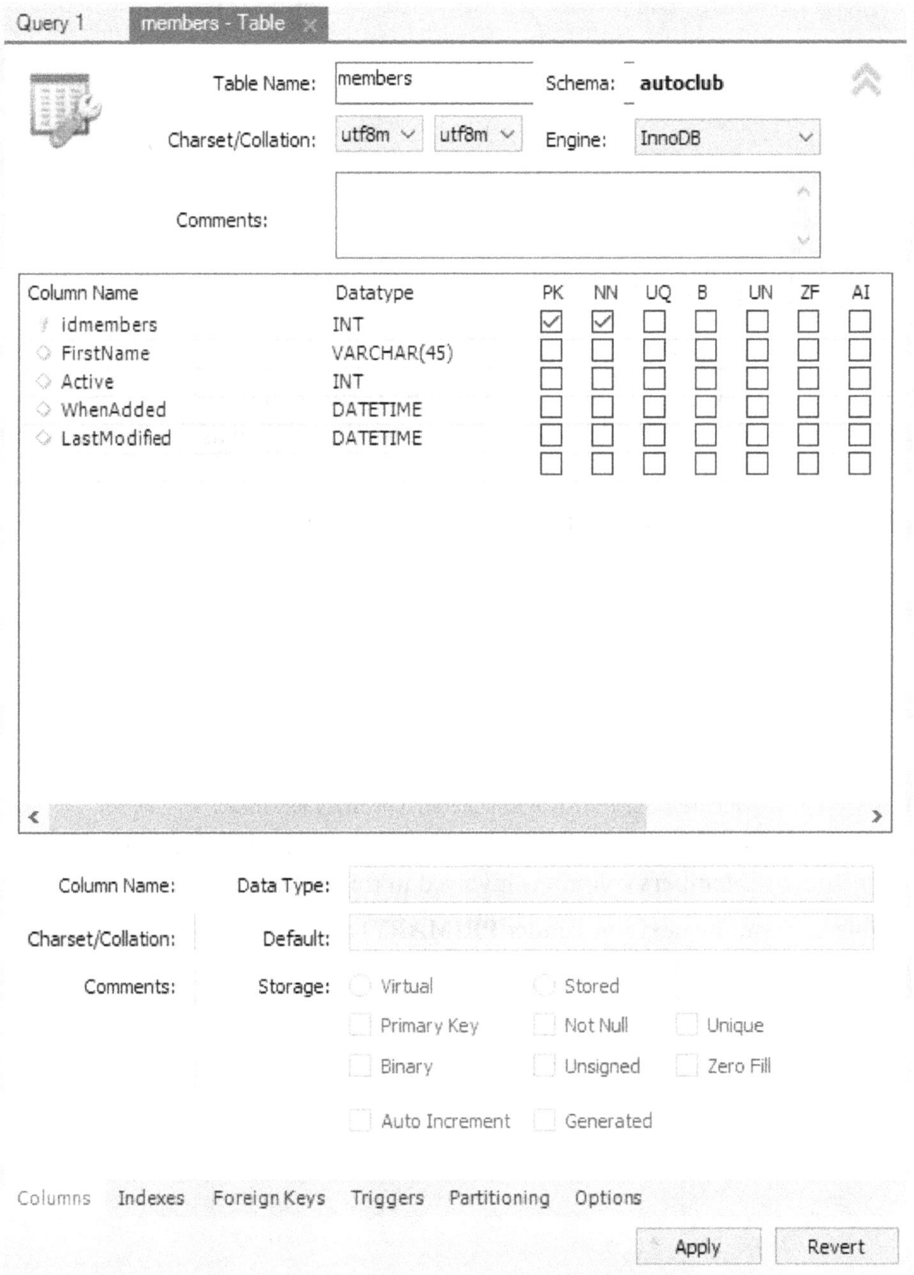

Figure 2.30: The Columns Design tab of the members table

4. Click on the **Indexes** tab at the bottom of the screen to open the **Indexes** tab, as shown in the following screenshot:

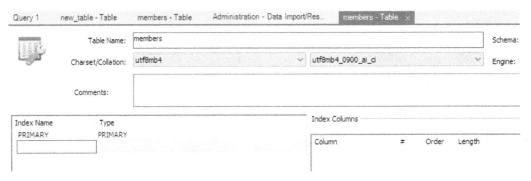

Figure 2.31: The Indexes tab for the members table

Notice that there is already an index named **PRIMARY** in the list. This has been automatically generated by MySQL when you set the column as the **Primary Key** type. This index will already be in all of your tables.

5. Click on the **Primary Index** type to view it, and you will see the following details:

Figure 2.32: The MySQL-generated primary key index

The panel on the right-hand side displays the columns used in the index. As you can see, only the **IdMembers** column is involved in this index.

6. Double-click on the next row (under **PRIMARY**) and enter a name for the new index; the FirstName type. The first column defines the name of the index you are creating, and the second column, **Type**, displays the types of indexes you can use:

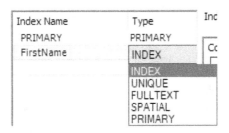

Figure 2.33: A list of index types, and the list of columns available for the index

7.   Select the default **INDEX** type as the new index type. In the **Index Columns** section of the window, you can now select the columns you wish to index. Since this index is defined to index the **FirstName** column, select the **FirstName** column from the column list:

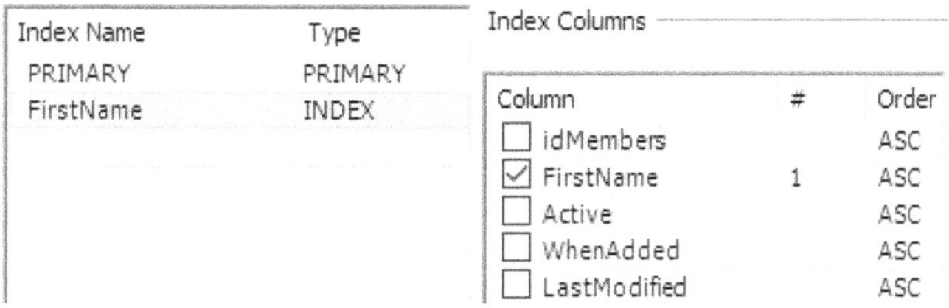

Index Name	Type
PRIMARY	PRIMARY
FirstName	INDEX

Index Columns

Column	#	Order
☐ idMembers		ASC
☑ FirstName	1	ASC
☐ Active		ASC
☐ WhenAdded		ASC
☐ LastModified		ASC

Figure 2.34: Selecting INDEX for the type and SURNAME for the column

8.   Click on **Apply** in the lower-right corner of the screen. The **Apply Script** popup will display. Click on **Apply** and then **Finish**.

In this exercise, you have learned how to successfully create an index. As the automobile club grows and the member table holds more and more members, searching for a member by their first name will be faster and more efficient within the index.

In the next section, you will learn how to apply indexes to multiple columns.

Indexes on multiple columns

Indexes can also be set in multiple columns. When an index is applied to multiple columns, it will follow the same principles as a single-column index. However, they will be applied to the combination of all the columns that make up the index. For example, let's say that you create an index of the Unique type on a single column name, called IDNumber. You might have the following data in the column:

ID	IDNumber
1	A123
2	A124
3	A125
4	A126

Figure 2.35: Multiple column indexing

However, if you tried to enter another record with an IDNumber column name of A125, the record would be rejected, as it does not meet the Unique index requirement.

> **Note**
>
> For the purposes of identification, the database will be able to accept any number of identification types such as a driver's license, passport, ID card, and more. Additionally, there is always a possibility that two different types of ID might have the same ID number.

Now, let's imagine that you had another column in the table, called IDType, and you created a Unique index using both the IDType column and the IDNumber column:

ID	IDType	IDNumber
1	Drivers License	A123
2	Drivers License	A124
3	Drivers License	A125
4	ID Card	A123
5	ID Card	A124

Figure 2.36: Additional column indexing

All of these records are valid because the combination of the IDType column and the IDNumber column is unique. However, if you try to add another record with Drivers License and A124, it would be rejected because there is already a record with that specific combination. However, you can add ID Card and A125.

Such an index can be used for the following purposes:

- To speed up searching

- To maintain data integrity

It is possible that any two different types of identification documents might have the same number. So, a unique index in the IDNumber column might only result in a valid identification document being rejected. However, you can be confident that a single type of identification will not have a duplicate number. So, setting the index on both columns is unlikely to reject valid records.

Foreign keys

Foreign keys are the links between your data tables. A value in a column that is defined as a foreign key refers to the primary key in another table. A column that is defined as a foreign key will hold the primary key value of the parent table. The table with the foreign key defined is considered the child of the parent table. Examine the following screenshot, which contains several tables in the `autoclub` database:

> **Note**
>
> The following screenshot is for demonstration purposes only. The `MemberID` foreign key column is not yet in the `identification`, `memberaddress`, and `vehicle` tables and will be added in the upcoming exercise and activity.

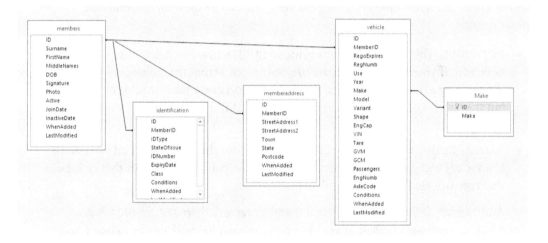

Figure 2.37: Tables with linking columns

The `identification`, `memberaddress`, and `vehicle` tables contain the `MemberID` columns of the `members` table. So, each of them is considered to be a child table of the `members` table.

Also, the `vehicle` table contains a column, named `Make`, which contains a value that matches the `ID` column in the `Make` table. So, it is also a child of the `Make` table.

When defined, foreign keys can help maintain data integrity by setting constraints; this is called **referential integrity**. In the preceding diagram, you might have several hundred members, all with their personal and vehicle details inside the database. Let's imagine that, for whatever reason, you remove the record of one member from the members table. If you did not also remove the matching records for each member of the child tables, identification, memberaddress, and vehicle, then those records will remain in the database and be orphaned, and they will be of no benefit to the database. Imagine that this has happened to many members. You created a report to find the average number of vehicles per member. A count of the total number of vehicles divided by a count of members would yield incorrect results. That's because if the records are removed to create orphaned vehicles, those vehicles will have no matching members, creating inaccurate results. Orphaned records can cause incorrect reporting, thereby impacting the integrity of the database.

Now, let's consider a few options of foreign keys that can be used for maintaining data integrity:

- SET NULL: This sets the column value to NULL when you delete the parent table row. If you were to delete the member record from the members table, the memberaddress record in the memberaddress table would remain, but the MemberID column would be set to NULL. The child tables would then be orphaned.

- CASCADE: This propagates the change throughout the database when the parent changes. If you delete a row, then the rows in the constrained tables that reference that row will also be deleted, and so on.

  Additionally, if you were to delete a member record, then ALL records in the memberaddress table that are linked to the member will also be deleted. You need to be very careful with this one.

  On the flip side, imagine that you had a Cascade on Update constraint set. You had a list of States in the States table, and the primary key was in the text column holding the state field. Let's suppose that this table has a foreign key relationship with a number of other tables, based on the States field. If we were to update the States table to change the States field from a value such as NSW to New South Wales, all tables with a foreign key relationship would also see NSW updated to New South Wales.

- RESTRICT: You cannot delete a given parent row if a child row exists that references the value for that parent row.

If you tried to delete a member record and there was a `memberaddress` record for that member, that is, the `members.ID` value was in `memberaddress.MemberID` in one or more records, the member record would not be deleted. The child records must all be deleted before the parent can be deleted (no orphans).

- `NO ACTION (default)`: `NO ACTION` is a keyword in standard SQL. In MySQL, it is the equivalent of `RESTRICT`. This is the default value. The behavior will be the same as `RESTRICT`.

Now that you have got a gist of the various options of foreign keys, you can get started by creating one in the next exercise.

Exercise 2.05 – creating a foreign key

Now you have been asked to link the `memberaddress` table to the `member` table. Both of these tables have a field to represent the member ID, meaning they share a similar unique identifier. In this exercise, you will create a foreign key to link the `memberaddress` table to the `members` table by linking the `memberaddress.MemberID` column to the `member.ID` column. To implement this exercise, perform the following steps:

1. Open MySQL Workbench and log in if required.
2. Open the tables list in the **autoclub** database.
3. Right-click on the **memberaddress** table, and select **Alter Table**.
4. Add a new column named `MemberID` with a datatype of `INT`, and save the table.
5. Select **Foreign Keys** from the tabs at the bottom of the table design screen. You will be presented with the foreign keys screen layout, as follows:

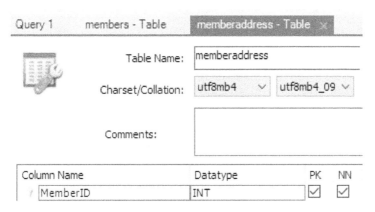

Figure 2.38: The foreign key screen

The upper section of the screen is standard across all of these tabs. The left-hand panel is where you enter the name of the foreign key and define the table a key it will be referencing. The middle panel is where you define the column that will be the foreign key and also what column in the parent table it will be referencing; usually, this is the primary key column. The right-hand panel is where you can set the options.

6. Starting from the left-hand panel, enter a name for the foreign key in the next available row in the **Foreign Key Name** column. Give it the name FK_MemberAddress_Members.

7. Select the **'autoclub'.'members'** table in the **Referenced Table** column, as shown in the following screenshot:

Foreign Key Name	Referenced Table
FK_MemberAddress_Members	`autoclub`.`members`

Figure 2.39: Foreign Key Name and Referenced Table

8. In the middle panel, MySQL might have selected the primary key for you in **Column** and **Referenced Column**. If not, select **MemberID** and **ID**, as shown in the following screenshot:

Figure 2.40: Column and Referenced Column for the FK_Member foreign key

9. In the right-hand panel, leave **On Update** as **No Action**. However, set **On Delete** to **Restrict**:

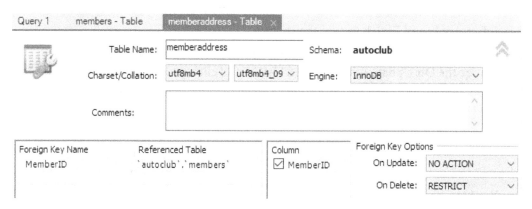

Figure 2.41: The foreign key options – On Update = No Action and On Delete = Restrict

10. Click on **Apply** to view the SQL script that the MySQL server generates:

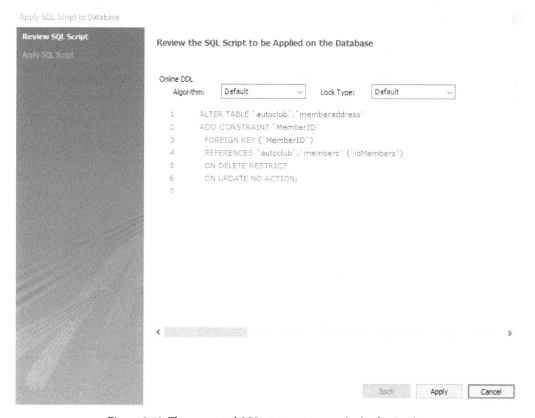

Figure 2.42: The generated SQL statement to apply the foreign key

> **Note**
> MySQL will also create an index in the new foreign key.

11. Select Apply and **Finish** to save the changes.

12. And, finally, just to prove that **No Action** and **Restrict** are the same in MySQL, change the **On Update** option to **RESTRICT** and click on **Apply**. The bottom line of the output panel will say **No Changes Detected**, as you can see in the following screenshot:

⬤   7 15:44:05 Apply changes to memberaddress   No changes detected

Figure 2.43: No changes detected when changing a foreign key option from No Action to Restrict

You have now created your first foreign key. You cannot test this until you insert some data into the tables. We will test this in the next chapter.

The `Cascade` option should be used with caution. If you wish to delete a parent record, first, you must delete all of the related child records and then remove the parent. This could be done by the code present in the application or by using a stored procedure in the database. The cascade option, if used on **On Delete**, will remove the child records automatically. This might be undesirable if the parents' deletion is accidental.

You can create foreign keys in several tables of the `autoclub` database. The following diagram will indicate which table should be joined with a foreign key. Go ahead and create the foreign key for the tables using the following diagram as your guide. Set all foreign key options to **RESTRICTED**:

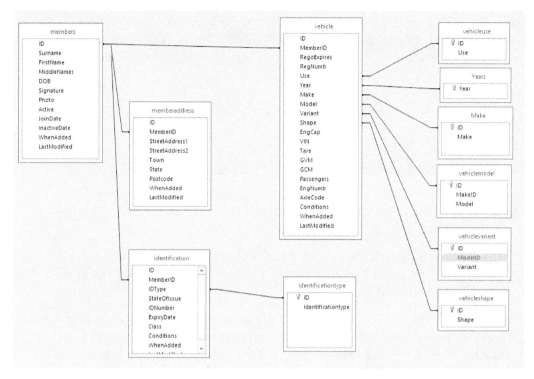

Figure 2.44: The ER diagram for the autoclub database; only tables required for this activity are included

Foreign keys are a handy way to help you keep your data in order and maintain the integrity of the database. When you are confident with the data stored in the database, you can be confident that the reports are accurate. The default value of **No Action** or **Restrict** for the options will be the most common constraint you will use. If you do use them, test them thoroughly.

> **Note**
>
> If you get an error when saving, check whether the primary key has been set in the referenced table. If one or more has not been set, set them and try. Ensure that you also set the auto-increment feature if required.

If you take the time to set the indexes and foreign keys where appropriate in your database, you will reap great benefits. Your database will become faster because of the indexing and more accurate with the foreign keys and constraints maintaining the data integrity for you.

Now, let's take a small detour to see how you can reverse engineer a database based on ER and EER.

Reverse engineering a database

You now have a small database complete with tables, indexes, and foreign keys. Let's imagine that you have a database with over a hundred tables. You will have to try and comprehend the data present if you do not have a database model to hand.

You can reverse engineer a database using MySQL Workbench so that you can create both the database model and the EER diagram, which, in return, will assist you greatly in coming to terms with the database.

An ER diagram is a snapshot of the database. It is an image of the tables in the database with lines connecting the tables to show the relationships as set by the foreign keys. There are a number of open source and proprietary software options that allow you to generate ER diagrams. Depending on the software used to create the diagram, you might be able to display information regarding indexes and foreign key columns. The lines connecting the tables might start and end at the actual connecting columns in the relationship, or they might just go from one table to the other without identifying the actual columns. ER diagrams provide a visual representation of the database. However, you cannot interact with the diagram:

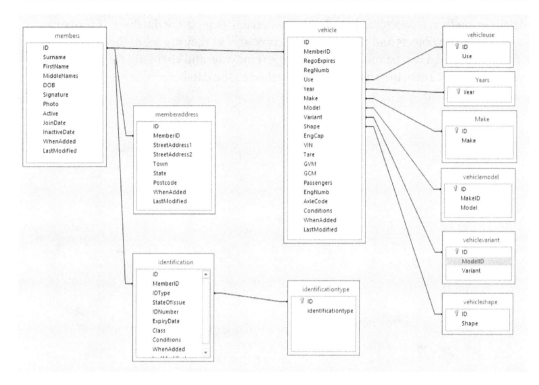

Figure 2.45: An ER diagram created in MS Access that limits interaction in the application

> **Note**
>
> For better viewing, you can find the preceding screenshot in full resolution at `https://github.com/PacktWorkshops/The-MySQL-Workshop/blob/master/Chapter02/Images/Image1.png`.

An EER diagram, as we find in MySQL servers, has all of the same features as an ER diagram, but you can interact with it. EER diagrams are software-based. This means that they are implemented through software tools such as SQL Workbench, and they are directly linked to the underlying conceptual model of the database so that you can select any object in the diagram and edit its properties. The changed properties are then saved to the model and retained. When the EER diagram is printed or exported to a PDF or other document type, it is no longer interactive while on paper or in the exported file format. Therefore, it becomes an ER diagram.

A database *model* is a conceptual model that accurately depicts the database. It contains all of the database objects and their properties precisely as they are set in the underlying database. The model can be modified by adding, removing, and changing the properties of various objects. Then, the changes can be pushed to the database:

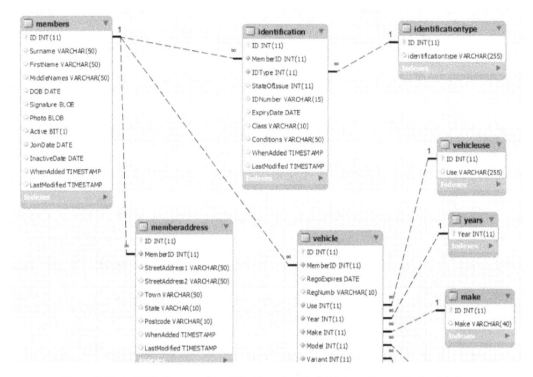

Figure 2.46: An EER diagram from Workbench with full interaction with the database

In the following exercise, you will create an EER diagram of the `autoclub` database.

Exercise 2.06 – creating an EER model from the autoclub database

Your boss needs you to verify a database's integrity as the documentation has been lost for the autoclub database. After some time has passed and you have moved on to developing other databases, your intimate knowledge of the autoclub database has waned a little. You are required to create an EER diagram from the autoclub database to visualize it.

> **Note**
>
> The model sits between the database and the EER diagram and is created for you when you create the EER diagram or reverse engineer the database. All changes made to the EER diagram will only affect the model. They will not modify the database until you forward engineer or synchronize the model to the database.

To create an EER diagram of the autoclub database, perform the following steps:

1. Open MySQL Workbench using **My First Connection**.

2. From the top-level tabs, select **Database** and click on **Reverse Engineer**:

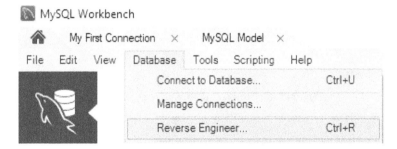

Figure 2.47: The Reverse Engineer menu

3.   In the wizard, select the connection, and click on **Next** at the bottom of the screen:

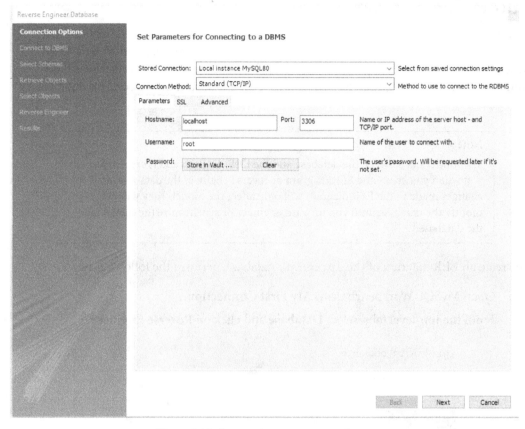

Figure 2.48: Connecting to reverse engineer

Upon successful connection, click on **Next**:

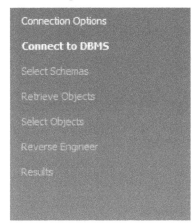

Figure 2.49: The connection is a success

4.   Now, select the database and click on **Next**:

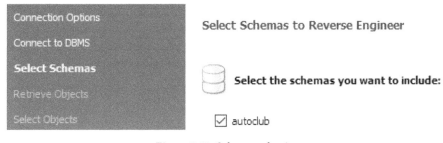

Figure 2.50: Schema selection

5. On the screen that follows, you can finalize the operation. To do this, select **Retrieve Objects from Selected Schemas**, along with **Check Results**, to retrieve all of the available data using the utility:

Figure 2.51: A successful connection to the server

6. Click on **Next**, and the schema list will be presented as follows:

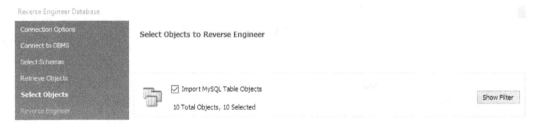

Figure 2.52: The list of schemas on the server, select autoclub

After completion, you will see the following window:

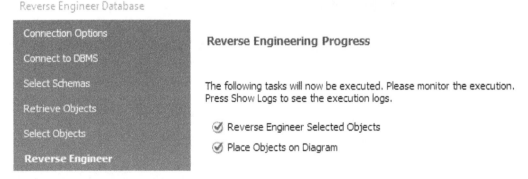

Figure 2.53: The successful retrieval of schema objects

In the ER diagram generated by the reverse engineering process, you should see the following:

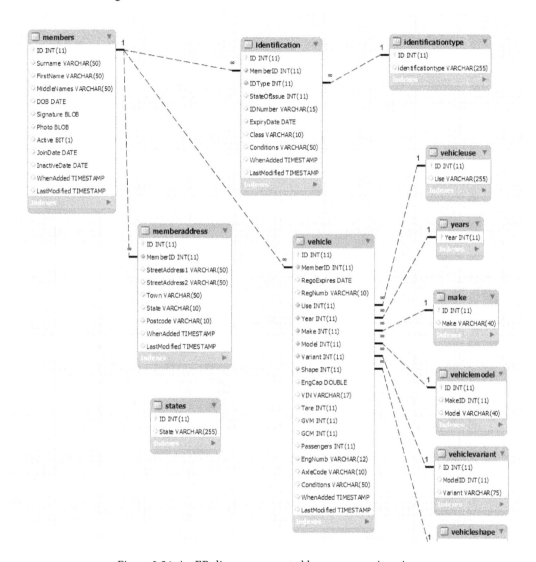

Figure 2.54: An ER diagram generated by reverse engineering

> **Note**
>
> For better viewing, you can find the preceding screenshot in full resolution at `https://github.com/PacktWorkshops/The-MySQL-Workshop/blob/master/Chapter02/Images/Image2.png`.

At this point, you can move the objects around to get a nice layout. However, it would probably be wise to rename and save it.

7. Click on the **MySQL Model** tab, and you will see the new EER diagram displayed:

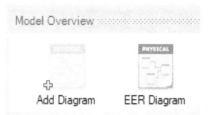

Figure 2.55: The MySQL Model window with the autoclub database

This contains the details of the `autoclub` database and its tables.

8. Right-click on the **EER Diagram** icon and select **Rename Diagram**. Enter the new name `autoclub`, and then click on **OK**:

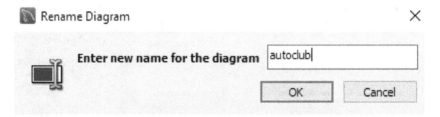

Figure 2.56: Rename the EER diagram

9.   The name will be changed, and you can see it in the following screenshot:

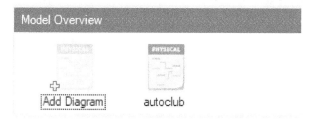

Figure 2.57: The EER Model icon in the Model Overview panel

So, you have created your EER diagram for the `autoclub` database. It is now much easier to visualize the database.

You can also interact with your database. As you hover over the tables, columns, lines, and more, an information window will pop up with all of the details about the item. The lines will highlight to see what they are linking, and the indexes can be displayed by clicking on the arrowhead at the bottom of the tables.

Some issues will really stand out, such as the `states` table, which is not linked to anything, as you never created the foreign key for it. In the upcoming exercise, you will fix it using the EER diagram.

Exercise 2.07 – using the EER diagram and forward engineering to manage the database model

Forward engineering will only add objects to an existing database; it will not remove them. If you change a table name, the table will be created with the new name, but the existing table will remain. If you delete a table from the model, it will not be deleted in the database. Usually, you will only use forward engineering to create a database from an EER diagram.

In this exercise, you will create a foreign key linking the `State` column of the `memberaddress` table with the `ID` column of the `states` table. This is needed because members can have multiple addresses in different states or even within the state:

1.   Open the `autoclub.mwb` file from `https://github.com/PacktWorkshops/The-MySQL-Workshop/blob/master/Chapter02/Excercise%202.07/autoclub.mwb`.

2. Double-click on the **memberaddress** table. The table design screen will open in the lower section of the screen, as shown in the following screenshot:

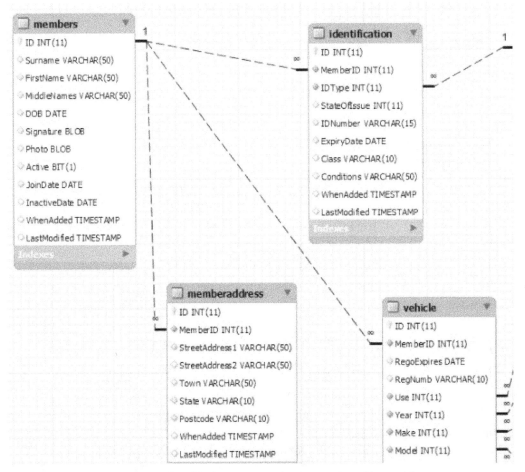

Figure 2.58: The table design screen for the memberaddress table

After examining the `State` column in the `memberaddress` table, you will observe that the column is of the `VARCHAR(10)` datatype:

Column Name	Datatype	PK	NN
MemberID	INT	☑	☑
StreetAddress1	VARCHAR(50)	☐	☑
StreetAddress2	VARCHAR(50)	☐	☑
Town	VARCHAR(50)	☐	☑
State	VARCHAR(10)	☐	☑
PostCode	VARCHAR(10)	☐	☑

Figure 2.59: The state is VARCHAR(10)

To make the foreign key, it needs to be an INT datatype in order to match the datatype of the State table. You will need to fix this first.

> **Note**
>
> If the table already contains data in the State column, you will not be able to change from the VarChar datatype to Int because an integer will not accept non-numeric characters. Fortunately, the table is still empty, so you can change it.

3. Select a datatype of **INT** for the State column, and you will notice this change taking immediate effect in the EER diagram:

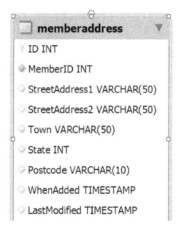

Figure 2.60: The type of State set to INT

4. Click on the **Foreign Keys** tab to view the work screen, as shown in the following screenshot:

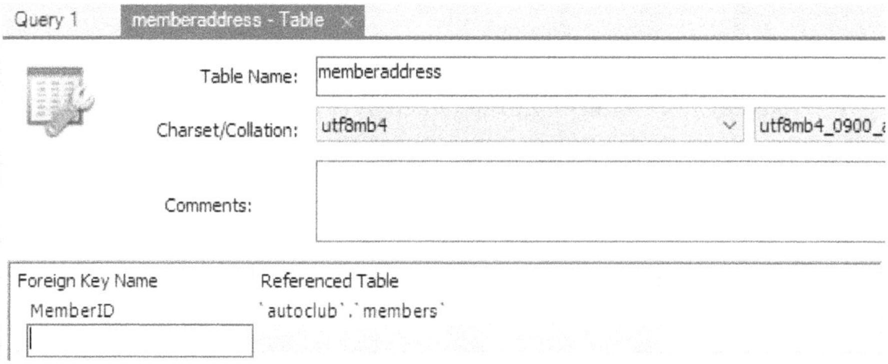

Figure 2.61: The Foreign Keys work screen

5. Enter the foreign key details just as you did in the earlier exercises. Enter **Foreign Key Name** as FK_MemberAddress_State and provide **Referenced Table** as 'autoclub'.'states'. Check the **State** type in the **Column** section and **ID** in the **Referenced Column** section. Also, set the foreign key options of **On Update** and **On Delete** to **RESTRICT**. As soon as these changes are applied, notice that the EER diagram is automatically changed. Please refer to the following screenshot:

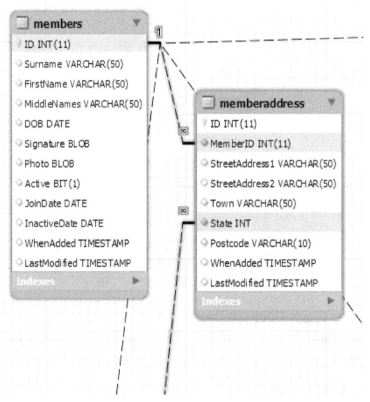

Figure 2.62: After the foreign key details are entered, the EER diagram changes immediately

6. Save the diagram by clicking on **File** and then **Save Model**. Use the **Save Model As...** if you wish to save it under a different name:

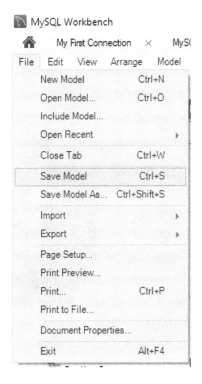

Figure 2.63: Using File and Save Model to save the EER diagram

7. The changes you have made in the EER diagram are not yet reflected in the database. To save the changes back to the database, select **Database** and then **Forward Engineer** from the top-level menu, as shown in the following screenshot:

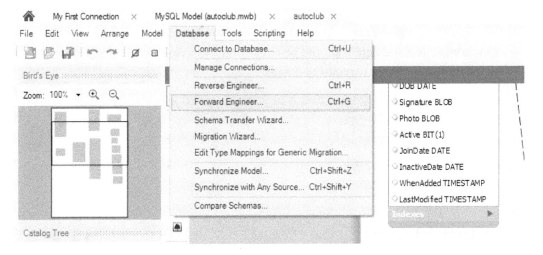

Figure 2.64: Selecting Database and Forward Engineer

This will open the **Forward Engineer to Database** screen, as shown in the following screenshot:

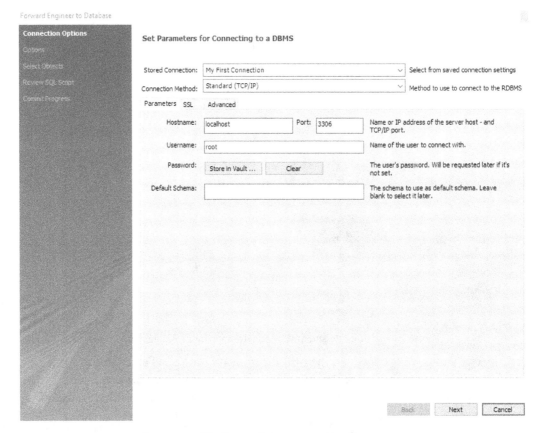

Figure 2.65: The Forward Engineer to Database screen

Your connection and details should be already on the screen. If not, select your connection and fill in any required details.

8. Click on **Next**, and the **Options** screen will open. Keep the screen options at their default settings, as shown in the following screenshot:

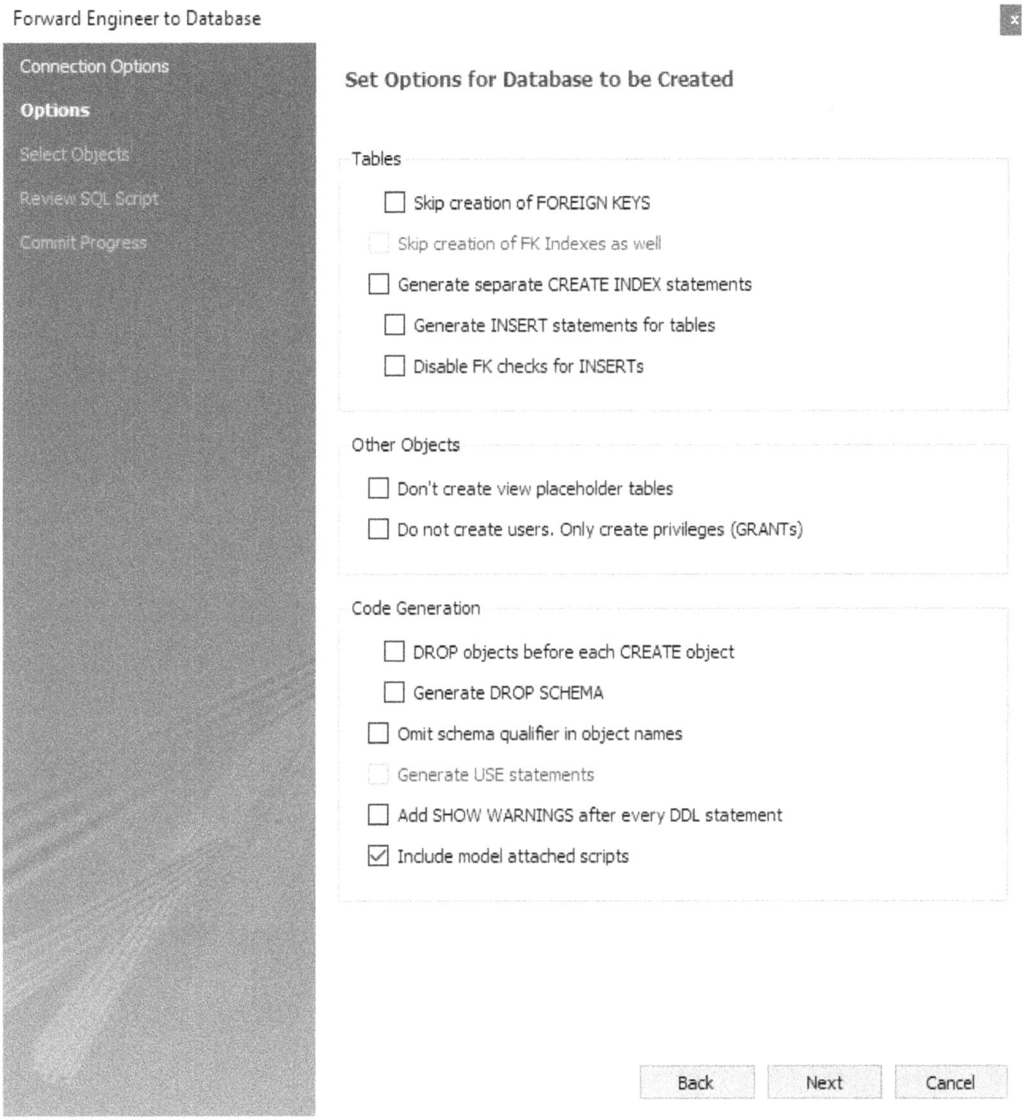

Forward Engineer to Database

Connection Options	Set Options for Database to be Created

Options

Select Objects

Review SQL Script

Commit Progress

Tables

- ☐ Skip creation of FOREIGN KEYS
- ☐ Skip creation of FK Indexes as well
- ☐ Generate separate CREATE INDEX statements
- ☐ Generate INSERT statements for tables
- ☐ Disable FK checks for INSERTs

Other Objects

- ☐ Don't create view placeholder tables
- ☐ Do not create users. Only create privileges (GRANTs)

Code Generation

- ☐ DROP objects before each CREATE object
- ☐ Generate DROP SCHEMA
- ☐ Omit schema qualifier in object names
- ☐ Generate USE statements
- ☐ Add SHOW WARNINGS after every DDL statement
- ☑ Include model attached scripts

[Back] [Next] [Cancel]

Figure 2.66: The Options screen

9. Click on **Next**, and the **Select Objects** screen will open. Click on the **Show Filter** button to view all of the objects. You have only changed the `memberaddress` table, so move all of the other objects to the right-hand panel, as shown in the following screenshot:

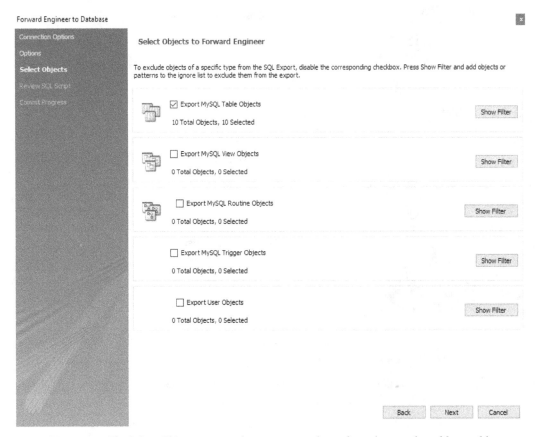

Figure 2.67: The Select Objects screen; changes were only made to the memberaddress table

10. Click on **Next** and the **Review SQL Script** screen will open, displaying the MySQL script that was generated to apply the changes. Review the script before executing it:

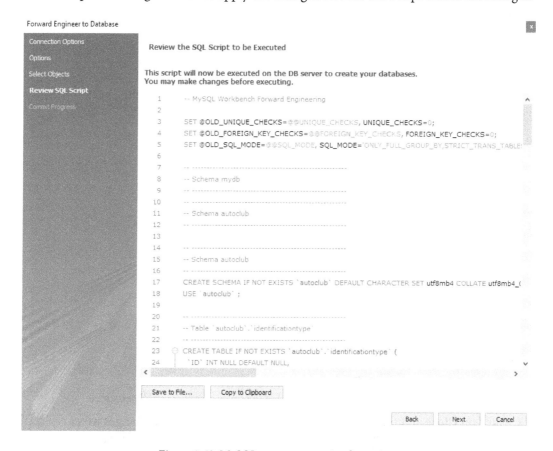

Figure 2.68: MySQL generates a script for review

11. Click on **Next** to execute the script. The progress screen will be displayed, and the script will run as follows:

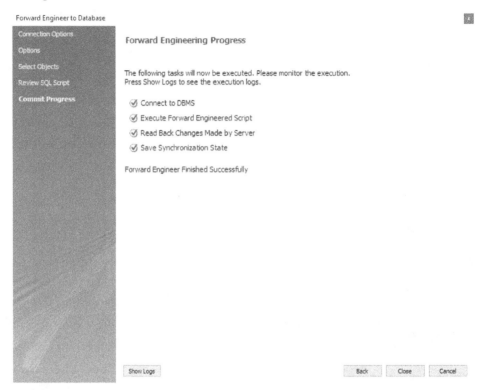

Figure 2.69: The progress screen showing that the script was successfully executed

12. Click on **Close** and the screen will close. Now, to check that the changes were applied, return to the **MySQL Model** (autoclub.mwb) tab. Double-click on the **memberaddress** table under **autoclub**. The **memberaddress** table definition will show that the **State** column is now an **INT** datatype:

Column Name	Datatype
MemberID	INT
State	INT

Figure 2.70: The State column is now an INT datatype

13. Click on the **Foreign Keys** tab at the bottom, and you will see new foreign keys:

Foreign Key Name	Referenced Table
MemberID	`autoclub`.`members`
FK_MemberAddress_State	`autoclub`.`states`

Figure 2.71: The foreign key for the State column exists

You have now successfully ensured that members with multiple addresses in different states can be registered.

During the course of development, changes could have been made either directly to the database or to the model. Due to this, they can become unsynchronized. You will need to use **Synchronize Model** to get everything in order or to apply your recent model changes to the database itself. In the following exercise, you will work through a synchronization task. You will take a look at how you can commit changes to the production database.

Exercise 2.08 – committing model changes to the production database with Synchronize Model

To be able to effectively use EER diagrams, it is important to understand how to synchronize them with a production database. In this exercise, you will update the autoclub database by committing the model through the synchronize model utility. This process will examine the provided EER model, process the relationships, and apply them to the existing autoclub database:

1. In the MySQL Model (that is, `autoclub.mwb` from `https://github.com/PacktWorkshops/The-MySQL-Workshop/blob/master/Chapter02/Excercise%202.08/autoclub.mwb`) window, select **Database** and then select **Synchronize Model...** to open the **Synchronize Model with Database** wizard:

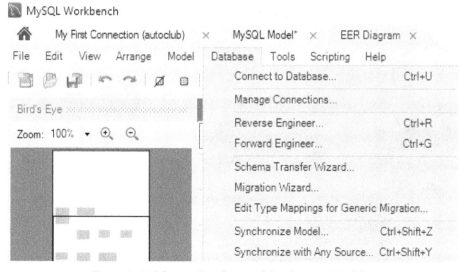

Figure 2.72: Selecting Database and Synchronize Model...

2. The initial screen is the same connection screen as shown in the previous exercises. Ensure the connection settings are correct as before. Then, click on **Next** to open the **Options** screen:

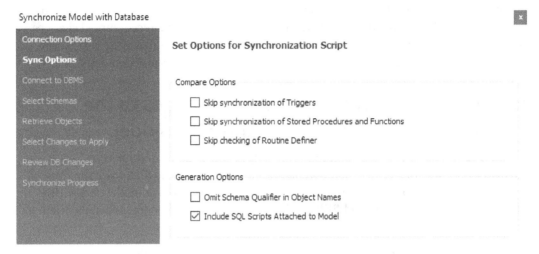

Figure 2.73: The options screen

3. Leave all options in their default settings, as shown in the preceding screenshot, and then click on **Next** to connect to the database.

4. As before, the wizard will connect to the database and collect the schema details. Click on **Next** when you are done to open the **Select Schemas** screen. The screen will open with **autoclub** already selected, as follows:

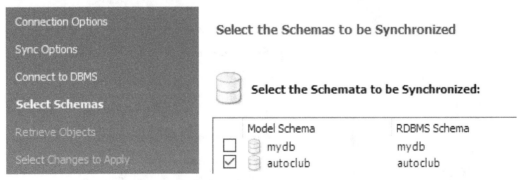

Figure 2.74: The schemata screen showing autoclub is already selected

5. Ensure **autoclub** is selected and click on **Next** to retrieve the database objects.

6. If the retrieval of objects was successful, click on **Next** to open the **selection** screen. When the screen opens, all arrows will be green. Double-click on the arrow of the object you do not want to update. You only modified the **memberaddress** table, so double-click on all the other table arrows. Your screen should now look similar to the following screenshot:

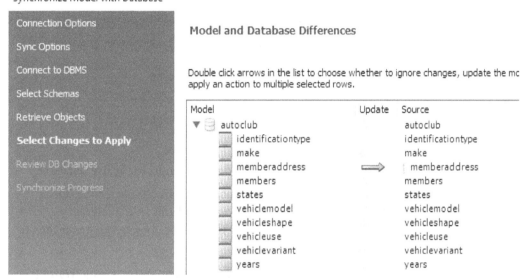

Figure 2.75: The object selection screen

You can click on the update arrow for each object to see what operation will be performed and decide whether to update the model, source, or ignore. In the preceding screenshot, you are ignoring all except the **memberaddress** table, and you are going to update the source from the model.

7. Click on **Next** to review the MySQL-generated SQL script to make the changes.

8. Click on **Execute** to run the script, and when finished, click on **Close**.

9. Return to the **My First Connection** tab. Right-click on the **autoclub** database and select **Refresh All** to refresh the list.

10. Right-click on the **memberaddress** table and select **Alter Table** to view the tables settings. You will notice that the datatype of the **State** column in the database has now changed to **INT**:

Column Name	Datatype
MemberID	INT
State	INT

Figure 2.76: The memberaddress.state column is now an INT datatype

11. Select the **Foreign Keys** tab, and you will see that two foreign keys have been created:

Foreign Key Name	Referenced Table
MemberID	`autoclub`.`members`
FK_MemberAddress_State	`autoclub`.`states`

Figure 2.77: The foreign key has been created

You have successfully made the changes defined in the model to the live database.

When you are developing a new database, after the initial analysis has been completed, take the approach of developing the model and EER diagram from your analysis documents first. Make sure everything is correct, and when done, forward engineer the model to the database. The database will be created for you.

When you are developing with an existing, undocumented system or migrating from MS Access, take the approach that you worked with here. For an MS Access migration, first, migrate the database into MySQL. Once the database is in MySQL (or you are working on an existing MySQL database), you can then reverse engineer to create the model and EER diagram. Then, you can make any changes to the EER diagram and model, and synchronize the model with the live database.

And now, for the final activity in this chapter, you will be modifying the EER diagram, the model, and the database as you add more objects when the business requirements change.

Activity 2.01 – modifying the EER diagram, the model, and the database

Your manager has asked you to include the ability to track Membership Fees in the autoclub database. Take a good long look at the EER diagram to see how you can insert this request into the database. You have decided to perform the following steps in order to implement this:

1. Insert a new table named membershipfees.

    The table will have the following columns and datatypes:

Column	Datatype	Options	Foreign key	Constraints
ID	Primary Key	Auto Increment		
MemberID	INT	Not Null	FK to members.ID	update Restricted delete Restricted
FeeAmount	Double	Not Null		
DatePaid	Date			
WhenAdded	TimeStamp			
LastModified	TimeStamp			

Figure 2.78: The membership fees table description

2. Save the EER diagram with the changes.

3. Synchronize the model with the database.

> **Note**
>
> The solution to this activity can be found in the *Appendix* section.

You have worked through updating the database via the EER diagram and model. This method is useful for situations where you want to plan a database structure and apply it directly to the SQL database. Often, EER diagrams are an easier way to visualize structures, so being able to directly apply them makes building databases easier.

> **Note**
>
> The CLI is another popular method that you can use to work with your database. The CLI will be covered in great detail in *Chapter 3*.

Summary

In this chapter, you learned how to work with the Workbench GUI to create a complete database with tables and columns, import new tables from an SQL file, create indexes and foreign keys, and create an EER model and diagram by reverse engineering the database. The ability to reverse engineer a database to create the model will make working with existing databases easier. Following this, you learned how to modify the EER diagram and forward engineer the changes to the model. Additionally, you explored how to synchronize the model with the production database.

In the next chapter, you will be using SQL statements to work with the database. You will learn how to back up and restore the database using MySQL Workbench and perform different operations using SQL statements.

3
Using SQL to Work with a Database

In this chapter and the next, you will be learning to use the SQL language to work with the database. There is much to learn, so the topic has been split into two chapters, with *Chapter 3, Using SQL to Work with a Database* (this chapter), concentrating on database creation, tables, fields, indexes, and foreign keys, the same topics that were covered in *Chapter 2, Creating a Database* (excluding EER) but using SQL statements and not a GUI such as Workbench. You will still be using the **Query** tabs in Workbench in which to write SQL. Learning to perform these functions in pure SQL will enhance your knowledge and skills. We will also cover some new topics, such as adding, modifying, and deleting data and records.

This chapter covers the following topics:

- Working with data
- Backing up databases
- Restoring databases
- Working with SQL code to maintain a database
- Creating a new database
- Creating and modifying tables

- SQL queries to create indexes and foreign keys
- Activity 3.1 – creating a table with indexes and foreign keys
- Altering table queries
- Adding data to a table
- Updating data in a record
- Deleting data from tables
- Drop queries
- Blobs, files, and file paths
- Files and file paths

An introduction to working with databases using SQL

In the last chapter, you learned about MySQL Workbench, and how to create a database, tables, and fields. You then learned how to import tables using an SQL script file and then set indexes and foreign keys. You learned how to create a database EER model and diagram by reverse engineering an existing database. You also learned how to modify the database structure using the EER diagram and forward engineering, the changes to the model, and finally, you learned how to synchronize the model with the live database.

In this chapter, you will learn the fundamentals of SQL queries, as well as the basics of creating backups for databases. Backups are valuable when you need to save data to prevent it from getting deleted or lost. It is important to keep backups of data; otherwise, data may become unrecoverable, creating a large amount of work to reconstruct a dataset.

To effectively work with MySQL, you will need to understand the fundamentals of SQL queries. If you want to use MySQL within an application, or query from anything external to MySQL Workbench, you will need to use SQL queries. These queries will not only allow you to query from outside MySQL Workbench but also build more complex queries, which MySQL Workbench is not able to do.

Before we start running queries, we will need to do a quick backup of the `autoclub` database. We will do this to ensure that there is always a copy of the data before any modifications occur. This way, if a modification causes issues in the data, we can recover to a previous copy, without any issues.

Working with data

The single most valuable component of any computerized system is data; without it, the system is meaningless. Over time, as the system is used, the data will build up to a point where it can provide valuable insights into a business and enable forecasting based on past trends, upon which business decisions will be made.

We will now start working with data, beginning with some simple additions, updating and removing records, through to more complex reading of the data from several joined tables for reporting purposes.

Types of SQL statements

SQL statements come under several main categories when working with MySQL:

- **System**: The statements will interact directly with the server to perform system-related tasks.

- **Database maintenance**: Statements that will work with the database, such as table and foreign key creation.

- **Data manipulation**: Statements that work directly with data, such as Insert and Update. We will be working with these, as these are the statement types you will be using more than any other.

- **Destructive**: A statement that removes database items such as records, tables, and entire databases. Always be careful of these as once something is gone, it may not be easy to recover it, if at all.

You will work with all of these types throughout this course; for now, we will work with a data manipulation language.

You will begin with adding data to the table in the following section.

> **Note**
>
> Before we start working with data, we are going to reset the database to ensure that it is in the proper state, with all the settings in place precisely as expected for the rest of the chapter to avoid possible issues. We will run a SQL script to do this. This script will remove the current `autoclub` database and then rebuild it. You can get the script at the following link: `https://github.com/PacktWorkshops/The-MySQL-Workshop/blob/master/Chapter02/Excercise%202.03/Chapter2%20Ancilliary%20Tables.sql`.

Backing up databases

Perhaps the most crucial task while working with data is that you should get into the practice of making regular backups of your database. Consideration should be given to the importance of the data and the problems it would cause if it became corrupt, hacked, or deleted. Most businesses demand daily backups at the very least. The backups can be created using a script scheduled to run overnight or at times when there is little demand for the database.

As a developer, backing up your database or even individual tables is essential because, while using SQL statements, you can easily mess up your database. A backup gives you a reset point.

You can create database backups in two different ways. The first way is through MySQL Workbench, and the second is through the command line. In MySQL Workbench, you can use the Data Export tool to create a backup of the data. Let's create a backup of our `autoclub` database using the Data Export tool.

To back up the `autoclub` database, we can use the following steps:

1. Open Workbench and **My First Connection**. Log in if required.

2. Open the **Server** menu from the top menu bar and select **Data Export**:

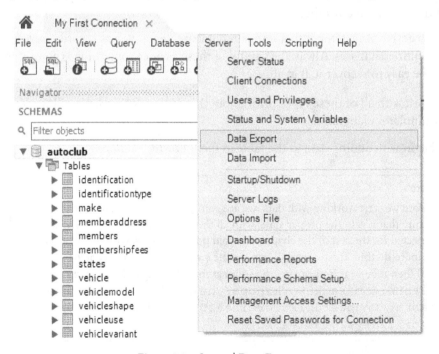

Figure 3.1 – Server | Data Export

3. Select autoclub from the database list and then ensure the following are set, as shown in the following screenshot:

- **Dump Structure** and **Data** are selected.

- **Export to Self-Contained File** is selected.

- The filename and path for the backup are entered.

- **Create Dump in a Single Transaction (self-contained file only)** is selected.

- **Include Create Schema** is selected:

Local instance MySQL80
Data Export

Object Selection Export Progress

Tables to Export

Exp...	Schema
☑	autoclub
☐	backuppractice
☐	chat
☐	coffee_data
☐	contactdb
☐	login
☐	ms_access_migration
☐	phpmyadmin
☐	shop
☐	test

Refresh

Objects to Export

☑ Dump Stored Procedures and Functions   ☑ Dump Events   ☑ Dump Triggers

Export Options

◉ Export to Dump Project Folder   C:\Users\scott\Documents\dumps\Dump20220310   ...

Each table will be exported into a separate file. This allows a selective restore, but may be slower.

○ Export to Self-Contained File   C:\Users\scott\Documents\dumps\Dump20220310.sql

All selected database objects will be exported into a single, self-contained file.

☐ Create Dump in a Single Transaction (self-contained file only)   ☑ Include Create Schema

Press [Start Export] to start...   Start Export

Figure 3.2 – The Data Export screen

4.   Click **Start Export**, and the database will be exported to a SQL file:

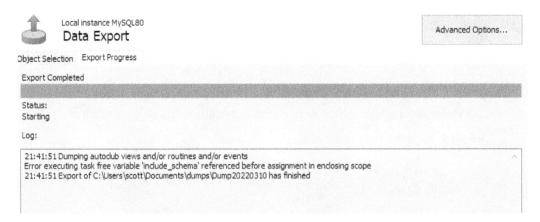

Figure 3.3 – Successful database export to a SQL file

The `autoclub` database has now been exported.

To back up a database through the command line, you can use the `mysqldump` command. This command will take in a username and password, as well as the databases that you wish to backup. The general format of the command will be `mysqldump --user [username] --p [database name] > [outputfile]`. This process will create a SQL file that can be used to recover the database. Let's try using this command on the `autoclub` database to create a backup.

Exercise 3.01 – Backing up the autoclub database

You have been asked by your manager to create a backup of the `autoclub` database. They would like the output to be a single file named `autoclub.sql`. To make the backup quickly, you have decided to use the `mysqldump` command-line tool. To create the backup successfully, take the following steps:

1.   Launch the command line on your computer.

2.   Run the following command:

```
mysqldump --user yourusernamehere --p autoclub >
autoclub.sql
```

Once the command is completed, you will have a single file named `autoclub.sql` that contains the backup of the database.

Of course, if something goes wrong, you will need to restore the database. In the next section, we will learn about restoring database backups.

Restoring databases

If you need to restore the `autoclub` database at any point during *Chapter 3* and *Chapter 4*, return to this section.

You are able to restore database backups using two different methods. The first is through MySQL Workbench, using the **Data Import** tool. To see how this works, let's try to restore our `autoclub` database.

To restore the `autoclub` database, follow the following steps:

1. Open Workbench and **My First Connection**. Log in if required.

2. From the top menu of the Workbench menu tab, select **Server** and then **Data Import**:

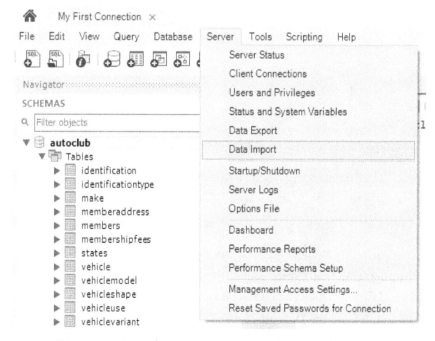

Figure 3.4 – Server | Data Import to restore the autoclub database

The following screen will appear:

Figure 3.5 – The Data Import screen

3.   Select **Import from Self-Contained File** and then locate the backup file you created in *Exercise 3.01* for the `autoclub` database.

4.   Select **autoclub** from the **Default Target Schema** drop-down menu.

5. Click **Start Import**. The following screen will display, and the database will be imported:

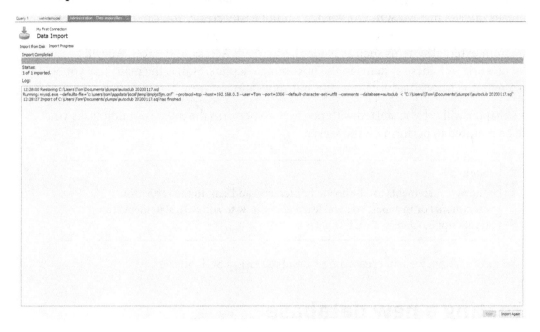

Figure 3.6 – A successful import

6. Now, close the screen.

The `autoclub` database is now restored to the last backup point.

To use the command line for restores, you can use the `mysql` command to run the SQL file created by the backup. The syntax for this command is `mysql -u [username] -p [database_name] < [filename]`. Let's try this command to recover our `autoclub` database.

Exercise 3.02 – restoring the autoclub database

A coworker has accidentally deleted one of the tables in the `autoclub` database! You need to recover the database to the latest backup so that the data is useable again. To do this, take the following steps:

1. Launch the command line on your computer.

2. Run the `mysql --u yourusernamehere --p autoclub < autoclub.sql` command.

Now that you have restored your previous work, we can continue working with SQL statements to modify the database.

Working with SQL code to maintain a database

Working with the MySQL server and SQL code provides flexibility, as you can construct complex queries that MySQL Workbench cannot easily create. You can not only run the SQL directly from Workbench but also send SQL statements to control the server and read data from external systems such as Node.js, Microsoft Access, and Excel. We will be doing a lot of work with these systems in the upcoming chapters. SQL is the main way you will work with the MySQL server from external applications.

This chapter will get you started with raw SQL to perform the most common tasks you will be required to perform on the server.

> **Note**
> The SQL statements can be created differently and sent to the server for execution. For instance, you will learn about how to run SQL statements in JavaScript in *Chapter 7* and *Chapter 8*.

In the next section, we will create a new database using SQL queries.

Creating a new database

The creation of an actual database is the first thing we are asked to do. Creating the database is, in itself, very simple in its most basic form. It can be done with a single line of code. The database in Workbench can be created using the following command:

```
CREATE SCHEMA '<database name>';
```

In addition, you can use the same syntax in the command line for MySQL to create a database.

> **Note**
> Some SQL statements have synonyms, which contain different syntax but have the same functionality. To create a database, you can use CREATE SCHEMA '<database name>'; or CREATE DATABASE '<database name>';. Both will create a database in the same way.

In the following exercises, we will create two new databases, just to go through the process. We will not use them any further, and all other work in the chapter will be with the autoclub database.

Exercise 3.03 – creating a new database

Your company is creating a new MySQL database in order to track shop orders. To do this, your manager has asked you to create a database called shoporder. To do so, follow the following steps:

1. Open Workbench and **My First Connection**. Log in if required.

2. Click the **Create a new SQL tab for executing queries** icon, as shown in *Figure 3.7*:

Figure 3.7 – Create a new SQL tab for executing queries

A new tab will be created.

3. Create a new database using the following command:

```
CREATE SCHEMA 'shoporder';
```

> **Note**
>
> The characters around the database name are backticks ('), which are located at the left of your keyboard. They are NOT single quotation marks.

4. Click **Execute the selected portion of the script or everything**, if there is no selection (the lightning bolt icon), to execute the preceding query. This will execute the statement and create the database.

> **Note**
>
> You can have multiple and distinct SQL statements in a single query tab. If you want to run only part of it, you can highlight the statement you wish to run and then click the lightning bolt icon. Only the highlighted section will execute then. If there is nothing highlighted, all of the statements will execute.

5. Right-click anywhere in the **SCHEMA** panel and select **Refresh All**. The list will be refreshed, and the new database will be displayed.

To complete this task through the command line instead of MySQL Workbench, follow the following steps:

1. Connect to your MySQL database by launching the command line and running `mysql -u [username] -p [password]`.

2. Once connected, run the command to create the database, `CREATE SCHEMA 'shoporder';`.

3. To verify that the database was created, run the `show databases` command, which shows all the databases in your current MySQL instance:

```
mysql> show databases;
+--------------------+
| Database           |
+--------------------+
| autoclub           |
| information_schema |
| mysql              |
| performance_schema |
| shoporder          |
| sys                |
+--------------------+
6 rows in set (0.02 sec)
```

Figure 3.8 – The result of the show databases query

Well, you can't get much simpler than that. The database has been created with a one-line statement and is ready to be filled with tables, data, and other objects.

In most databases, you will use the default collation as defined in the server during installation. However, in some circumstances, you may need to define a specific collation for the database. For instance, if you plan to connect to the database using Microsoft Access, it has a specific requirement regarding MySQL and collation. This will be discussed in further detail in *Chapter 6, Stored Procedures and Other Objects*. For now, we will use the default collation.

That completes creating a database using SQL statements. We will continue working with the `autoclub` database for the remainder of this chapter.

In the next section, we will learn how to create and modify a table using SQL statements.

Creating and modifying tables

Once the database is created, you want to start adding tables to it. You can, at any time, add new tables to the database and even add new fields to the tables. However, once applications are using your database, you should be very careful about removing or renaming procedures, views, tables, and fields because applications or MySQL views and procedures using these objects will stop working.

You can create a new table using the following command:

```
CREATE TABLE [IF NOT EXISTS] tableName (FieldName1 Datatype,
FieldName2 Datatype, …)
```

There are a number of properties we can set when we add a field to a table. Before we move on to an example, let's briefly discuss the properties available for our fields. The first common type of property is to set controls for whether a field can be null or not. If a field should never be null, you can add NOT NULL after the field data type. Otherwise, you can place NULL after the data type to allow for null values.

AUTO_INCREMENT is another common property that is set in table creations. This property can be set for integer values and allows for a field to automatically increment each time a record is added. So, for example, when you add your first record, the AUTO_INCREMENT field will be set to 1. The next record will get 2, then 3, and so on.

With the DEFAULT property, we can specify the default value for a field if one is not provided. Finally, we can set ON UPDATE to change a field to a specific value when the record is updated. This is typically used for timestamps to keep track of when a record was last changed.

In the next exercise, you will create a new table using the SQL statements.

Exercise 3.04 – creating a new table

The Automobile Club has several staff members who access a database. It is important to ensure that each user has their own user ID and password to gain access to the database. You have to create a user table to control who has access to the database. This table will contain the following fields and properties:

Field name	Data type	Properties
ID	INT	NOT NULL, AUTO_INCREMENT
Username	VARCHAR(16)	NOT NULL
Email	VARCHAR(255)	NULL

Field name	Data type	Properties
Password	VARCHAR(32)	NOT NULL
Active	Bit	NOT NULL, DEFAULT 1
WhenAdded	TIMESTAMP	NOT NULL, DEFAULT CURRENT_TIMESTAMP
LastModified	TIMESTAMP	NULL, DEFAULT, CURRENT_TIMESTAMP, ON UPDATE CURRENT_TIMESTAMP

Figure 3.9 – The user table with values

To create the user table, perform the following steps:

1. Open Workbench and select **My First Connection**. Log in if required.

2. Click the **Create a new SQL tab for executing queries** icon:

Figure 3.10 – Create a new SQL tab for executing queries

3. Enter the following SQL statement to create a new table, user, in the autoclub database:

```
-- -----------------------------------------------------------
-- Table 'autoclub'.'user'
-- -----------------------------------------------------------
CREATE TABLE IF NOT EXISTS 'autoclub'.'user' (
    'ID' INT NOT NULL AUTO_INCREMENT,
    'username' VARCHAR(16) NOT NULL,
    'email' VARCHAR(255) NULL,
    'password' VARCHAR(32) NOT NULL,
    'Active' BIT NOT NULL DEFAULT 1,
    'WhenAdded' TIMESTAMP NULL DEFAULT CURRENT_TIMESTAMP,
    'LastModified' TIMESTAMP NULL DEFAULT CURRENT_TIMESTAMP
ON UPDATE CURRENT_TIMESTAMP,
    PRIMARY KEY ('ID'));
```

The first line of this query defines the name of the table and specifies that it should be created only if it does not currently exist. The next set of lines defines the fields in the table, their data types, and any properties that are required for them.

4. Execute the SQL query by clicking the **Execute SQL** (lightning bolt) icon.

5. Right-click anywhere in the **SCHEMA** panel and select **Refresh All**. The list will be refreshed, and the new **user** table will be visible in the **autoclub** table list:

Figure 3.11 – The new user table

6. Right-click on the table in the schema list and select **Alter Table** to view the table in design mode:

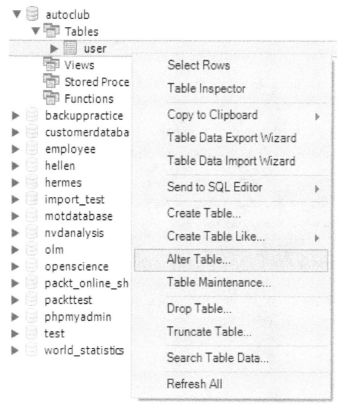

Figure 3.12 – Right-click the table and select Alter Table

You will get the following screen:

Column Name	Datatype	PK	NN	UQ	B	UN	ZF	AI	G	Default/Expression
ID	INT(11)	☑	☑	☐	☐	☐	☐	☑	☐	
username	VARCHAR(16)	☐	☑	☐	☐	☐	☐	☐	☐	
email	VARCHAR(255)	☐	☐	☐	☐	☐	☐	☐	☐	NULL
password	VARCHAR(32)	☐	☑	☐	☐	☐	☐	☐	☐	
Active	BIT(1)	☐	☑	☐	☐	☐	☐	☐	☐	b'1'
WhenAdded	TIMESTAMP	☐	☐	☐	☐	☐	☐	☐	☐	CURRENT_TIMESTAMP
LastModified	TIMESTAMP	☐	☐	☐	☐	☐	☐	☐	☐	CURRENT_TIMESTAMP ON...
		☐	☐	☐	☐	☐	☐	☐	☐	

Figure 3.13 – The user table in design view, with the settings as defined in SQL

This shows us that each of the fields was added as expected. The output also shows the data type, as well as any properties that were set at the time of creating the table.

If you want to complete these steps using the command line, the same syntax applies. The steps are as follows:

1. Connect to MySQL using `mysql -u [user] -p [password]`.

2. Once connected, type or copy the query we used for MySQL Workbench:

```
CREATE TABLE IF NOT EXISTS 'autoclub'.'user' (
    'ID' INT NOT NULL AUTO_INCREMENT,
    'username' VARCHAR(16) NOT NULL,
    'email' VARCHAR(255) NULL,
    'password' VARCHAR(32) NOT NULL,
    'Active' BIT NOT NULL DEFAULT 1,
    'WhenAdded' TIMESTAMP NULL DEFAULT CURRENT_TIMESTAMP,
    'LastModified' TIMESTAMP NULL DEFAULT CURRENT_TIMESTAMP
ON UPDATE CURRENT_TIMESTAMP,
    PRIMARY KEY ('ID'));
```

3. Once the query has been executed, you can verify it was successful using the following queries:

- Use `autoclub` to set the database to the `autoclub` database.

- Run the `show tables;` query to display all of the tables in the `autoclub` database:

```
mysql> show tables;
+-------------------+
| Tables_in_autoclub |
+-------------------+
| user              |
+-------------------+
1 row in set (0.01 sec)
```

Figure 3.14 – The result of the show tables query showing the user table

In this exercise, you have created a simple table with several fields, a primary key, and a few default values; you also enforced that some value must be entered into the required fields by setting the NOT NULL field.

> **Note**
>
> You may have noticed that there is nothing to ensure that the value entered in the `username` field is unique while creating the table. This is done intentionally so that you can correct it in *Exercise 3.05*.

In the next exercise, you will create a new table with an index and a foreign key included. Primary keys aside, indexes and foreign keys are likely to be the most common settings you should include when creating a new table. Assuming you have already worked through the initial analysis and design stage of the database, you will know what fields these are to be applied to.

SQL queries to create indexes and foreign keys

When working in MySQL, we will often have multiple tables, containing multiple datasets. These datasets are often related to each other in some way, typically with a common field between them. For example, if we had a table of customers, each customer might have a customer ID. From here, we may have a table of orders that contains the customer ID of the person who ordered it. We can relate these two tables using the customer ID field that they both share.

This type of relationship is called a foreign key relationship. To help to define these relationships, MySQL allows us to specify them at the time of creating a table. This creates a relationship between the two tables. The main advantage of this is that we can enforce policies for the foreign keys. For example, suppose we wanted to change a customer ID in our customer table. It will, in turn, make sense that we want to update the same customer ID in every other table that it appears in. To achieve this, we can set a property for our foreign key so that if the customer ID in the customer table changes, it will also change in the other tables that have it as a foreign key. This allows for our data integrity to be easily maintained.

To define a foreign key in a create query, we will use the following syntax:

```
CONSTRAINT '[NameOfForeignKey]'
    FOREIGN KEY ('FieldName')
    REFERENCES 'OtherTable' ('FieldName')
    [Additional Properties])
```

In addition to defining foreign keys, we can also define indices on our database tables. Indices define how data is stored in a database system. When we index a table, we order the data within it in a way that is easier to search through. For example, if you index a field that contains customer IDs, they will typically be sorted, allowing for faster searching through the values.

We will define an index using the following syntax:

```
[UNIQUE] INDEX 'IndexName' ('FieldName' [ASC|DESC])
```

We use the UNIQUE keyword when the field being indexed does not contain duplicates. With this understanding, we can now look at how these queries work in an example.

Exercise 3.05 – creating tables with indexes and foreign keys

The Automobile Club holds regular events for its members and their families. You need to include tables to hold the data of the events. In this exercise, you will create three new tables in the autoclub database for the events and assign the indexes and foreign keys at the time of creation.

To add the new tables, perform the following steps:

1. Open Workbench and click **My First Connection**. Log in if required.

2. Click the **Create a new SQL tab for executing queries** icon:

Figure 3.15 – Create a new SQL tab for executing queries

3. In the tab, create the first table with the following query:

```sql
-- -----------------------------------------------------
-- Table 'autoclub'.'eventvenues'
-- -----------------------------------------------------
CREATE TABLE IF NOT EXISTS 'autoclub'.'eventvenues' (
  'ID' INT NOT NULL AUTO_INCREMENT,
  'VenueName' VARCHAR(100) NOT NULL,
  'VenueAddress1' VARCHAR(255) NULL,
  'VenueAddress2' VARCHAR(255) NULL,
  'VenueTown' VARCHAR(30) NULL,
  'VenueState' INT NULL,
  'VenuePostcode' VARCHAR(10) NULL,
  'VenueContactName' VARCHAR(20) NULL,
  'VenuePhone' VARCHAR(15) NULL,
  'VenueEmail' VARCHAR(255) NULL,
  'VenueWebsite' VARCHAR(255) NULL,
  PRIMARY KEY ('ID'),
  INDEX 'FK_EventVenue_States_idx' ('VenueState' ASC),
  UNIQUE INDEX 'Idx_VenueName' ('VenueName' ASC),
  CONSTRAINT 'FK_EventVenue_States'
    FOREIGN KEY ('VenueState')
    REFERENCES 'autoclub'.'states' ('ID')
    ON DELETE RESTRICT
    ON UPDATE RESTRICT)
ENGINE = InnoDB;
```

This will create the eventvenues table, which will contain details about the event venues that exist for the autoclub. This table contains an index on the venue state, as well as a unique index on the venue name. There is additionally a foreign key, linking the venue state field to the states ID table. This foreign key restricts update and delete, meaning that these operations cannot be completed on the table, in order to keep the integrity of the key. Next, we can create the second table. The code to create this table can be found at https://github.com/PacktWorkshops/ The-MySQL-Workshop/blob/master/Chapter03/Exercise05/ Exercise%205%20%E2%80%93%20Creating%20a%20new%20table%20 with%20Indexes%20and%20Foreign%20Keys.txt.

This query creates the club events table, which contains all the events for the autoclub. This table has an index on the venue start, venue end, and event date fields. This table also contains two foreign keys, one for the venue start field and one for the venue end field. Finally, we can create the last table:

```
-- ------------------------------------------------------
-- Table 'autoclub'.'eventtype'
-- ------------------------------------------------------
CREATE TABLE IF NOT EXISTS 'autoclub'.'eventtype' (
    'ID' INT NOT NULL AUTO_INCREMENT,
    'EventType' VARCHAR(45) NULL,
    PRIMARY KEY ('ID'))
ENGINE = InnoDB;
```

This query creates the eventtype table, which stores the event types that exist for the autoclub.

4. Execute the SQL code by clicking the **Execute SQL** (lightning bolt) icon:

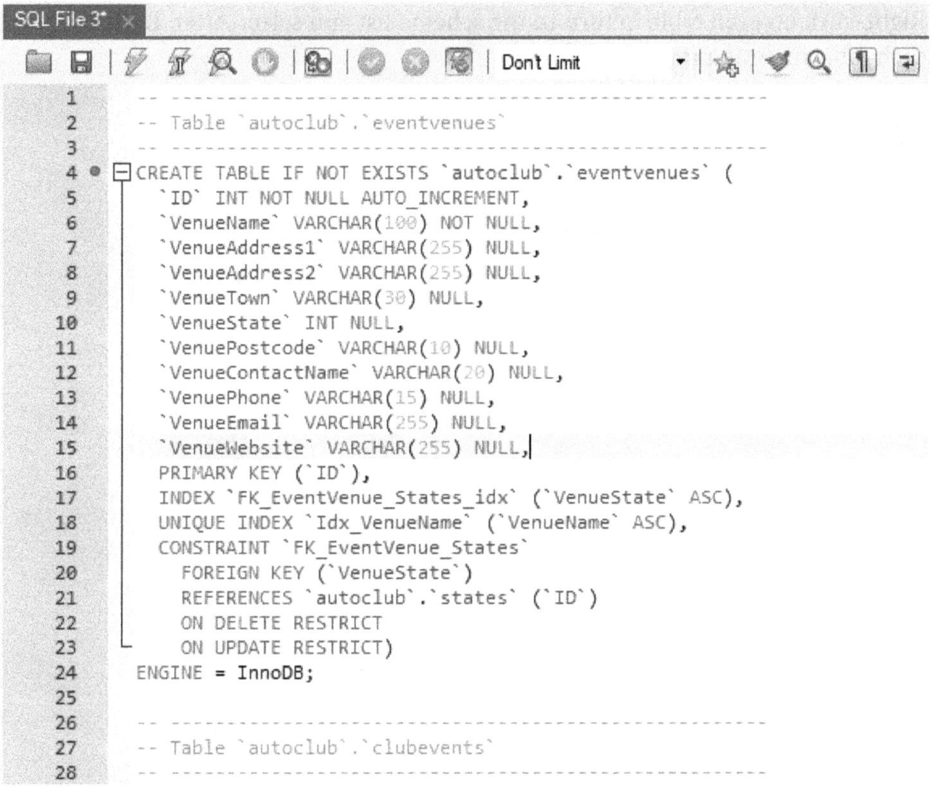

```
   SQL File 3*  ×

      📁 💾 | �-  🔅  🔍 ⏱ | 🔲 | ✓  ⊗  🔳 | Don't Limit          ▾ | 🎯 | 🧹 | 🔍 📘 🔁
      1       -- -------------------------------------------------------
      2       -- Table `autoclub`.`eventvenues`
      3       -- -------------------------------------------------------
      4  ●  ⊟ CREATE TABLE IF NOT EXISTS `autoclub`.`eventvenues` (
      5          `ID` INT NOT NULL AUTO_INCREMENT,
      6          `VenueName` VARCHAR(100) NOT NULL,
      7          `VenueAddress1` VARCHAR(255) NULL,
      8          `VenueAddress2` VARCHAR(255) NULL,
      9          `VenueTown` VARCHAR(30) NULL,
     10          `VenueState` INT NULL,
     11          `VenuePostcode` VARCHAR(10) NULL,
     12          `VenueContactName` VARCHAR(20) NULL,
     13          `VenuePhone` VARCHAR(15) NULL,
     14          `VenueEmail` VARCHAR(255) NULL,
     15          `VenueWebsite` VARCHAR(255) NULL,
     16          PRIMARY KEY (`ID`),
     17          INDEX `FK_EventVenue_States_idx` (`VenueState` ASC),
     18          UNIQUE INDEX `Idx_VenueName` (`VenueName` ASC),
     19          CONSTRAINT `FK_EventVenue_States`
     20            FOREIGN KEY (`VenueState`)
     21            REFERENCES `autoclub`.`states` (`ID`)
     22            ON DELETE RESTRICT
     23            ON UPDATE RESTRICT)
     24       ENGINE = InnoDB;
     25
     26       -- -------------------------------------------------------
     27       -- Table `autoclub`.`clubevents`
     28       -- -------------------------------------------------------
```

Figure 3.16 – The SQL code and the lightning bolt icon to run it

5. Right-click anywhere in the **SCHEMA** panel and select **Refresh All**. The list will be refreshed, and the new tables should be visible in the `autoclub` table list:

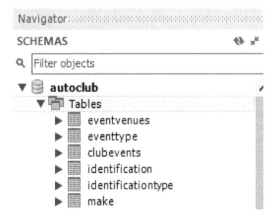

Figure 3.17 – New eventvenues, eventtype, and clubevents tables are visible

6. Right-click on each table in turn in the schema list and select **Alter Table** to view each table in design mode:

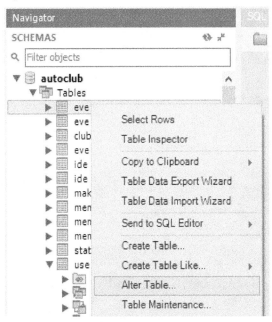

Figure 3.18 – Right-click the tables in turn and select Alter Table

The eventvenues table in design mode should look like the following:

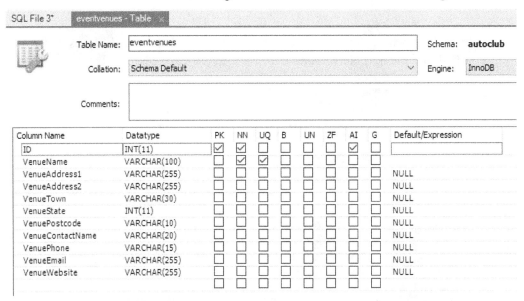

Figure 3.19 – The eventvenues table in design view with settings as defined in SQL

In this exercise, you have created three new tables with foreign keys and indexes. You will have noticed that when you start creating indexes and foreign keys, it can get a little more complicated, but certainly, with some practice, you will master the concept of creating foreign keys and indexes.

In the upcoming activity, you will add one more table to the `autoclub` database along with indexes and foreign keys.

Activity 3.1 – creating a table with indexes and foreign keys

The autoclub now wants members to be able to register for events. To do this, they would like you to create a table named `EventMemberRegistration`. This table will contain details about the members registered for particular events.

In this activity, perform the following steps:

1. Add a new table to the `autoclub` database and name it `EventMemberRegistration`.

2. Add the following fields to the table:

    - `'ID' INT NOT NULL AUTO_INCREMENT,`
    - `'ClubEventID' INT NOT NULL,`
    - `'MemberID' INT NOT NULL,`
    - `'ExpectedGuestCount' INT NOT NULL DEFAULT 0,`
    - `'RegistrationDate' DATE NOT NULL,`
    - `'FeesPaid' BIT NOT NULL DEFAULT 0,`
    - `'TotalFees' DOUBLE NOT NULL DEFAULT 0,`
    - `'MemberAttended' BIT NOT NULL DEFAULT 0,`
    - `'ActualGuestCount' INT NOT NULL DEFAULT 0,`
    - `'Notes' MEDIUMTEXT NULL,`
    - `'WhenAdded' TIMESTAMP NULL DEFAULT CURRENT_TIMESTAMP,`
    - `'LastModified' TIMESTAMP NULL DEFAULT CURRENT_TIMESTAMP ON UPDATE CURRENT_TIMESTAMP,`

3. Set the ID field as the primary key.

4. Create a standard INDEX named 'Idx_EventID' on the ClubEventID field; set its sort order to descending.

5. Create a standard INDEX named 'FK_EventReg_Members_idx' on the MemberID field; set its sort order to ascending.

6. Create a foreign key named FK_EventReg_ClubEvents for the ClubEventID field that references the ID field in 'autoclub'.'clubevents'; both UPDATE and DELETE constraints should be NO ACTION.

7. Create a foreign key named FK_EventReg_Members for the MemberID field that references the ID field in 'autoclub'.'members'; both UPDATE and DELETE constraints should be NO ACTION.

8. Set the table to use the InnoDB database engine.

> **Note**
> The solution to this activity can be found in the *Appendix*.

Altering table queries

In addition to creating tables, it is also possible to modify existing tables. This can be done using an ALTER query. An ALTER query uses the following syntax:

```
ALTER TABLE [table_name] [alter_options]
```

ALTER queries can be used for a number of purposes. One common reason is to change how a field in a table is defined. For example, suppose we have a customer table that contains a field for username. Currently, it allows for a VARCHAR value of size 15, but we want to extend this to be size 30. To do this, we can use an ALTER query, as follows:

```
ALTER TABLE customer MODIFY username VARCHAR(30);
```

We can also use the ALTER query to add an index to our table. To do this, we first need to create the index using a CREATE query. So, for example, suppose we now wanted to add an index to the username of our customer table. First, we create the index for username:

```
CREATE UNIQUE INDEX 'idx_username' ON customer('username')
```

The next exercise will show further applications of modifying tables with an index.

Exercise 3.06 – modifying an existing table

In most database applications, usernames are unique. Although they do have a unique numerical user ID, the text name should be unique as well. You are required to make the username field unique in the user table by creating a unique index. Also, you are asked to add a foreign key named ('EventType') that references the 'EventType' tables ('ID') field.

Perform the following steps:

1. Open a new SQL panel in Workbench and enter the following command:

   ```
   CREATE UNIQUE INDEX 'idx_username' ON 'autoclub'.'user'
   ('username')
   ```

2. Execute the SQL to create the index.

   A new index will be created on the username field by making it a unique index, which means that no two records can have the same name.

   > **Note**
   >
   > The preceding query will work fine on an empty table. If you already have data in the table and any names are the same, the index will not be created, and the corresponding SQL statement will fail.

3. Enter the following SQL statement into the query tab to add a foreign key on the EventType field that references the ID field of the EventType table:

   ```
   ALTER TABLE 'clubevents'
   ADD CONSTRAINT 'FK_Clubevents_EventType'
   FOREIGN KEY ('EventType') REFERENCES 'EventType'('ID');
   ```

4. Execute the SQL query by clicking the lightning bolt icon:

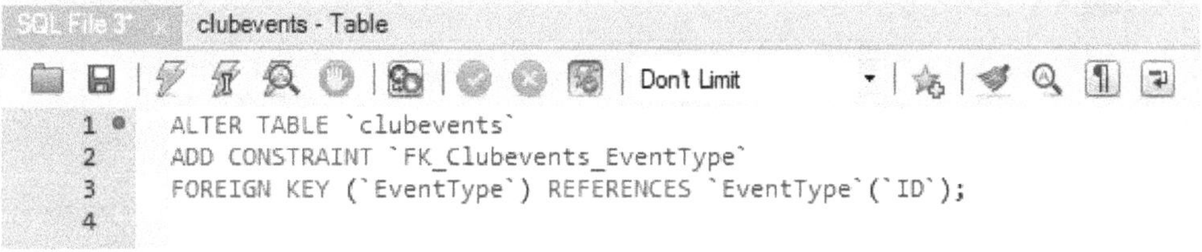

Figure 3.20 – The SQL code and the lightning bolt icon to run it

5. Refresh the **SCHEMA** panel, and you will see that the table now has a foreign key added to it.

6.   Open the **clubevents** table in design view to examine the table design, indexes, and foreign keys:

Figure 3.21 – Right-click the clubevents table and select Alter Table

The new foreign key will be visible, and the default values for the UPDATE and DELETE options of RESTRICT have been included automatically:

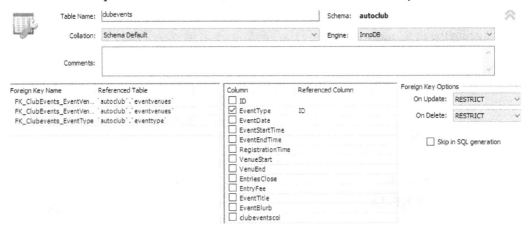

Figure 3.22 – The design view of the clubevents table

You will also see that the index is automatically created on the new foreign key field:

Index Name	Type
PRIMARY	PRIMARY
FK_ClubEvents_EventVenue_idx	INDEX
FK_ClubEvents_EventVenue_End_idx	INDEX
Idx_EventDate	INDEX
FK_Clubevents_EventType	INDEX

Index Columns

Column	#	Order	Length
☐ ID		ASC	
☑ EventType	1	ASC	
☐ EventDate		ASC	
☐ EventStartTime		ASC	
☐ EventEndTime		ASC	
☐ RegistrationTime		ASC	
☐ VenueStart		ASC	
☐ VenueEnd		ASC	
☐ EntriesClose		ASC	
☐ EntryFee		ASC	
☐ EventTitle		ASC	
☐ EventBlurb		ASC	
☐ clubeventscol		ASC	

Figure 3.23 – An index is created automatically on the new foreign key field

This exercise demonstrates how we can add indices and alter existing tables. This allows us to change tables if we ever need to accommodate different datasets.

Creating a database can be a large undertaking, and mistakes or omissions in the design can creep in. In some cases, you may not find them until you are much further into development or the database has gone live. Approaching the task carefully and systemically during the initial design and creation can allow you to identify issues early in development when it is simplest to correct them. However, when database objects, applications, or websites are developed using the database, fixing them may prove far more complex, as alterations can quickly stop applications from working or render database objects (such as views and stored procedures) inoperative.

The two issues we fixed here will be among the most common changes you need to make in your database or when migrating an existing one. However, always thoroughly investigate the possible effects of any change. Sometimes, changes can cause more issues, so don't just dive in and change something; investigate it first.

So far in this chapter, we have backed up our database due to impending changes, added some tables and fields, and corrected some issues, all with SQL commands only.

Now, we get to work with some data; after all, that's what tables are for, right? In the next section, we will be adding, modifying, and removing records from the database tables.

Adding data to a table

When adding data to a database through an application, you will usually add one record at a time, although you may string several additions together in a single script. The data can come from a system user, sensors on a production line, be scraped from a webpage, another computer, or any other method where it is possible to extract data for recording, and the computer applications behind all of these possible data entry sources will record it in a similar way.

You can add a record in the table using the following command:

```
INSERT INTO [TableName] ([field1], [field2],…, [fieldn]) VALUES
(Value1, Value2,…, Valuen)
```

It is also possible to insert data from another table into the current target. These types of queries will use a SELECT statement and work as shown here:

```
INSERT INTO [TableName] ([field1], [field2],…, [fieldn]) SELECT
[field1, field2,…, fieldn] FROM [tablename]
```

In the next exercise, you will add a record to a database table.

Exercise 3.07 – adding a single record to a members table

The autoclub table has its first official member and would like to add them to the database! In order to add the member, you will need to run an insert query into the members table. The following steps show how this can be done:

1. Open Workbench and click **My First Connection**. Log in if required.

2. Click the **Create a new SQL tab for executing queries** icon:

Figure 3.24 – Create a new SQL tab for executing queries

3. In the tab, enter the following SQL statement to add a record:

```
INSERT INTO members
('Surname','FirstName','DOB','JoinDate')
      SELECT "Bloggs","Frederick","1990/06/15","2020/01/15";
```

Let's break down the preceding SQL command:

```
INSERT INTO members
```

It tells the server that you want to insert record/s into the `members` table:

```
('Surname','FirstName','DOB','JoinDate')
```

This is a list of the fields to enter data into. The field names are enclosed in backticks and are separated by commas. Also, the entire field name list is enclosed in brackets:

```
SELECT "Bloggs","Frederick","1990/06/15","2020/01/15";
```

The `SELECT` commands tell the server to use the following data; each data item in this sample is passed in as a string, so they are enclosed in quotes. They are also separated by commas, and they are in the **exact** same order as the field names are listed.

4. Execute the SQL query by clicking the **Execute SQL** (lightning bolt) icon. You should get the following output:

Figure 3.25 – The Output pane at the bottom of the screen will display the query result

5.   To view the data in the table, right-click on the **members** table and click
Select Rows:

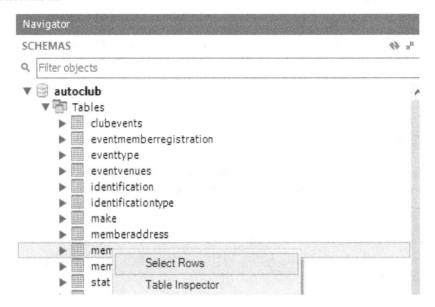

Figure 3.26 – Right-click on members and click Select Rows

You should get the following screen:

Figure 3.27 – The record has been added to the table

In the preceding figure, note that there are some fields that we have not included in
our SQL statement. This is because they are either set to default or can accept null
values or are incremented by default:

- `ID`: This is the primary key field. The Auto Increment option we have set will
cause this number to increase every time a record is inserted.

- `MiddleNames`: Not everyone has a middle name, so this field is set to allow nulls.

- `Signature and Photo`: We may not have these immediately when the records
are added, so they can accept a null value too.

- `Active`: This is 1 (`True`) by default. We expect that when a member is initially added to the database, they will be active.

- `InactiveDate`: This can accept a null value because this field may be used at a later date.

- `WhenAdded` and `LastModified`: When the record is first added, they will have the same date/time value set by their respective defaults. `WhenAdded` will never change from this value; however, `LastModified` will change each and every time the record is modified in any way.

Note that this same syntax can be used in exactly the same way when working with the MySQL command line.

Now it is your turn; in the next activity, you will insert a record into the `members` table.

Records in a database are often updated, as information often needs to be added or updated. In the next section, we will learn how to update a record.

Updating data in a record

A database is not meant to be totally static, unless, of course, it is an archive. Some information in the database will need to be changed at times or perhaps added to.

You can update a record using the following command:

```
UPDATE [tablename] SET [field1] = [Value1], … , [fieldn] =
[Valuen];
```

For example, if you had a table named `customers` and you wanted to set the active field to 0, you could use the following:

```
UPDATE customers SET active = 0;
```

In the next exercise, we will update a single record of a table.

Exercise 3.08 – updating a record

Fred Bloggs has informed the Automobile Club that he will no longer be retaining his membership. You are required to make him inactive in the database so that he doesn't receive invitations to club events.

To make Fred inactive in the database, perform the following steps:

1. Open Workbench and click **My First Connection**.

2. Click the **Create a new SQL tab for executing queries** icon:

Figure 3.28 – Create a new SQL tab for executing queries

A new tab will be created.

3. In the tab, enter the following SQL statement to make Fred inactive:

```
UPDATE members
SET
    active = 0,
    InactiveDate = CURRENT_DATE()
WHERE
    ID=1;
```

Let's break down the preceding SQL command:

- `UPDATE members` instructs the server to modify the `members` table.

- `SET` updates the following fields to the indicated values:

    - `active= 0` sets the active field to `0`.

    - `InactiveDate = CURRENT_DATE()` sets the `InactiveDate` field to the current date, and `CURRENT_DATE()` is a MySQL function that returns the current date.

- Multiple fields are separated by commas.

- `WHERE ID=1;` sets the preceding values only to records whose ID field has a value of `1`.

The `WHERE` clause in an SQL statement allows us to filter the records to specific criteria so that the actions will only affect those/that record(s). In this case, Fred's ID value is `1`, so only that record was affected. If `WHERE` was left out, all records would have been changed.

4. Execute the SQL query by clicking the **Execute SQL** (lightning bolt) icon:

```
1 ●    UPDATE members
2      SET
3          active  = 0,
4          InactiveDate = CURRENT_DATE()
5      WHERE
6          ID=1;
```

Figure 3.29 – The SQL code and the lightning bolt icon to run it

5. To view the data in the table, right-click on the **members** table and click **Select Rows**:

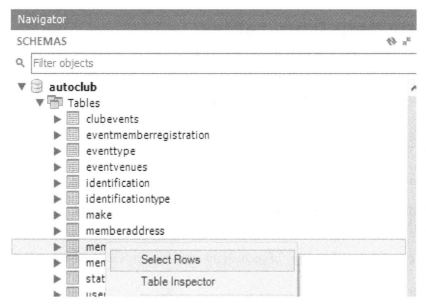

Figure 3.30 – Right-click on members and click Select Rows

You should get the following screen:

	ID	Surname	FirstName	MiddleNames	DOB	Signature	Photo	Active	JoinDate	InactiveDate	WhenAdded	LastModified
▶	1	Bloggs	Frederick	NULL	1990-06-16	NULL	NULL	0	2020-01-15	2020-01-21	2020-01-21 20:02:30	2020-01-21 21:41:03
	2	Pettit	Thomas	William	1960-10-15	NULL	NULL	1	2020-01-20	NULL	2020-01-21 21:01:56	2020-01-21 21:01:56
*	NULL	NULL	NULL	NULL	NULL	NULL	NULL	NULL	NULL	NULL	NULL	NULL

Figure 3.31 – Fred's record has been marked inactive and the inactive date has been updated

Note that the update query syntax is the same for the MySQL command line and can be used exactly as demonstrated in the exercise.

The most important part of updating data is to know which specific record or set of records you want to update, so you use the WHERE clause to limit the records to only those you need to perform the action on. The rest of the command is straightforward in that you specify the table to perform the action on, the fields, their new values, and the commas to separate them if there are more than one, and then you include the WHERE statement to filter. Using the ID field is often the safest approach, as it identifies a single record; however, you can use any other field or combination of fields you like in the WHERE statement, as long as you separate those commas. We will use multiple fields in the WHERE statement when we get to remove records from the database.

In the next section, we will learn about queries for deleting data from tables.

Deleting data from tables

As mentioned earlier, the DELETE statement removes rows from a table. This looks similar to the SELECT statement, but you don't specify a list of columns to return.

Consider the following example in which you first create a table named fruits using the following query:

```
CREATE TABLE fruits (id int primary key, fruit varchar(255));
```

Then, you insert 4 records into it:

```
INSERT INTO fruits VALUES (1, 'Apple'), (2, 'Pear'), (3,
'Orange'), (4, 'Carrot');
```

In order to check the total number of records inserted into the table, use the SELECT command:

```
SELECT * FROM fruits;
```

This will produce the following output:

```
+----+--------+
| id | fruit  |
+----+--------+
|  1 | Apple  |
|  2 | Pear   |
|  3 | Orange |
|  4 | Carrot |
+----+--------+
4 rows in set (0.01 sec)
```

Figure 3.32 – Records stored in the fruits table

Now, in order to delete a single record from a table, use the following command:

```
DELETE FROM fruits WHERE fruit='Carrot';
```

Here, you are deleting the record containing the Carrot fruit. To check whether the record has been successfully removed from the table, you use the SELECT command once again:

```
SELECT * FROM fruits;
```

This produces the following output:

```
+----+--------+
| id | fruit  |
+----+--------+
|  1 | Apple  |
|  2 | Pear   |
|  3 | Orange |
+----+--------+
3 rows in set (0.00 sec)
```

Figure 3.33 – The updated fruits table after deleting a single record

In the preceding example, you created a table called fruits and populated it with 4 items. Then, you used DELETE to remove one item.

Another very useful statement related to deleting data is TRUNCATE, which allows you to remove all data from a table. This is a very powerful command and should be used carefully. The statement looks like this:

```
TRUNCATE <table name>;
```

TRUNCATE will always delete all of the data from a table, so when using it, make sure to verify that all data should be deleted.

Drop queries

In addition to deleting data, it is also possible to delete tables and databases using the DROP query. The DROP query syntax is shown here:

```
DROP [database|table] [name]
```

It is important to note that this query will delete all data associated with the table or database it is targeted at. Only use this query if you are absolutely sure you want to delete the data.

In the next section, we will continue with updating records, but this time, we will be working with images.

Blobs, files, and file paths

When it comes to storing images and files in databases, there are two ways you can achieve this:

1. MySQL offers four `blob` data types of varying sizes that will store files and images in the database.

    This method is okay if you have small files and not too many records. Too many or large images can impact database performance, and developers often tend to avoid this method. Just because you can store an image in the database doesn't mean that you should.

2. You can set up a `VARCHAR(255)` field, in which you can store a file path and name pointing to a file or image stored somewhere on your network.

    This is the preferred method, especially for many or large files, and requires no messing around with the server settings. The application can read the path and name from the database and then load the file and do what it needs to with it – display it, transfer it, or whatever. However, it does require that the file storage should be organized. Losing the links or changing a file server's address can be challenging to recover from, so keep the backups happening, regularly.

> **Note**
> MySQL cannot display the image or file. It can only serve the image or file path to the client application.

In order to load files into MySQL, we will use a function called `LoadFile`. This function takes in the directory of a file, parses it, and uploads it for use as a blob.

In the next exercise, we will attempt to load a file into a blob field; the success will depend on your access to the upload folder and your specific server settings in relation to file access.

> **Note**
>
> If you find you cannot access the file, then simply move to *Exercise 3.11* where we will use the VARCHAR file path method.

Exercise 3.09 – files and blobs

Fred Bloggs has changed his mind and wishes to retain his membership and has had his photo taken and his signature digitized. Both are now image files to be put on file.

His photograph can be found at https://github.com/PacktWorkshops/ The-MySQL-Workshop/blob/master/Chapter03/Exercise09/ FredBloggs_Phtoto.jpg and his signature can be found at https://github. com/PacktWorkshops/The-MySQL-Workshop/blob/master/Chapter03/ Exercise09/FredBloggs_Signature.JPG.

You are required to update the members table with Fred's details. In this exercise, you will add an image to a blob field and also reinstate Fred's active status.

Perform the following steps:

1. Open a SQL tab.

2. Enter the following code to determine your file upload directory:

   ```
   SHOW VARIABLES LIKE "secure_file_priv";
   ```

   The server will return the secure file path if it is activated:

Figure 3.34 – The Value column is the MySQL server's secure file path

3. Remember, this location is on the computer where the MySQL server is installed. Test whether you have access to it by navigating to it in Windows Explorer.

4. Copy both the image files to the folder path found in secure_file_priv.

5. Now that you have determined you can access the folder and have saved the files, let's get on with it. Open an SQL tab and enter the following SQL statements. If the file path you received in *step 1* is different from that shown here, enter your path:

   ```
   UPDATE 'members'
   SET
           active  = 1,
   ```

```
            InactiveDate = NULL,
            Signature = LOAD_File('C:/ProgramData/MySQL/
MySQL Server 8.0/Uploads/Fred Bloggs_Signature.JPG'),
            Photo = LOAD_File('C:\\ProgramData\\MySQL\\
MySQL Server 8.0\\Uploads\\Fred Bloggs_Photo.JPG')
    WHERE   'ID'=1;
```

This updates the members table to set the signature of the user Fred equal to the image uploaded for their signature.

> **Note**
>
> MySQL will not accept a file path delimiter of \ as used in Windows. You need to change it to either \\ or /. Both are used in the preceding script for demonstration.

6. Execute the script using the lightning bolt icon:

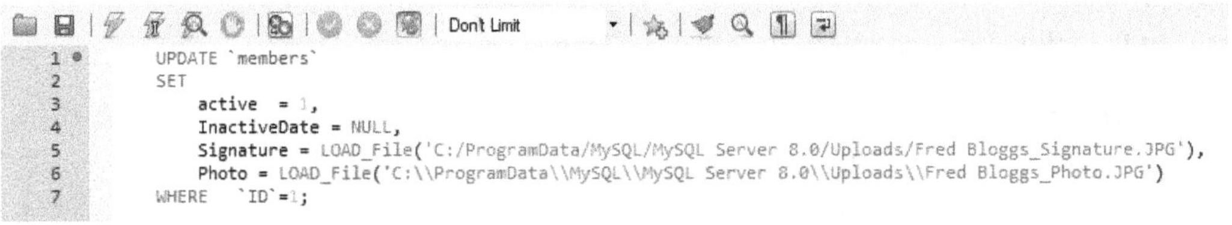

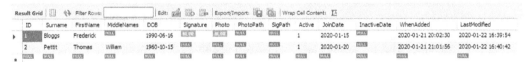

Figure 3.35 – Execute the script

7. To view the data in the table, right-click on the **members** table and click **Select Rows**.

   You should get the following output:

ID	Surname	FirstName	MiddleNames	DOB	Signature	Photo	PhotoPath	SigPath	Active	JoinDate	InactiveDate	WhenAdded	LastModified
1	Bloggs	Frederick	NULL	1990-06-16	BLOB	BLOB	NULL	NULL	1	2020-01-15	NULL	2020-01-21 20:02:30	2020-01-22 16:39:54
2	Pettit	Thomas	William	1960-10-15	NULL	NULL	NULL	NULL	1	2020-01-20	NULL	2020-01-21 21:01:56	2020-01-22 16:40:42
NULL	NULL	NULL	NULL	NULL	NULL	NULL	NULL	NULL	NULL	NULL	NULL	NULL	NULL

Figure 3.36 – View the results

   You should see the word **BLOB** in the **Signature** and **Photo** fields, indicating that there is a file stored in the field. Fred has been reactivated, and there is no inactive date.

The goal of this exercise was to introduce you to the blob and images or files. There are several things that make this method a little challenging to use, starting with the server setting, MySQL's access to folders, and more. If you do not have access to the server settings and your database administrator won't change them for security reasons, then you may not be able to use blobs to store files.

In the next section, we will look at a better method for file uploads, which is working with file paths directly in the MySQL database. This method will allow us to customize where a file is uploaded, which allows for better success and fewer permission-based issues.

Files and file paths

You can store an entire file path and name in a field; however, if the file repository's drive mappings or IP address changes, all files will need to be updated to reflect the new address. A popular method is to store the root address of the file repository in a lookup table in the database, which an application can look up and concatenate the path with the value stored in the table. A considerable advantage of this method is that the files can be as large as you like; you will only be limited by the capacity of your storage media.

The lookup table looks like the following:

Key	Value
ImageRepository	D:\FileRepository\

Figure 3.37 – The lookup table

The key indicates what type of data is stored in the folder. In this case, the directory is used to store images. The value is the location of the image store, which in this case is the `D:\FileRepository\` path.

And within the `FileRepository` folder, the files can be separated into folders, as shown in the following screenshot:

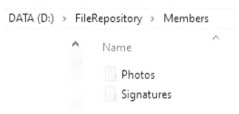

Figure 3.38 – FileRepository with the Members folder and subfolders

The images can then be stored in the subfolders, using a generic name with the relevant ID:

DATA (D:) › FileRepository › Members › Photos

MemberPhoto_1.
jpg

Figure 3.39 – A photo for member ID #1

And in the members table, you can store the value as shown here:

ID	PhotoPath
1	Members\Photos\MemberPhoto_1.jpg

Figure 3.40 – The members table

The application or an SQL query will then join the root with the stored path and name to get the full D:\FileRepository\Members\Photos\MemberPhoto_1.jpg file path, and should the repository ever need to move, maybe to a faster and bigger computer, the entire repository structure can be copied across. The ImageRepository value in the lookup table will then change to the new location.

To be able to use this path, we will need to take advantage of concatenation. This can be done using the CONCAT function in MySQL.

CONCAT is the MySQL command to concatenate or join two or more character strings, and the strings can be typed in directly to the SQL or retrieved from the database. The basic syntax is as follows:

```
CONCAT(String1, String2, String3,…)
```

You can include multiple strings. In the examples, we included two.

In our sample, String1 was extracted from the lookups table with an embedded SQL statement:

```
(SELECT 'Value' FROM 'lookups' WHERE 'Key'="ImageRepository")
```

When you embed an SQL statement within another, the embedded SQL must be enclosed in brackets; it will be executed separately and the result passed back to the primary SQL to be used in whatever the context is (based on its position in the primary SQL). In this case, the embedded SQL extracted the **image root folder**, and the primary SQL took that as `String1` in the `CONCAT` command.

In our sample, `String2` is the `'members'.'PhotoPath'` field; the value is extracted as part of the main query and used as `String2`.

At the end of the `CONCAT` command, we assigned it a field name to display the results with, so our command was as follows:

```
CONCAT(String 1,String 2) AS FullPhotoPath
```

In the next exercise, you will work with the file path method.

Exercise 3.10 – files and file paths

You are now asked to update Fred's details using the file path method. His photograph can be found at `https://github.com/PacktWorkshops/The-MySQL-Workshop/blob/master/Chapter03/Exercise10/MemberPhoto_1.jpg`, and his signature can be found at `https://github.com/PacktWorkshops/The-MySQL-Workshop/blob/master/Chapter03/Exercise10/MemberSignature_1.JPG`

You will first set up the file repository and then store the root path. Once the root path is stored, you will store the image path of Fred. The full path will then be extracted at the end of this exercise:

> **Note**
>
> You are dealing with images in this exercise. The files could easily be any other type of document.

1. Create the following file structure on your computer. You can place this on any disk drive on your computer. You will need to adjust the path accordingly in these steps:

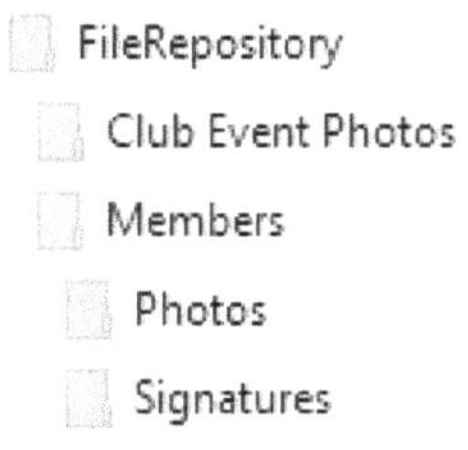

FileRepository
Club Event Photos
Members
Photos
Signatures

Figure 3.41 – The file repository structure

> **Note**
>
> Under the `FileRepository` folder, you can set up any number of subfolders to group your images and files accordingly.

Download the images from GitHub and copy `MemberPhoto_1.jpg` into the `Photos` folder and `MemberSignature_1.JPG` into the `Signatures` folder.

We are now set up to write SQL code to insert the files into the database table.

> **Note**
>
> In the real world, your application will deal with copying and renaming the images appropriately and generating the SQL to store the images, or calling a stored procedure that will store them.

2. Open an SQL tab to run the following query. If your `FileRepository` folder is in a different location to that shown here, enter your location:

```
INSERT INTO lookups
('Key','Value','Descriptions')
SELECT
"ImageRepository",
"D:\\FileRepository\\",
"Automobile Club images";
```

> **Note**
>
> The backslash (\) is an escape character in MySQL that tells it to ignore the meaning of the next character. So, if you want to include the backslash in the text, you need to enter it twice, as shown in the preceding query. If you want to include a double quote in the text, you will need to precede each quotation mark that is part of the text with a backslash in the SQL statement.
>
> Alternatively, you can replace the backslash (\) with a single forward slash (/).

This query has created a lookup that stores the image repository location. We can then use this to store and retrieve images.

3. Execute the SQL statement with the lightning bolt icon:

```
1 ●   INSERT INTO lookups
2     (`Key`,`Value`,`Descriptions`)
3     SELECT
4     "ImageRepository",
5     "D:\\FileRepository\\",
6     "Automobile Club images";
```

Figure 3.42 – Execute the code

4. Then, view the results in the lookups table. You should get the following output:

Figure 3.43 – The key, value, and description for the repository root folder

This shows that the values were successfully inserted into the lookups table.

5. Now, open an SQL tab to run a query and type the following SQL command to update the members table with Fred's details:

```
UPDATE 'members'
SET
SigPath = "Members\\Signatures\\MemberSignature_1.jpg",
PhotoPath = "Members\\Photos\\MemberPhoto_1.jpg"
WHERE    'ID'=1;
```

6. Execute the SQL query and view the results. You should get the following screen:

Figure 3.44 – The Photo and Signature text fields hold the path and image name
within the repository root folder

This shows that we now have values for the photo and signature path in the members table.

7. Now, extract the details from the database for Fred, including the image paths. Open a SQL tab to run the following query:

```
SELECT 'FirstName','Surname',
```

```
CONCAT((SELECT 'Value' FROM 'lookups' WHERE
'Key'="ImageRepository") , 'PhotoPath') AS FullPhotoPath,
CONCAT((SELECT 'Value' FROM 'lookups' WHERE
'Key'="ImageRepository") , 'SigPath') AS
FullSignaturePath
```

```
FROM 'members' WHERE 'members'.'ID'=1
```

8. Execute the SQL query, and the results should be as follows:

FirstName	Surname	FullPhotoPath	FullSignaturePath
Frederick	Bloggs	D:\FileRepository\Members\Photos\MemberPhoto_1.jpg	D:\FileRepository\Members\Signatures\MemberSignature_1.jpg

Figure 3.45 – Fred's names and both image paths

The application using this data can now access Fred's images and display them where it needs to.

Working with files, images, and your MySQL database can be a little tricky and requires some thought in setting up – that is, what approach you should take, and whether you should store them in the database or use file path pointers. That decision will be yours as the developer, although you may be required to choose one method over another due to business infrastructure and rules.

Where possible, try to avoid storing files in a database if you suspect they will be large or plentiful. If you elect to use the file pointer method used in *Exercise 3.11*, then put some thought into how you want to structure your repository.

In the next activity, you will add an image to your repository and update the database with the image path.

Activity 3.2 – adding image file paths to the database

One of the members with the surname Pettit has added himself as a member of the Automobile Club. You are asked to add the image of the new member in the members table using the file path method. You are also asked to fetch the full file path of the new image that is added.

Perform the following steps to achieve the goal of this activity:

1. Determine the ID of the member with the surname Pettit.

2. Download the image from https://github.com/PacktWorkshops/ The-MySQL-Workshop/tree/master/Chapter03/Activity02 and save it in the Member/Photos folder.

3. Now that you have your image in place, open another SQL tab and create a script to place the path and photo name in your member record.

4. Create and run another SQL query to extract the full file path for your image.

On successful completion of the activity, you should get the following output:

Figure 3.46 – The full file path for the image

> **Note**
> The solution to this activity can be found in the *Appendix*.

In this section, you learned about blobs, updating records with images, and file path pointers. You learned how to organize image and file storage and how to work with them with SQL statements; you also learned about embedded SQL statements and the CONCAT command to join strings together.

Dealing with images and files with a database is a widespread practice and a valuable skill to get your head around. It isn't difficult, and a little practice will serve you well.

Summary

In this chapter, you have learned how to back up your database and run an SQL script to restore the database. You have learned how to use SQL statements and scripts to create a database and tables, as well as how to modify tables, create indexes and foreign keys, and insert, update, and delete data. You also worked with images and files with your database.

In the next chapter, you will continue working with SQL statements. You will learn some more about SQL queries and how to create and use SQL queries with stored procedures, functions, and views.

4

Selecting, Aggregating, and Applying Functions

In this chapter, we cover different ways to get the information we need out of MySQL data. We will learn how to filter out the records and apply functions on the data – for example, to only return the first 15 characters of a field. We will then start to use GROUP BY to group rows and calculate results built on the groups. This is often used to sum all the values in a group or count how many items they are in one.

This chapter covers the following topics:

- An introduction to querying data
- Querying tables in MySQL
- Exercise 4.01 – simple queries
- Filtering results
- Exercise 4.02 – filtering results

- Using functions on data

- Exercise 4.03 – using functions

- Aggregating data

- Exercise 4.04 – aggregating data

- Case statements

- Exercise 4.05 – writing case statements

- Activity 4.01 – collecting information for a travel article

An introduction to querying data

In the previous chapter, we covered multiple ways of getting data into MySQL. We imported data in CSV, JSON, and SQL formats into tables and collections. Now, we want to use MySQL to get information out of the data. The main benefit of having data in a MySQL database is that you can query it, combine multiple tables, and aggregate and filter results. This makes it easy to create reports on the data. This is not limited to data stored in tables; it is still possible to do this if the data resides in a collection of JSON documents.

An example of this is having a database that stores an inventory of the laptops that the company has and then producing reports based on the different types of laptops and the different warranty periods.

In this chapter, you will learn how to filter the results – for example, filtering for only one brand of laptop. Then, you will learn how to use functions – for example, to calculate the days remaining in a warranty. Then, you will learn to summarize data by aggregating multiple rows.

Querying tables in MySQL

To get data out of MySQL, we use a SELECT query. A basic SELECT query has the following format:

```
SELECT <items> FROM <table>
```

Here, <items> can be many different kinds of things. It can be a wildcard (*) character, which returns all columns from a table, a list of columns, or even something that's not in the table at all but should still be in resultset – for example, a constant such as production or number. The FROM <table> part is optional, but it is there in most cases.

The SQL language is a declarative language, which means that the focus is more on the results that are obtained rather than how they are obtained. This is why we describe in a SELECT statement what the returned data should look like (for example, what fields it should have). We don't instruct the database to open the data file and navigate through the data structures. Based on the instructions we give for what the results should look like, the database server will figure out the best way to get this data to you.

Consider the following query:

```
SELECT * FROM city;
```

This query returns all the rows and columns from the city table.

Now, consider the following query:

```
SELECT * FROM city LIMIT 5;
```

This query limits the result set to 5 records from the city table. Note that the results will not be in any order because we did not specify one. This can return any 5 or fewer records from the cities table. If you test this out, you might notice that the records come up ordered by their primary key, but this can easily change with bigger tables and more complex tables, so we cannot rely on it.

A SELECT query with all columns and only a few rows is often a good way to see what data looks like if you are not familiar with a table. Consider the following query:

```
SELECT Name, Capital FROM country;
```

This query returns the Name and Capital columns from the country table.

Now, you will complete an exercise with some simple queries before continuing to practice filtering out rows in which you are interested.

Exercise 4.01 – working with simple queries

In this exercise, you will be using the `world` database. As a developer, you will often need to use languages in your applications. You can download the world database here: `https://downloads.mysql.com/docs/world-db.zip/`. You are told that languages are stored in the `countrylanguage` table of the database, but you are not aware whether the language is stored as a name or in the form of code. You will first inspect the table definition and then get a sample of the table. You need to make sure that you have the world database available; refer to the *Loading data from a SQL file* section in *Chapter 11, MS Excel VBA and MySQL* if you have questions. Follow these steps to complete this exercise:

1. Connect to MySQL with the CLI and the appropriate user.

2. Select the world database to be used:

   ```
   USE world;
   ```

   The current database will be changed to the `world` database, as you can see in the following figure:

   ```
   mysql> USE world
   Database changed
   ```

   Figure 4.1 – The USE output

3. Inspect the `countrylanguage` table definition by using the `DESCRIBE` command:

   ```
   DESCRIBE countrylanguage;
   ```

   This query returns the following table definition:

   ```
   mysql> DESCRIBE countrylanguage;
   +-------------+---------------+------+-----+---------+-------+
   | Field       | Type          | Null | Key | Default | Extra |
   +-------------+---------------+------+-----+---------+-------+
   | CountryCode | char(3)       | NO   | PRI |         |       |
   | Language    | char(30)      | NO   | PRI |         |       |
   | IsOfficial  | enum('T','F') | NO   |     | F       |       |
   | Percentage  | decimal(4,1)  | NO   |     | 0.0     |       |
   +-------------+---------------+------+-----+---------+-------+
   4 rows in set (0.01 sec)
   ```

   Figure 4.2 – The DESCRIBE output

The DESCRIBE command will display the available columns. To get languages, you need the Language column. However, this does not specify whether this is a language code or the name of the language, nor does it specify whether the name is in English or the native language (for example Spanish versus Español).

4. Obtain a sample of the table by writing the following query:

```
SELECT Language FROM countrylanguage LIMIT 5;
```

This produces the following output:

```
mysql> SELECT Language FROM countrylanguage LIMIT 5;
+------------+
| Language   |
+------------+
| Dutch      |
| English    |
| Papiamento |
| Spanish    |
| Balochi    |
+------------+
5 rows in set (0.01 sec)
```

Figure 4.3 – The SELECT output, limited to five records and the Language column

With this output, you now know that the languages are stored in name form (in English).

In this exercise, we inspected the countrylanguage table and used a sample of the Language column to learn what the data in this column looks like. In the next section, we will explore how to filter out the fetched results.

Filtering results

Often, the table or tables you are querying have many more rows than you are interested in. Filtering is done in two ways; the first way is only selecting the columns we need. This is what we did in the previous section. The second way is to filter out the rows; this is done with a WHERE clause in the SELECT statement. Besides only returning the data you need, this also allows the database server to use a more efficient way of retrieving the data, which translates to faster queries.

Consider the following query:

```
SELECT * FROM city WHERE CountryCode='CHE';
```

This query will return the following results:

```
mysql> SELECT * FROM city WHERE CountryCode='CHE';
+------+----------+-------------+-------------+------------+
| ID   | Name     | CountryCode | District    | Population |
+------+----------+-------------+-------------+------------+
| 3245 | Zürich   | CHE         | Zürich      |     336800 |
| 3246 | Geneve   | CHE         | Geneve      |     173500 |
| 3247 | Basel    | CHE         | Basel-Stadt |     166700 |
| 3248 | Bern     | CHE         | Bern        |     122700 |
| 3249 | Lausanne | CHE         | Vaud        |     114500 |
+------+----------+-------------+-------------+------------+
5 rows in set (0.00 sec)
```

Figure 4.4 – The SELECT output, filtered by CountryCode CHE for Switzerland

Here, you return all columns for rows that have CHE as CountryCode. Every row is a city in Switzerland. Now, consider the following example:

```
SELECT Name, Population FROM country
WHERE Continent='Oceania' AND Population > 1000000;
```

Here, you filter out countries in the continent of Oceania that have a population of more than 1000000. The > operator checks whether the value on the left is bigger than the value on the right. Other similar operators are inequalities, such as < for less than, > for greater than, <= for less than or equal, and >= for greater than or equal. We also have operations such as = for equality, * for multiply, and / for divide.

This query produces the following output:

```
mysql> SELECT Name, Population FROM country WHERE Continent='Oceania'
    -> AND Population > 1000000;
+------------------+------------+
| Name             | Population |
+------------------+------------+
| Australia        |   18886000 |
| New Zealand      |    3862000 |
| Papua New Guinea |    4807000 |
+------------------+------------+
3 rows in set (0.01 sec)
```

Figure 4.5 – The SELECT output, filtered on Oceania and > 1000000 population

As you can see in the preceding screenshot, the query returns only the name and population columns from the country table that are both in the `Oceania` continent and have a population of more than `1000000`. Here, you can see that we combine two filters with the `AND` keyword. It is also possible to use the `OR` keyword to match multiple filters – for example, `Continent='Oceania' OR Continent='Europe'`. Consider the following query:

```
SELECT Name FROM country WHERE Name LIKE 'United %';
```

This query will return the following results:

```
mysql> SELECT Name FROM country WHERE Name LIKE 'United %';
+----------------------------------------+
| Name                                   |
+----------------------------------------+
| United Arab Emirates                   |
| United Kingdom                         |
| United States                          |
| United States Minor Outlying Islands   |
+----------------------------------------+
4 rows in set (0.01 sec)
```

Figure 4.6 – The SELECT output for countries that start with United

The result set from this query will have only one column. It returns all countries from the `country` table that start with the word `United`. In SQL, `%` is a wildcard for one or more characters and `_` is a wildcard for a single character. Other languages often use `*` and `.` for this.

Note

MySQL also has a whole range of features and syntax for more advanced text matching and full-text indexing, but we won't cover that here.

If you are combining `OR` and the `WHERE` clause of the query, then you might need to group operations. Consider the following examples:

```
SELECT * FROM city WHERE District='New York' OR District='New
Jersey'
AND Population>100000;
SELECT * FROM city WHERE (District='New York' OR District='New
Jersey')
AND Population>100000;
```

The preceding queries will return the following results:

```
mysql> SELECT * FROM city WHERE District='New York' OR District='New Jersey'
    -> AND Population>100000;
+------+-------------+-------------+------------+------------+
| ID   | Name        | CountryCode | District   | Population |
+------+-------------+-------------+------------+------------+
| 3793 | New York    | USA         | New York   |    8008278 |
| 3850 | Buffalo     | USA         | New York   |     292648 |
| 3855 | Newark      | USA         | New Jersey |     273546 |
| 3864 | Jersey City | USA         | New Jersey |     240055 |
| 3871 | Rochester   | USA         | New York   |     219773 |
| 3888 | Yonkers     | USA         | New York   |     196086 |
| 3932 | Paterson    | USA         | New Jersey |     149222 |
| 3935 | Syracuse    | USA         | New York   |     147306 |
| 3978 | Elizabeth   | USA         | New Jersey |     120568 |
| 4048 | Albany      | USA         | New York   |      93994 |
+------+-------------+-------------+------------+------------+
10 rows in set (0.01 sec)

mysql> SELECT * FROM city WHERE (District='New York' OR District='New Jersey')
    -> AND Population>100000;
+------+-------------+-------------+------------+------------+
| ID   | Name        | CountryCode | District   | Population |
+------+-------------+-------------+------------+------------+
| 3793 | New York    | USA         | New York   |    8008278 |
| 3850 | Buffalo     | USA         | New York   |     292648 |
| 3855 | Newark      | USA         | New Jersey |     273546 |
| 3864 | Jersey City | USA         | New Jersey |     240055 |
| 3871 | Rochester   | USA         | New York   |     219773 |
| 3888 | Yonkers     | USA         | New York   |     196086 |
| 3932 | Paterson    | USA         | New Jersey |     149222 |
| 3935 | Syracuse    | USA         | New York   |     147306 |
| 3978 | Elizabeth   | USA         | New Jersey |     120568 |
+------+-------------+-------------+------------+------------+
9 rows in set (0.01 sec)
```

Figure 4.7 – The SELECT output, demonstrating group filters

Note the parenthesis in the preceding figure. Without this, you would return all cities in New York and all cities in New Jersey that have a population of more than 100000. With the parenthesis, you return cities in both New York and New Jersey that have a population of more than 100000. In the next section, you will solve an exercise in which you will be filtering the results.

Exercise 4.02 – filtering results

Imagine that you are working for a TV station. For an item about Western Europe, you need to get the surface area from the database. In this exercise, you will connect to the world database, get the table definition, and filter on the Western Europe region. Follow these steps to complete this exercise:

1. Connect to MySQL with the CLI and the appropriate user.

2. Select the world database to work with:

```
USE world;
```

The preceding query connects to the `world` database:

```
mysql> USE world
Database changed
```

Figure 4.8 – The USE output

3. Inspect the table definition with the following query:

```
DESCRIBE country;
```

The preceding query produces the following output:

```
+----------------+------------------------------------------------------------------------------------------+------+-----+
| Field          | Type                                                                                     | Null | Key |
| Default | Extra |
+----------------+------------------------------------------------------------------------------------------+------+-----+
| Code           | char(3)                                                                                  | NO   | PRI |
|         |       |
| Name           | char(52)                                                                                 | NO   | MUL |
|         |       |
| Continent      | enum('Asia','Europe','North America','Africa','Oceania','Antarctica','South America')    | NO   |     |
| Asia    |       |
| Region         | char(26)                                                                                 | NO   |     |
|         |       |
| SurfaceArea    | decimal(10,2)                                                                            | NO   |     |
| 0.00    |       |
| IndepYear      | smallint                                                                                 | YES  |     |
| NULL    |       |
| Population     | int                                                                                      | NO   |     |
| 0       |       |
| LifeExpectancy | decimal(3,1)                                                                             | YES  |     |
| NULL    |       |
| GNP            | decimal(10,2)                                                                            | YES  |     |
| NULL    |       |
| GNPOld         | decimal(10,2)                                                                            | YES  |     |
| NULL    |       |
| LocalName      | char(45)                                                                                 | NO   |     |
|         |       |
| GovernmentForm | char(45)                                                                                 | NO   |     |
|         |       |
| HeadOfState    | char(60)                                                                                 | YES  |     |
| NULL    |       |
| Capital        | int                                                                                      | YES  |     |
| NULL    |       |
| Code2          | char(2)                                                                                  | NO   |     |
|         |       |
+----------------+------------------------------------------------------------------------------------------+------+-----+
```

Figure 4.9 – The DESCRIBE output for the country table

> **Note**
>
> The output is very wide because of the definition of the `Continent` column. You can use `\G` instead of `;` at the end of the statement to return output in a vertical format.

4. Filter on the `Western Europe` region with the following query:

```
SELECT Name, SurfaceArea FROM country WHERE
Region='Western Europe';
```

The preceding query produces the following results:

```
mysql> SELECT Name, SurfaceArea FROM country WHERE Region='Western Europe';
+---------------+-------------+
| Name          | SurfaceArea |
+---------------+-------------+
| Austria       |    83859.00 |
| Belgium       |    30518.00 |
| Switzerland   |    41284.00 |
| Germany       |   357022.00 |
| France        |   551500.00 |
| Liechtenstein |      160.00 |
| Luxembourg    |     2586.00 |
| Monaco        |        1.50 |
| Netherlands   |    41526.00 |
+---------------+-------------+
9 rows in set (0.00 sec)
```

Figure 4.10 – The SELECT output for countries in Western Europe

In this exercise, you filtered out two columns of the `country` table and only returned records that match the `Western Europe` region. In the next section, we will explore functions.

Using functions on data

MySQL comes with a big list of functions to work with all the common data types. In addition to this, it also allows you to create your own functions in SQL, C, or C++. This can help you to filter data based on specific conditions and format it. The following sections will detail some of these functions.

Math functions

These are +, -, and / to add, subtract, and divide. In addition to that, there are also functions such as `FLOOR()`, `CEILING()`, `POWER()`, `ROUND()`, and quite a few more to help you do calculations on numerical data that you have in the database. Consider the following query:

```
SELECT 1 + 2, 10 - 11, 1 / 3, POW(2, 3), ROUND(1/3, 1),
CEILING(0.9);
```

This query produces the following results:

```
mysql> SELECT 1 + 2, 10 - 11, 1 / 3, POW(2, 3), ROUND(1/3, 1), CEILING(0.9);
+-------+---------+--------+-----------+---------------+--------------+
| 1 + 2 | 10 - 11 | 1 / 3  | POW(2, 3) | ROUND(1/3, 1) | CEILING(0.9) |
+-------+---------+--------+-----------+---------------+--------------+
|     3 |      -1 | 0.3333 |         8 |           0.3 |            1 |
+-------+---------+--------+-----------+---------------+--------------+
1 row in set (0.02 sec)
```

Figure 4.11 – The SELECT output with a demonstration of mathematical functions in MySQL

In the figure, we see the following observations:

- `1 + 2` is an addition and will return 3.

- `10 - 11` is a subtraction and returns `-1`.

- `POW(2, 3)` is 2 to the power of 3, which will return 8.

- `ROUND(1/3, 1)` is `0.333333` but rounded down to one number after the period(`.`).

- `CEILING(0.9)` rounds up to the next integer, which is 1.

This query doesn't use any table or collections and returns a single row with all the results of the calculations. Now, let's look at another query:

```
SELECT
    Name,
    ROUND(Population/1000000,1) AS 'Population (Million)'
FROM city
WHERE CountryCode='MEX' AND Population>1000000;
```

This query returns the following results:

```
mysql> SELECT
    ->    Name,
    ->    ROUND(Population/1000000,1) AS 'Population (Million)'
    -> FROM city
    -> WHERE CountryCode='MEX' AND Population>1000000;
+---------------------+----------------------+
| Name                | Population (Million) |
+---------------------+----------------------+
| Ciudad de México    |                  8.6 |
| Guadalajara         |                  1.6 |
| Ecatepec de Morelos |                  1.6 |
| Puebla              |                  1.3 |
| Nezahualcóyotl      |                  1.2 |
| Juárez              |                  1.2 |
| Tijuana             |                  1.2 |
| León                |                  1.1 |
| Monterrey           |                  1.1 |
| Zapopan             |                  1.0 |
+---------------------+----------------------+
10 rows in set (0.01 sec)
```

Figure 4.12 – The SELECT output with a demonstration of using a function on data from a table

Here, you are listing big cities in Mexico and showing the population number in millions, formatted to only show one digit after the decimal point. Here, the calculation is done for every row returned by the query. Let's now look at string functions.

String functions

To cut a string at a specific character, you can use the LEFT() function. Consider the following query:

```
use world;
SELECT Name FROM city WHERE LEFT(Name, 3) = 'New';
```

This returns all the cities that have New as the first three letters of their name:

```
mysql> SELECT Name FROM city WHERE LEFT(Name, 3) = 'New';
+---------------------+
| Name                |
+---------------------+
| Newcastle           |
| Newcastle upon Tyne |
| Newport             |
| Newcastle           |
| New Bombay          |
| New Delhi           |
| New York            |
| New Orleans         |
| Newark              |
| Newport News        |
| New Haven           |
| New Bedford         |
+---------------------+
12 rows in set (0.00 sec)
```

Figure 4.13 – The SELECT output with LEFT()

Another way of splitting a string is by using the SUBSTRING_INDEX() function, which you can see in the next example:

```
use sakila;
SELECT
  email,
  SUBSTRING_INDEX(email, "@", 1),
  SUBSTRING_INDEX(email, "@", -1)
FROM customer
WHERE store_id=1 AND active=0;
```

This splits the email address on @ and returns the user and domain parts in different columns:

```
mysql> SELECT
    -> email,
    -> SUBSTRING_INDEX(email, "@", 1),
    -> SUBSTRING_INDEX(email, "@", -1)
    -> FROM customer
    -> WHERE store_id=1 AND active=0;
+---------------------------------+--------------------------------+---------------------------------+
| email                           | SUBSTRING_INDEX(email, "@", 1) | SUBSTRING_INDEX(email, "@", -1) |
+---------------------------------+--------------------------------+---------------------------------+
| SHEILA.WELLS@sakilacustomer.org | SHEILA.WELLS                   | sakilacustomer.org              |
| PENNY.NEAL@sakilacustomer.org   | PENNY.NEAL                     | sakilacustomer.org              |
| HARRY.ARCE@sakilacustomer.org   | HARRY.ARCE                     | sakilacustomer.org              |
| NATHAN.RUNYON@sakilacustomer.org| NATHAN.RUNYON                  | sakilacustomer.org              |
| MAURICE.CRAWLEY@sakilacustomer.org | MAURICE.CRAWLEY             | sakilacustomer.org              |
| CHRISTIAN.JUNG@sakilacustomer.org | CHRISTIAN.JUNG               | sakilacustomer.org              |
| JIMMIE.EGGLESTON@sakilacustomer.org | JIMMIE.EGGLESTON           | sakilacustomer.org              |
| TERRANCE.ROUSH@sakilacustomer.org | TERRANCE.ROUSH               | sakilacustomer.org              |
+---------------------------------+--------------------------------+---------------------------------+
8 rows in set (0.00 sec)
```

Figure 4.14 – The SELECT output with SUBSTRING_INDEX()

To determine the length of a string, you can use either LENGTH() or CHAR_LENGTH(). They often return the same value, as LENGTH returns the length in bytes and CHAR_LENGTH() returns the length in characters. Characters can vary in length based on encoding. For example, characters such as English alphabet characters are 1 byte long, and Unicode characters are 2 bytes long.

Consider another query:

```
SELECT LENGTH('Café'), CHAR_LENGTH('Café');
```

This will return 5 for the length in bytes and 4 for the length in characters. This is because é is two bytes:

```
mysql> SELECT LENGTH('Café'), CHAR_LENGTH('Café');
+-----------------+----------------------+
| LENGTH('Café')  | CHAR_LENGTH('Café')  |
+-----------------+----------------------+
|               5 |                    4 |
+-----------------+----------------------+
1 row in set (0.00 sec)
```

Figure 4.15 – The SELECT output with LENGTH() and CHAR_LENGTH()

Other useful functions are UPPER() and LOWER(), which make a string uppercase or lowercase respectively, and CONCAT(), which concatenates strings together. In other databases, you might have used || to concatenate data, but that doesn't work in MySQL because, by default, it is a synonym for OR. In the next section, we will explore date and time functions.

Date and time functions

The first set of functions gets the current time, date, or timestamp:

- CURRENT_TIME
- CURRENT_DATE
- CURRENT_TIMESTAMP
- NOW()

For NOW(), parentheses are required; for the other functions, this is optional. The functions that deal with time accept fractional seconds part (FSP) as an argument. This allows you to specify the precision of the time. By default, FSP is 0, which means precision in seconds. The maximum is 6, which means microsecond precision with 6 digits.

Consider the following examples:

```
SELECT CURRENT_TIME(), CURRENT_DATE(), CURRENT_TIMESTAMP(),
NOW();
```

```
SELECT CURRENT_TIME(6), CURRENT_DATE(), CURRENT_TIMESTAMP(6),
NOW(6);
```

These queries return the following results:

```
mysql> SELECT CURRENT_TIME(), CURRENT_DATE(), CURRENT_TIMESTAMP(), NOW();
+----------------+----------------+---------------------+---------------------+
| CURRENT_TIME() | CURRENT_DATE() | CURRENT_TIMESTAMP() | NOW()               |
+----------------+----------------+---------------------+---------------------+
| 15:30:03       | 2020-02-01     | 2020-02-01 15:30:03 | 2020-02-01 15:30:03 |
+----------------+----------------+---------------------+---------------------+
1 row in set (0.00 sec)

mysql> SELECT CURRENT_TIME(6), CURRENT_DATE(), CURRENT_TIMESTAMP(6), NOW(6);
+-----------------+----------------+----------------------------+----------------------------+
| CURRENT_TIME(6) | CURRENT_DATE() | CURRENT_TIMESTAMP(6)       | NOW(6)                     |
+-----------------+----------------+----------------------------+----------------------------+
| 15:30:04.506387 | 2020-02-01     | 2020-02-01 15:30:04.506387 | 2020-02-01 15:30:04.506387 |
+-----------------+----------------+----------------------------+----------------------------+
1 row in set (0.00 sec)
```

Figure 4.16 – The SELECT output with date/time functions

The second set of functions is for adding and subtracting dates and times. To do this, you use DATE_ADD() and DATE_SUB(). Both take a date as the first argument and INTERVAL as the second argument. An interval looks like INTERVAL <number> <unit>. The unit is always singular, even when the number is more than 1 – for example, INTERVAL 5 DAY.

Consider the following query:

```
SELECT DATE_ADD('2010-01-01', INTERVAL 1 YEAR);
```

This returns 2011-01-01 because that's one year after 2010-01-01:

```
mysql> SELECT DATE_ADD('2010-01-01', INTERVAL 1 YEAR);
+-----------------------------------------+
| DATE_ADD('2010-01-01', INTERVAL 1 YEAR) |
+-----------------------------------------+
| 2011-01-01                              |
+-----------------------------------------+
1 row in set (0.00 sec)
```

Figure 4.17 – The SELECT output with a calculated date field

Some systems use Unix timestamps (the number of seconds since January 1, 1970). This is often done to prevent timezone-related issues, as the Unix timestamp is always stored in the same timezone. With `FROM_UNIXTIME()`, you can convert a Unix timestamp to a timestamp, and with `UNIX_TIMESTAMP()`, you can do the opposite.

Consider the following example:

```
SELECT UNIX_TIMESTAMP('2030-01-01 00:00:00'), FROM_
UNIXTIME(1573846979);
```

This query outputs the following results:

```
mysql> SELECT UNIX_TIMESTAMP('2030-01-01 00:00:00'), FROM_UNIXTIME(1573846979);
+-------------------------------------+---------------------------+
| UNIX_TIMESTAMP('2030-01-01 00:00:00') | FROM_UNIXTIME(1573846979) |
+-------------------------------------+---------------------------+
|                          1893452400 | 2019-11-15 20:42:59       |
+-------------------------------------+---------------------------+
1 row in set (0.00 sec)
```

Figure 4.18 – The SELECT output with the Unix timestamp conversion

We use various functions to modify the data returned from tables or things such as the current time. For example, you can convert timestamps to a human-readable format if they are stored as Unix timestamps, which then also allows you to calculate how far in the future or past that timestamp is. In the next section, you will complete an exercise based on these functions.

Exercise 4.03 – using functions

In this exercise, you will use the `world` database again. For a news article related to countries' independence, you want to compile a list of countries that have been independent for more than 1,000 years. For this, you need the following:

- The independence year
- The number of years since independence
- The population in millions
- The average population per square km (rounded to integers)

You will connect to the `world` database, select the columns you need, and apply a condition to find countries that are more than 1,000 years old. Then, you will add calculated columns and convert values where needed. Follow these steps to complete this exercise:

1. Connect to MySQL with Workbench and the appropriate user.

2. Make sure that you are using the `world` database:

   ```
   USE world;
   ```

   This query provides the following results:

   ```
   mysql> USE world
   Database changed
   ```

   Figure 4.19 – The USE output

3. Select the columns we need and apply the condition for countries that have been independent for more than 1,000 years by writing the following query:

   ```
   SELECT Name, IndepYear, Population, SurfaceArea FROM
   country
   WHERE YEAR(NOW()) - IndepYear > 1000;
   ```

   This outputs the following results:

   ```
   mysql> SELECT Name, IndepYear, Population, SurfaceArea FROM country
       -> WHERE YEAR(NOW()) - IndepYear > 1000;
   +------------+-----------+------------+-------------+
   | Name       | IndepYear | Population | SurfaceArea |
   +------------+-----------+------------+-------------+
   | China      |     -1523 | 1277558000 |  9572900.00 |
   | Denmark    |       800 |    5330000 |    43094.00 |
   | Ethiopia   |     -1000 |   62565000 |  1104300.00 |
   | France     |       843 |   59225700 |   551500.00 |
   | Japan      |      -660 |  126714000 |   377829.00 |
   | San Marino |       885 |      27000 |       61.00 |
   | Sweden     |       836 |    8861400 |   449964.00 |
   +------------+-----------+------------+-------------+
   7 rows in set (0.00 sec)
   ```

Figure 4.20 – The SELECT output with countries that have been independent for more than 1,000 years

You now have the data, but you need to do some calculations and transformations.

4. Add calculated columns. A calculated column is a column where you take the raw data from MySQL and transform it by using a function. Here, you divide `Population` by `SurfaceArea` and then use the `ROUND()` function to round it down to 0 numbers after `period` (.). You also use the `YEAR()` function on the value returned from `NOW()` to get the year out of the current timestamp, and then you subtract the independence year of the country to reach the value you need. Write the following query to achieve this:

```
SELECT
    Name,
    IndepYear,
    YEAR(NOW()) - IndepYear,
    Population,
    ROUND(Population/SurfaceArea,0)
FROM country
WHERE YEAR(NOW()) - IndepYear > 1000;
```

This outputs the following results:

```
mysql> SELECT
    ->     Name,
    ->     IndepYear,
    ->     YEAR(NOW()) - IndepYear,
    ->     Population,
    ->     ROUND(Population/SurfaceArea,0)
    -> FROM country
    -> WHERE YEAR(NOW()) - IndepYear > 1000;
+------------+-----------+-------------------------+------------+--------------------------------+
| Name       | IndepYear | YEAR(NOW()) - IndepYear | Population | ROUND(Population/SurfaceArea,0) |
+------------+-----------+-------------------------+------------+--------------------------------+
| China      |     -1523 |                    3543 | 1277558000 |                            133 |
| Denmark    |       800 |                    1220 |    5330000 |                            124 |
| Ethiopia   |     -1000 |                    3020 |   62565000 |                             57 |
| France     |       843 |                    1177 |   59225700 |                            107 |
| Japan      |      -660 |                    2680 |  126714000 |                            335 |
| San Marino |       885 |                    1135 |      27000 |                            443 |
| Sweden     |       836 |                    1184 |    8861400 |                             20 |
+------------+-----------+-------------------------+------------+--------------------------------+
7 rows in set (0.00 sec)
```

Figure 4.21 – The SELECT output with calculated columns

You now have the number of years since independence and the average population per square km, but you need to convert the population to millions.

5. Convert the values where needed. Divide `Population` by `1000000` and round it down to 0 decimals:

```
SELECT
    Name,
    IndepYear,
```

```
    YEAR(NOW()) - IndepYear,
    ROUND(Population / 1000000, 0),
    ROUND(Population/SurfaceArea,0)
FROM country
WHERE YEAR(NOW()) - IndepYear > 1000;
```

This outputs the following results:

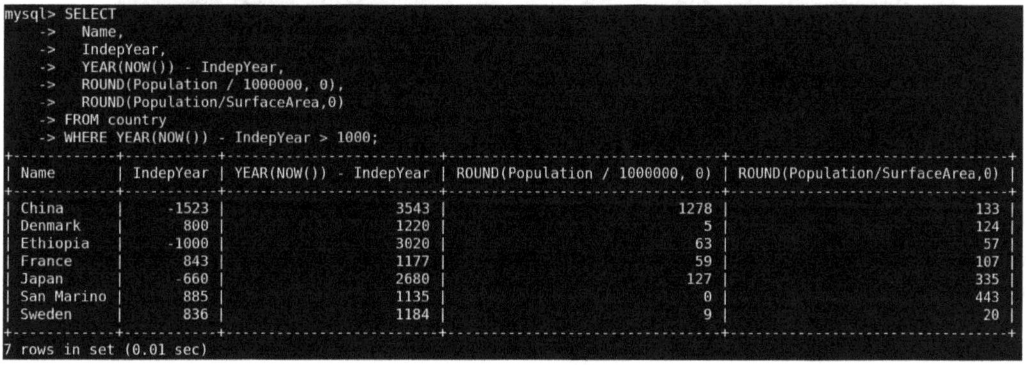

```
mysql> SELECT
    ->    Name,
    ->    IndepYear,
    ->    YEAR(NOW()) - IndepYear,
    ->    ROUND(Population / 1000000, 0),
    ->    ROUND(Population/SurfaceArea,0)
    -> FROM country
    -> WHERE YEAR(NOW()) - IndepYear > 1000;
+------------+-----------+------------------------+-------------------------------+---------------------------------+
| Name       | IndepYear | YEAR(NOW()) - IndepYear | ROUND(Population / 1000000, 0) | ROUND(Population/SurfaceArea,0) |
+------------+-----------+------------------------+-------------------------------+---------------------------------+
| China      |     -1523 |                   3543 |                          1278 |                             133 |
| Denmark    |       800 |                   1220 |                             5 |                             124 |
| Ethiopia   |     -1000 |                   3020 |                            63 |                              57 |
| France     |       843 |                   1177 |                            59 |                             107 |
| Japan      |      -660 |                   2680 |                           127 |                             335 |
| San Marino |       885 |                   1135 |                             0 |                             443 |
| Sweden     |       836 |                   1184 |                             9 |                              20 |
+------------+-----------+------------------------+-------------------------------+---------------------------------+
7 rows in set (0.01 sec)
```

Figure 4.22 – The SELECT output with the final result

In this exercise, you performed calculations and transformations on MySQL data and filtered rows based on this. The exercise used YEAR(), NOW(), ROUND(), and / to do this. In the next section, you will learn about aggregating data.

Aggregating data

This is one of the most powerful aspects of the SQL language. To do this, we use the GROUP BY clause in a SELECT statement. This groups one or more rows together and reports values based on this group. MySQL has many functions that operate on a group of rows, one of which is MAX(), which gets the maximum value from the group. It is important to only ever use the columns on which you are grouping by and/or other columns with an aggregate function.

Consider this data in the following table:

employee_id	region	city	sales
1	EMEA	London	300,000
2	EMEA	Milan	250,000
3	APAC	Singapore	350,000
4	APAC	Jakarta	100,000

Figure 4.23 – The sales table

Consider the following query:

```
SELECT region, SUM(sales) FROM sales GROUP BY region;
```

This outputs the following results:

```
mysql> SELECT region, SUM(sales) FROM sales GROUP BY region;
+--------+------------+
| region | SUM(sales) |
+--------+------------+
| EMEA   |     550000 |
| APAC   |     450000 |
+--------+------------+
2 rows in set (0.00 sec)
```

Figure 4.24 – The SELECT output, demonstrating GROUP BY

This groups the rows by region, creating two groups, and then it sums the rows in each group.

Now, consider this query:

```
SELECT city, SUM(sales) FROM sales GROUP BY region;
```

This outputs the following results:

```
mysql> SELECT city, SUM(sales) FROM sales GROUP BY region;
ERROR 1055 (42000): Expression #1 of SELECT list is not in GROUP BY clause and c
ontains nonaggregated column 'test.sales.city' which is not functionally depende
nt on columns in GROUP BY clause; this is incompatible with sql_mode=only_full_g
roup_by
```

Figure 4.25 – The SELECT output, demonstrating GROUP BY with ERROR 1055

This is similar to the previous query, but here, we select the `city` column, while grouping on the region. Older versions of MySQL will be performing this by default, thereby allowing you to do this. The result is two groups, one for each region. Then, it picks a more or less random city from the group to give you a result for the region column you asked for. Newer versions of MySQL, by default, set `ONLY_FULL_GROUP_BY`, and this will cause the query to end with this error:

```
ERROR: 1055: Expression #1 of SELECT list is not in GROUP BY
clause and contains nonaggregated column 'test.sales.city'
which is not functionally dependent on columns in GROUP BY
clause; this is incompatible with sql_mode=only_full_group_by
```

Almost always, this is done by accident. Here, you probably wanted to group by `city` instead of region, like this:

```
SELECT city, SUM(sales) FROM sales GROUP BY city;
```

This outputs the following results:

```
mysql> SELECT city, SUM(sales) FROM sales GROUP BY city;
+-----------+------------+
| city      | SUM(sales) |
+-----------+------------+
| London    |     300000 |
| Milan     |     250000 |
| Singapore |     350000 |
| Jakarta   |     100000 |
+-----------+------------+
4 rows in set (0.00 sec)
```

Figure 4.26 – The SELECT output, GROUP BY with SUM()

This creates four groups with one record in each group for this data. But if the table was bigger, it might have had groups of multiple records.

Now, let's go over a few commonly used functions' aggregations:

- `SUM()`: Sum all rows in the group.
- `MAX()` and `MIN()`: Pick the highest or lowest value.
- `COUNT()`: Return how many records we have in the group.
- `AVG()`: Return the average of the values in the group.
- `GROUP_CONCAT()`: Concatenate (join) all values from the group together.

It is also possible to filter which groups you want in your result with the `HAVING` keyword. Consider the following query:

```
SELECT region, AVG(sales)
FROM sales
GROUP BY region
HAVING AVG(sales) > 230000;
```

This outputs the following results:

```
mysql> SELECT region, AVG(sales)
    -> FROM sales
    -> GROUP BY region
    -> HAVING AVG(sales) > 230000;
+--------+------------+
| region | AVG(sales) |
+--------+------------+
| EMEA   | 275000.0000 |
+--------+------------+
1 row in set (0.00 sec)
```

Figure 4.27 – The SELECT output, showing GROUP BY with a HAVING clause

Here, you again have two groups for the two regions, but only one region matches the filter on the average sales numbers, resulting in only one result from this query. In this case, you need to use WHERE to filter on non-aggregated data and use HAVING to filter on data that is aggregated.

The COUNT function is usually used with * as an argument to work on the whole group. Another common thing to do is to use the optional DISTINCT keyword, such as COUNT(DISTINCT city). This will return the number of unique cities in the group. In the next section, you will solve an exercise based on aggregating data.

Exercise 4.04 – aggregating data

In this exercise, you will utilize the world database once again. You need some data about continents and regions for another news article, including the following information for each continent:

- The surface area for all the countries in that continent combined

- The average GNP per continent

- The total surface area per region for Asia

You will first connect to the world database, get the per continent data, and then get the per region data. Perform the following steps to complete this exercise:

1. Connect to the MySQL shell with Workbench and the appropriate user.

2. Make sure that you are using the world database:

```
USE world;
```

This outputs the following results:

```
mysql> USE world
Database changed
```

Figure 4.28 – The USE output

3. Obtain the data about each continent by writing the following query:

```
SELECT Continent, AVG(GNP), SUM(SurfaceArea)
FROM country GROUP BY Continent;
```

This outputs the following results:

```
mysql> SELECT Continent, AVG(GNP), SUM(SurfaceArea)
    -> FROM country GROUP BY Continent;
+---------------+---------------+-----------------+
| Continent     | AVG(GNP)      | SUM(SurfaceArea) |
+---------------+---------------+-----------------+
| North America | 261854.789189 |     24214470.00 |
| Asia          | 150105.725490 |     31881005.00 |
| Africa        |  10149.322034 |     29010887.00 |
| Europe        | 206497.065217 |     23049133.90 |
| South America | 107991.000000 |     17864926.00 |
| Oceania       |  14991.953571 |      8564294.00 |
| Antarctica    |      0.000000 |     13132101.00 |
+---------------+---------------+-----------------+
7 rows in set (0.01 sec)
```

Figure 4.29 – The SELECT output, grouped by continent

Here, you use `GROUP BY` on the `Continent` column. Then, use `AVG()` on `GNP` to calculate the average GNP for that continent and use `SUM()` on `SurfaceArea` to sum the surface areas of all the countries in that continent.

Get the per region data by writing the following query:

```
SELECT Region, SUM(SurfaceArea) FROM country
WHERE Continent='Asia' GROUP BY Region;
```

This outputs the following results:

```
mysql> SELECT Region, SUM(SurfaceArea) FROM country
    -> WHERE Continent='Asia' GROUP BY Region;
+----------------------------+------------------+
| Region                     | SUM(SurfaceArea) |
+----------------------------+------------------+
| Southern and Central Asia  |      10791130.00 |
| Middle East                |       4820592.00 |
| Southeast Asia             |       4494801.00 |
| Eastern Asia               |      11774482.00 |
+----------------------------+------------------+
4 rows in set (0.00 sec)
```

Figure 4.30 – The SELECT output grouped by region

In this exercise, you filtered by the continent of Asia and then used GROUP BY on region. For each group, you summed SurfaceArea. You did not need to use HAVING, as the filter is not on aggregated data. In the next section, we will explore how to write output directly to a file.

Case statements

Often, we want to display data based on some sort of condition. In these situations, a case statement can be used to display data relative to a condition. The case statement syntax is shown here:

```
CASE WHEN [condition 1] THEN [result1]
WHEN [condition 2] THEN [result2]
...
 [ELSE] [resultn]
END
```

For example, suppose that you had a table of users named userTable, which contained users of varying ages. If a user is age 18 or older, you want to show them as an adult. If a user is younger than age 18, you want to show them as a youth. To achieve this, you can use a case statement, like so:

```
SELECT CASE WHEN age >= 18 THEN 'adult'
ELSE 'youth' END AS isadult FROM user;
```

The next exercise demonstrates a practical example of case statements.

Exercise 4.05 – writing case statements

Your company wants to run analysis on the country data of the world database to determine the size of the countries. They have asked you to create a query that categorizes countries based on the following criteria:

- If a country has a population under 100,000, it is small.
- If a country has a population between 100,000 and 500,000, it is medium.
- In all other cases, the country is large.

To achieve this, we can use a case statement. Here are the steps to write the query:

1. Open MySQL Workbench and create a new query window.
2. First, it is helpful to determine the cases that our query has. There are three cases to consider:

   - WHEN population < 100,000, THEN 'small'
   - WHEN population < 500,000, THEN 'medium'
   - ELSE 'large'

3. Next, we will put these cases into a formal case statement. This will give us the following query:

```
SELECT Name, CASE WHEN population < 100000 THEN 'small'
WHEN population < 500000 then 'medium'
ELSE 'large' END AS countrySize
FROM world.country;
```

4. Run the query to get the following result:

Name	countrySize
Aruba	medium
Afghanistan	large
Angola	large
Anguilla	small
Albania	large
Andorra	small
Netherlands Antilles	medium
United Arab Emirates	large
Argentina	large
Armenia	large
American Samoa	small
Antarctica	small

Figure 4.31 – The result of the case statement query

With this, you now have a query that successfully categorizes the sizes of the countries.

Activity 4.01 – collecting information for a travel article

For a travel magazine, you need to collect some information from the world schema to add bits of trivia to some of the articles in next month's edition. The information that is requested is this:

- What is the population size of the smallest city in the database?

- How many languages are spoken in India?

- Which languages are spoken in more than 20 countries?

- What are the five biggest cities in the "Southern and Central Asia" region?

- How many cities have a name that ends with "ester"?

Follow these steps to complete this activity:

1. Connect to the `world` schema with a MySQL client.

2. For each question, follow a few basic steps:

    - Select the tables that we need.

    - Filter out the rows that we need.

    - Aggregate the rows if needed.

    - Select the fields that we need.

> **Note**
>
> The solution for this activity can be found in the *Appendix*.

In this activity, you collected the required information from the world schema that can be used to add bits of trivia to some of the articles in next month's edition of a travel magazine.

Summary

In this chapter, you learned how to select databases and query their tables. You also learned how to apply different filters to the results using `WHERE`. You got hands-on practice with popular built-in functions that help you manipulate data, such as `ROUND()`, `POW()`, and `CEILING()`, string functions to slice and dice output, and used date and time functions to enable you to capture different points in time when a record was inserted or manipulated. Finally, you got to practice aggregating data, which is a must-have skill for any database admin.

In the next chapter, we will continue our journey and cover using joins to correlate related data.

Section 2: Managing Your Database

This section covers the various ways that you can manage and analyze your MySQL data. We will discuss different ways of joining tables, creating objects to analyze data, and creating basic database clients through Node.js.

This section consists of the following chapters:

- *Chapter 5, Correlating Data across Tables*
- *Chapter 6, Stored Procedures and Other Objects*
- *Chapter 7, Creating Database Clients in Node.js*
- *Chapter 8, Working with Data using Node.js*

5

Correlating Data across Tables

In this chapter, you will learn multiple ways to query data that is spread over more than one table. You will then use **Common Table Expressions (CTEs)** to build easy-to-follow queries where parts of the main query are abstracted out into separate parts. In addition, you will see how to work with a CTE to query for hierarchical data and generate ranges of numbers, dates, and more. Finally, you will learn how to use EXPLAIN to see how MySQL would execute a query.

This chapter covers the following topics:

- Introduction to processing data across tables
- Joining two tables
- Analyzing subqueries
- Common table expressions
- Analyzing query performance with EXPLAIN
- Activity 5.01: The Sakila video store
- Activity 5.02: Generating a list of years

Introduction to processing data across tables

In the previous chapter, we covered querying a single table. We used WHERE to filter out the rows we were interested in, and we used GROUP BY to aggregate rows into groups of rows to then use aggregate functions such as COUNT() and SUM(). We also learned about working with JSON data.

In a relational database such as MySQL, data is stored across multiple tables. The reason for doing this is that it avoids storing the same piece of information multiple times.

An example of this is a database for a simple website with comments. It probably has a table of users consisting of values such as username, display name, and password hash. Then it has a table named posts that stores all the posts, and then there is a table with comments. The comment table stores a reference to the post the comment is linked to and a reference to the user commenting.

If the user changes their password, then only one table has to be updated. And if a comment gets edited, then one table also gets updated. If you want to get a list of display names that commented on a post, you now have to use multiple tables to get this information.

In the next section, we will learn the basics of joining tables together. This allows us to query a set of tables with one statement and get one coherent result set.

Joining two tables

If there is related data in two tables, you often need to query both to get the information you want. You can do this with two queries, but often it is easier and more efficient to query the two tables with a single query. An example of related data is the city and country tables in the world database.

Let's use a simplified version of the city and country tables to learn how to join two tables. Here is the city table, consisting of ID, Name, and CountryCode columns:

ID	Name	CountryCode
539	Sofija	BGR
540	Plovdiv	BGR
3018	Bucuresti	ROM
3019	Iasi	ROM

Figure 5.1 – The city table

Here is the `country` table, consisting of `Code` and `Name` columns:

Code	Name
BGR	Bulgaria
ROM	Romania

Figure 5.2 – The country table

As you can see, the values in the `CountryCode` column in the `city` table match the values in the `Code` column in the country table. Now, let's see what happens if we join the two tables:

1. First, write the following query to join both tables:

    ```
    SELECT * FROM city JOIN country;
    ```

    This query produces the following output in the compiler:

    ```
    mysql> SELECT * FROM city JOIN country;
    +------+-----------+-------------+------+----------+
    | ID   | Name      | CountryCode | Code | Name     |
    +------+-----------+-------------+------+----------+
    |  539 | Sofija    | BGR         | BGR  | Bulgaria |
    |  539 | Sofija    | BGR         | ROM  | Romania  |
    |  540 | Plovdiv   | BGR         | BGR  | Bulgaria |
    |  540 | Plovdiv   | BGR         | ROM  | Romania  |
    | 3018 | Bucuresti | ROM         | BGR  | Bulgaria |
    | 3018 | Bucuresti | ROM         | ROM  | Romania  |
    | 3019 | Iasi      | ROM         | BGR  | Bulgaria |
    | 3019 | Iasi      | ROM         | ROM  | Romania  |
    | 2460 | Skopje    | MKD         | BGR  | Bulgaria |
    | 2460 | Skopje    | MKD         | ROM  | Romania  |
    +------+-----------+-------------+------+----------+
    10 rows in set (0.01 sec)
    ```

    Figure 5.3 – SELECT output with JOIN

This is probably not the output you expected as it doesn't contain information. MySQL combined all the rows in the `country` table with all the records in the `city` table. As there are 4 records in the first table and 2 in the latter, this resulted in 2 x 4 = 8 records. In this case, you should tell MySQL how the tables are related to each other.

2. To tell MySQL how the tables are related to each other, use the query shown here:

```
SELECT * FROM city JOIN country ON city.
CountryCode=country.Code;
```

This query produces the following output:

```
mysql> SELECT * FROM city JOIN country ON city.CountryCode=country.Code;
+------+-----------+-------------+------+-----------+
| ID   | Name      | CountryCode | Code | Name      |
+------+-----------+-------------+------+-----------+
|  539 | Sofija    | BGR         | BGR  | Bulgaria  |
|  540 | Plovdiv   | BGR         | BGR  | Bulgaria  |
| 3018 | Bucuresti | ROM         | ROM  | Romania   |
| 3019 | Iasi      | ROM         | ROM  | Romania   |
+------+-----------+-------------+------+-----------+
4 rows in set (0.00 sec)
```

Figure 5.4 – SELECT output for a join describing the relation between the two tables

This looks much better. Now the information makes sense, and we no longer see rows where city and country don't match. But we do see two columns that have the same information: the CountryCode column, which comes from the city table, and the Code column, which comes from the country table. For now, we are not interested in the ID column.

3. To specify the columns we want to see, we can write a query that modifies the field list in the select:

```
SELECT city.Name, country.Code, country.Name FROM city
JOIN country ON city.CountryCode=country.Code;
```

This results in the following output:

```
mysql> SELECT city.Name, country.Code, country.Name FROM city
    -> JOIN country ON city.CountryCode=country.Code;
+-----------+------+-----------+
| Name      | Code | Name      |
+-----------+------+-----------+
| Sofija    | BGR  | Bulgaria  |
| Plovdiv   | BGR  | Bulgaria  |
| Bucuresti | ROM  | Romania   |
| Iasi      | ROM  | Romania   |
+-----------+------+-----------+
4 rows in set (0.00 sec)
```

Figure 5.5 – SELECT output, joined with a specified set of columns

There is, once again, some improvement. We only see the three columns we want to see, and `city` and `country` are still combined correctly. However, it is not without problems. There are two columns called `Name`, which is confusing, and we have to specify the table names quite often.

4. To fix the duplicate naming, we can use aliases for the table names and the column names:

```
SELECT ci.Name, co.Code AS CountryCode, co.Name  AS
CountryName
FROM city ci JOIN country co ON ci.CountryCode=co.Code;
```

This query produces the following output:

```
mysql> SELECT ci.Name, co.Code AS CountryCode, co.Name  AS CountryName
    -> FROM city ci JOIN country co ON ci.CountryCode=co.Code;
+-----------+-------------+-------------+
| Name      | CountryCode | CountryName |
+-----------+-------------+-------------+
| Sofija    | BGR         | Bulgaria    |
| Plovdiv   | BGR         | Bulgaria    |
| Bucuresti | ROM         | Romania     |
| Iasi      | ROM         | Romania     |
+-----------+-------------+-------------+
4 rows in set (0.01 sec)
```

Figure 5.6 – SELECT output with table and column aliases

The column names are now unique, and we aliased the city table as `ci` and the country table as `co`. Aliasing table names is especially useful if the table names are long.

> **Joins and Collections**
>
> While data in tables is usually spread out over multiple tables, this is not common for collections. In the case of cities and countries, this would be combined in a single collection, probably with nested data. In X DevAPI, there is no support for joining tables or collections. In SQL mode, you can join collections with other collections or with tables.

Accidental cross joins

If you don't specify the relation between two tables, then the database server will join every record in the first table with every record in the second table. This is often not what you want and can produce very big result sets.

A regular query will look like the following:

```
SELECT ci.name, co.name
FROM city ci
JOIN country co ON ci.CountryCode=co.Code;
```

The same query, but with the ON part forgotten, might look like this:

```
SELECT ci.name, co.name
FROM city ci
JOIN country co
```

The first query returns 4,079 rows and the second returns 978,960 rows.

The city table has 4,079 rows, while the country table has 240 rows, and 4,079 x 240 = 978,960. So, you can see that it matches every record in the country table with every record in the city table. Therefore, be careful to not forget the ON part of the join.

LEFT JOIN versus INNER JOIN

Now, add a new city to the table with the insert query:

```
INSERT INTO city VALUES(2460, 'Skopje', 'MKD');
```

And run the same query again:

```
SELECT ci.Name, co.Code AS CountryCode, co.Name  AS CountryName
FROM city ci JOIN country co ON ci.CountryCode=co.Code;
```

This produces the following output:

```
mysql> INSERT INTO city VALUES(2460, 'Skopje', 'MKD');
Query OK, 1 row affected (0.01 sec)

mysql> SELECT ci.Name, co.Code AS CountryCode, co.Name  AS CountryName
    -> FROM city ci JOIN country co ON ci.CountryCode=co.Code;
+-----------+-------------+-------------+
| Name      | CountryCode | CountryName |
+-----------+-------------+-------------+
| Sofija    | BGR         | Bulgaria    |
| Plovdiv   | BGR         | Bulgaria    |
| Bucuresti | ROM         | Romania     |
| Iasi      | ROM         | Romania     |
+-----------+-------------+-------------+
4 rows in set (0.00 sec)
```

Figure 5.7 – SELECT output after adding a city

This is not showing the new city. This is because there are multiple types of joins. JOIN in MySQL means INNER JOIN. With INNER JOIN, MySQL will only show results if there is a matching record in both tables. In this example, there is no country with the code MKD in the country table, so it is not showing the new city. One of the other options is LEFT JOIN. With LEFT JOIN, all the records from the first table are shown even if there is no matching record in the second table. In that case, the columns from that table will have NULL as a value.

Now, try a LEFT JOIN:

```
SELECT ci.Name, co.Code AS CountryCode, co.Name  AS CountryName
 FROM city ci LEFT JOIN country co ON ci.CountryCode=co.Code;
```

This produces the following output:

```
mysql> SELECT ci.Name, co.Code AS CountryCode, co.Name  AS CountryName
    -> FROM city ci LEFT JOIN country co ON ci.CountryCode=co.Code;
+-----------+-------------+-------------+
| Name      | CountryCode | CountryName |
+-----------+-------------+-------------+
| Sofija    | BGR         | Bulgaria    |
| Plovdiv   | BGR         | Bulgaria    |
| Bucuresti | ROM         | Romania     |
| Iasi      | ROM         | Romania     |
| Skopje    | NULL        | NULL        |
+-----------+-------------+-------------+
5 rows in set (0.00 sec)
```

Figure 5.8 – SELECT with LEFT JOIN

It now shows all five records from the city table and shows code and name from the country table if there is a matching record.

Say you have a city table with the following values:

ID	Name	CountryCode
539	Sofija	BGR
540	Plovdiv	BGR
2460	Skopje	MKD
3018	Bucuresti	ROM
3019	Iasi	ROM

Figure 5.9 – The city table

And you have a `country` table with the following values:

Code	Name
BGR	Bulgaria
ROM	Romania

Figure 5.10 – The country table

Then, an `INNER JOIN` looks like this:

ID	Name	CountryCode	Code	Name
539	Sofija	BGR	BGR	Bulgaria
540	Plovdiv	BGR	BGR	Bulgaria
3018	Bucuresti	ROM	ROM	Romania
3019	Iasi	ROM	ROM	Romania

Figure 5.11 – INNER JOIN of the city and country tables

And a `LEFT JOIN` looks like this:

ID	Name	CountryCode	Code	Name
539	Sofija	BGR	BGR	Bulgaria
540	Plovdiv	BGR	BGR	Bulgaria
2460	Skopje	MKD	NULL	NULL
3018	Bucuresti	ROM	ROM	Romania
3019	Iasi	ROM	ROM	Romania

Figure 5.12 – LEFT JOIN of the city and country tables

So, the difference is that `INNER JOIN` requires matching rows in both tables, whereas `LEFT JOIN` will show rows that have a match only in the left table.

You can combine joining tables with aggregations, which you learned about in the previous chapter, by writing the following query:

```
SELECT co.Name, COUNT(*) FROM country co
LEFT JOIN city ci ON ci.CountryCode=co.Code
GROUP BY co.Name;
```

This produces the following output:

```
mysql> SELECT co.Name, COUNT(*) FROM country co
    -> LEFT JOIN city ci ON ci.CountryCode=co.Code
    -> GROUP BY co.Name;
+-----------+----------+
| Name      | COUNT(*) |
+-----------+----------+
| Bulgaria  |        2 |
| Romania   |        2 |
+-----------+----------+
2 rows in set (0.00 sec)
```

Figure 5.13 – SELECT with JOIN and GROUP BY

So, you can see that, for every record in the country table, there are two matches in the city table. This is the number of cities per country.

But what happened to the city we just added? It is not shown, as the base of this query is the country table, and with that information, it goes to look for matching cities. However, we can change that by starting with the city table and then looking for matching countries. In this case, we can use RIGHT JOIN, which does the same as LEFT JOIN, but with the order of the tables reversed.

Consider the following query:

```
SELECT co.Name, COUNT(*) FROM country co
RIGHT JOIN city ci ON ci.CountryCode=co.Code GROUP BY co.Name;
```

This produces the following output:

```
mysql> SELECT co.Name, COUNT(*) FROM country co
    -> RIGHT JOIN city ci ON ci.CountryCode=co.Code GROUP BY co.Name;
+-----------+----------+
| Name      | COUNT(*) |
+-----------+----------+
| Bulgaria  |        2 |
| Romania   |        2 |
| NULL      |        1 |
+-----------+----------+
3 rows in set (0.01 sec)
```

Figure 5.14 – SELECT with RIGHT JOIN

So, it now starts with the five cities and then, for each of them, looks for a matching country and uses NULL if there is no match. The preceding output shows that Bulgaria and Romania both have two cities and that there is one city for which we don't know the country.

Now that you have learned about the joins, the next section will take you through an exercise wherein you will be joining two tables.

Exercise 5.01: Joining two tables

This exercise assumes you have the world database available from the previous chapter. In the country table in the world database, you have the Region column to store the region that country is in. In the city table, you store the population of the cities. You now want to get the five biggest cities in the Middle East region. For this, you need to query both tables. In this exercise, you will try to join the tables by selecting the fields you need, and then apply filtering, sorting, and limit options. Follow these steps to complete this exercise:

1. Connect to the MySQL client with Workbench and the appropriate user.

2. Select the world database for the execution:

    ```
    USE world;
    ```

    This produces the following output:

    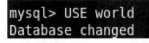

    Figure 5.15 – USE output

3. Join the two tables. Here, you, know that the CountryCode column of the city table stores a reference to the Code column of the country table. So, the JOIN part of your query will be this:

    ```
    FROM city ci JOIN country co ON ci.CountryCode=co.Code
    ```

4. Select the fields you need. You want the city name and population, so the SELECT part of your query will be the following:

    ```
    SELECT ci.Name, ci.Population
    ```

5. Add filtering, sorting, and the limit. You want to filter Region to get the top five cities by population. So, the last part of your query will be as follows:

    ```
    WHERE co.Region='Middle East' ORDER BY ci.Population DESC
    LIMIT 5
    ```

6. Now, combine and execute the query:

```
SELECT ci.Name, ci.Population
FROM city ci JOIN country co ON ci.CountryCode=co.Code
WHERE co.Region='Middle East' ORDER BY ci.Population DESC
LIMIT 5;
```

The preceding query produces the following output:

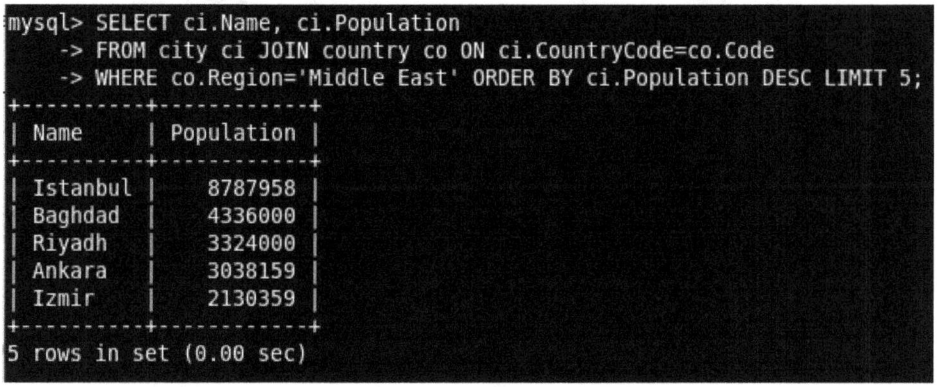

```
mysql> SELECT ci.Name, ci.Population
    -> FROM city ci JOIN country co ON ci.CountryCode=co.Code
    -> WHERE co.Region='Middle East' ORDER BY ci.Population DESC LIMIT 5;
+----------+------------+
| Name     | Population |
+----------+------------+
| Istanbul |    8787958 |
| Baghdad  |    4336000 |
| Riyadh   |    3324000 |
| Ankara   |    3038159 |
| Izmir    |    2130359 |
+----------+------------+
5 rows in set (0.00 sec)
```

Figure 5.16 – SELECT output with the top five cities in the Middle East region

Finally, you joined two tables and got the five biggest cities in the Middle East region (namely, Istanbul, Baghdad, Riyadh, Ankara, and Izmir). You also applied filtering, sorting, and limit options to get the desired result. In the next section, we will learn about subqueries.

Analyzing subqueries

Another way of joining the tables is available in MySQL. It consists of using the output of a query directly in another query.

Use the world_simple table as an example again and look at the following query:

```
SELECT Name FROM city WHERE CountryCode=(
  SELECT Code FROM country WHERE Name='Romania'
);
```

This query produces the following output:

```
mysql> SELECT Name FROM city WHERE CountryCode=(
    ->    SELECT Code FROM country WHERE Name='Romania'
    -> );
+-----------+
| Name      |
+-----------+
| Bucuresti |
| Iasi      |
+-----------+
2 rows in set (0.00 sec)
```

Figure 5.17 – SELECT with a subquery

The preceding query is essentially running this command:

```
SELECT Code FROM country WHERE Name='Romania';
```

It then saves the result and runs the following query:

```
SELECT Name FROM city WHERE CountryCode=<saved_result>
```

So, to use a subquery, you place the query inside (and) and place it where you want to see the output. This can be in the WHERE part of the query, but also the SELECT part and most other places.

Dependent subqueries

In the previous example, the two queries were independent, but in some cases, you can make the subquery depend on the main query. That looks like this:

```
SELECT
  Name,
  CountryCode,
  (SELECT Name FROM country WHERE Code=city.CountryCode) AS
CountryName
FROM city;
```

This query produces the following output:

```
mysql> SELECT
    ->    Name,
    ->    CountryCode,
    ->    (SELECT Name FROM country WHERE Code=city.CountryCode) AS CountryName
    -> FROM city;
+-----------+-------------+-------------+
| Name      | CountryCode | CountryName |
+-----------+-------------+-------------+
| Sofija    | BGR         | Bulgaria    |
| Plovdiv   | BGR         | Bulgaria    |
| Bucuresti | ROM         | Romania     |
| Iasi      | ROM         | Romania     |
| Skopje    | MKD         | NULL        |
+-----------+-------------+-------------+
5 rows in set (0.00 sec)
```

Figure 5.18 – SELECT with a dependent subquery

Here, the subquery refers to `city.CountryCode`. In MySQL, this is run like this: it runs the subquery for every row in the `city` table. Now that we have gained a good understanding of subqueries, let's do an exercise on them.

Exercise 5.02: Using a subquery

In the `countrylanguage` table, you have a list of languages. You want to get a list of countries where Portuguese is the official language. To do this, you will first be filtering out the rows you need from the `countrylanguage` table. Later, you will be adding a subquery to look up the name of the `country`. Follow these steps to accomplish this:

1. Connect to the MySQL client with Workbench and the appropriate user.

2. Select the `world` database to be used:

```
USE world;
```

This produces the following output:

```
mysql> USE world
Database changed
```

Figure 5.19 – USE output

3. Filter out the rows you need from the `countrylanguage` table by writing the following query:

```
SELECT * FROM countrylanguage
WHERE Language='Portuguese' AND IsOfficial='T';
```

This produces the following output:

```
mysql> SELECT * FROM countrylanguage
    -> WHERE Language='Portuguese' AND IsOfficial='T';
+-------------+------------+-----------+------------+
| CountryCode | Language   | IsOfficial | Percentage |
+-------------+------------+-----------+------------+
| BRA         | Portuguese | T         |       97.5 |
| CPV         | Portuguese | T         |        0.0 |
| GNB         | Portuguese | T         |        8.1 |
| MAC         | Portuguese | T         |        2.3 |
| PRT         | Portuguese | T         |       99.0 |
| TMP         | Portuguese | T         |        0.0 |
+-------------+------------+-----------+------------+
6 rows in set (0.00 sec)
```

Figure 5.20 – SELECT output for countries with Portuguese as the official language

So, you now have a list of `CountryCodes` for countries that have `Portuguese` as one of the official languages. This is close to what you need, but you need to look up those `CountryCodes` in the `country` table to get the actual names of those countries.

4. Add a subquery to look up the name of the country:

```
SELECT (
    SELECT Name FROM country
    WHERE Code=CountryCode
) AS CountryName FROM countrylanguage
WHERE Language='Portuguese' AND IsOfficial='T';
```

This produces the following output:

```
mysql> SELECT (
    ->    SELECT Name FROM country
    ->    WHERE Code=CountryCode
    -> ) AS CountryName FROM countrylanguage
    -> WHERE Language='Portuguese' AND IsOfficial='T';
+---------------+
| CountryName   |
+---------------+
| Brazil        |
| Cape Verde    |
| Guinea-Bissau |
| Macao         |
| Portugal      |
| East Timor    |
+---------------+
6 rows in set (0.00 sec)
```

Figure 5.21 – SELECT output with a subquery for countries that have Portuguese as an official language

You now have the list you wanted; the names of the countries that have `Portuguese` as one of their official languages. You did this in two steps, where you first filtered the right entries from the `countrylanguage` table and verified the result. Later, you added the subquery to look up the names of the countries. In the next section, you will learn about common table expressions.

Common table expressions

If a query is joining multiple tables and also has subqueries, then things might start to look a little complex. But luckily, there is a way to do this that's easier to understand. This is called **Common Table Expressions** (**CTEs**). This is also known as `WITH` because that's the keyword we have to use for this. Consider the following expression:

```
WITH city_in_romania AS (
   SELECT ci.Name, ci.CountryCode, co.Name AS CountryName
   FROM city ci INNER JOIN country co ON ci.CountryCode=co.Code
AND co.Name='Romania'
)
SELECT * FROM city_in_romania;
```

This produces the following output:

```
mysql> WITH city_in_romania AS (
    ->    SELECT ci.Name, ci.CountryCode, co.Name AS CountryName
    ->    FROM city ci INNER JOIN country co ON ci.CountryCode=co.Code AND co.Name='Romania'
    -> )
    -> SELECT * FROM city_in_romania;
+-----------+-------------+-------------+
| Name      | CountryCode | CountryName |
+-----------+-------------+-------------+
| Bucuresti | ROM         | Romania     |
| Iasi      | ROM         | Romania     |
+-----------+-------------+-------------+
2 rows in set (0.00 sec)
```

Figure 5.22 – SELECT with CTE

In the first few lines, we define `city_in_romania` as a new table that's only available for this query. It is made by a join of the `country` and `city` tables and then filtered on the `country` name. Then we can use this new table in the second part of the query. We can define multiple virtual tables in this way.

Without the CTE, the query would look like this:

```
SELECT * FROM (
  SELECT ci.Name, ci.CountryCode, co.Name AS CountryName
  FROM city ci INNER JOIN country co ON ci.CountryCode=co.Code
  AND co.Name='Romania'
) AS city_in_romania;
```

In this example, the query that uses the `city_in_romania` table is very simple. If that part of the query becomes more complex, the usefulness of the CTE is more obvious. Let's look at recursive CTE in the next section.

Recursive CTE

There is a variation of normal CTE called **recursive CTE**. It can be useful in quite a few situations. One example is an `employee` table wherein the direct manager for every employee is recorded. Then, a recursive CTE can find out which employees are under the same manager even if there are multiple levels between them. Another category of problems in which recursive CTEs are useful is when generating ranges of data.

To understand CTEs, we first need to cover `UNION`. `UNION` combines the output of multiple queries into a single result set. This requires the queries to have exactly the same column order and types. Consider the following query:

```
SELECT Name FROM country WHERE Code='ROM'
UNION
SELECT Name FROM country WHERE Code='BGR';
```

This produces the following output:

```
mysql> SELECT Name FROM country WHERE Code='ROM'
    -> UNION
    -> SELECT Name FROM country WHERE Code='BGR';
+----------+
| Name     |
+----------+
| Romania  |
| Bulgaria |
+----------+
2 rows in set (0.00 sec)
```

Figure 5.23 – SELECT output with UNION

This output is combining the results of the two cases, the first case is where the code is equal to ROM, and the second case is where the code is equal to BGR. Note that the duplicates that exist in the output of both queries are not shown. If you want all rows from both queries to show even if there are duplicates, we then need to use UNION ALL, which would look something like this:

```
SELECT Name FROM country WHERE Code='ROM'
UNION ALL
SELECT Name FROM country WHERE Code='BGR';
```

In our country table, there are no duplicate records, so the result of UNION and UNION ALL are the same. There are many cases where the results of these queries will be different. One such case is recursive CTEs, which will potentially have duplicate results in their recursive calls.

As an example of a recursive CTE, consider a simple example of recursively counting from 1 to 12. To do this, we start with the number 1, and on each recursive call, we add 1 to the previous number, until we reach our target value. To achieve this in a CTE, we start with our initial condition in a SELECT operation. We then union that initial condition with a recursive call, using a WHERE operation. clause to determine when the recursive call ends. The resulting query is shown here:

```
WITH RECURSIVE numbers AS (
    SELECT 1 AS n
    UNION ALL
    SELECT n+1 FROM numbers WHERE n<12
)
SELECT
```

```
  n,
  monthname(CONCAT("2019-",n,"-01"))
 FROM numbers;
```

This produces the following output:

```
mysql> WITH RECURSIVE numbers AS (
    ->    SELECT 1 AS n
    ->    UNION ALL
    ->    SELECT n+1 FROM numbers WHERE n<12
    -> )
    -> SELECT
    ->   n,
    ->   monthname(CONCAT("2019-",n,"-01"))
    -> FROM numbers;
+-------+------------------------------------+
| n     | monthname(CONCAT("2019-",n,"-01")) |
+-------+------------------------------------+
|     1 | January                            |
|     2 | February                           |
|     3 | March                              |
|     4 | April                              |
|     5 | May                                |
|     6 | June                               |
|     7 | July                               |
|     8 | August                             |
|     9 | September                          |
|    10 | October                            |
|    11 | November                           |
|    12 | December                           |
+-------+------------------------------------+
12 rows in set (0.01 sec)
```

Figure 5.24 – SELECT query with recursive CTE

On the first line, we see the RECURSIVE keyword, which is needed if you want to use a recursive CTE. Then, on the next line, we have SELECT 1 AS n, which generates the first result and is used to initialize the recursion. Then, on the next line, we have UNION ALL, which is required in order for the CTE to work. It combines the first result with every next result.

Then, on the next line, we have a SELECT query that returns the value of n+1 but refers to the numbers table, which is the recursive CTE. This is the second part of the CTE that is called for every recursion and uses the data from the previous recursion. This also has the WHERE condition, n<12, to stop the recursion. If the query after the union is no longer returning results, then the recursion is done.

With CTE, we have created a virtual table that has numbers from 1 to 12. Then we use these numbers to get the names of the 12 months. As `monthname()` requires data, we create some data that has this month.

We could also move the `monthname` query to a CTE by writing the following query:

```
WITH RECURSIVE numbers AS (
    SELECT 1 AS n
    UNION ALL
    SELECT n+1 FROM numbers WHERE n<12
),
months AS (
    SELECT n, monthname(CONCAT("2019-",n,"-01"))
    FROM numbers
)
SELECT * FROM months;
```

This produces the following output:

Figure 5.25 – SELECT output with recursive CTE and regular CTE combined

Here, you can see that you can combine recursive CTEs with regular CTEs. Having a way to generate a range of numbers or months can be very useful, for example, for a report of sales per day of the week. There might be a holiday or some other reason why there are no sales on a particular day. Then, the report will only have the days on which there are sales. If you join against a virtual table with days of the week, you can ensure that the days for which there were no sales are still in the result. In the next section, we will solve an exercise based on CTE.

Exercise 5.03: Using a CTE

In the previous exercise, you got a list of countries that have Portuguese as an official language. Now you will go a step further. You want to get the total list of languages spoken in each of the countries that have Portuguese as an official language. To do this, you will create a CTE with the list of `CountryCodes` that have Portuguese as an official language. Next, you will join with the `countrylanguage` table to get the other languages spoken in that list of countries. Finally, you will join with the country table to get the names of each of the countries. Follow these steps to complete this exercise:

1. Connect to the MySQL client with Workbench and the appropriate user.

2. Select the `world` database to be used:

```
USE world;
```

This produces the following output:

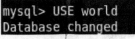

Figure 5.26 – USE output

3. Create a CTE with the list of `CountryCodes` that have `Portuguese` as an official language:

```
WITH country_portuguese AS (
    SELECT CountryCode FROM countrylanguage
    WHERE Language='Portuguese' AND IsOfficial='T'
)
SELECT * FROM country_portuguese;
```

This produces the following output:

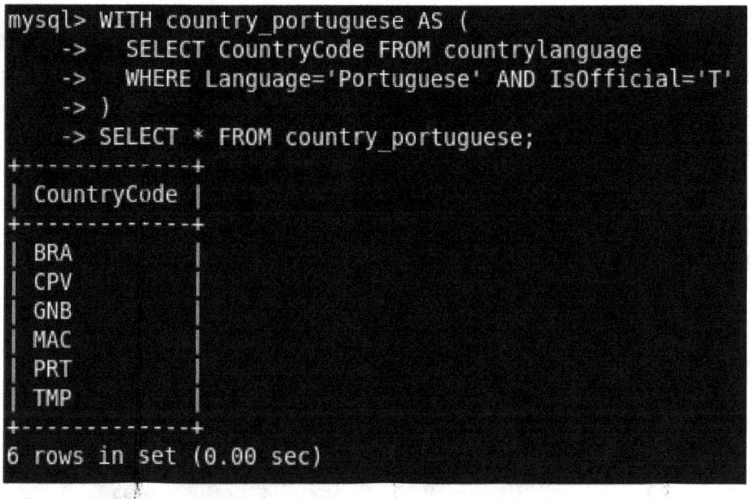

```
mysql> WITH country_portuguese AS (
    ->    SELECT CountryCode FROM countrylanguage
    ->    WHERE Language='Portuguese' AND IsOfficial='T'
    -> )
    -> SELECT * FROM country_portuguese;
+-------------+
| CountryCode |
+-------------+
| BRA         |
| CPV         |
| GNB         |
| MAC         |
| PRT         |
| TMP         |
+-------------+
6 rows in set (0.00 sec)
```

Figure 5.27 – SELECT output with CTE

This might not look like much, but you can now use this as a table to join in the next step.

4. Join the CTE from the previous step with the countrylanguage table to get the other languages spoken in that list of countries:

```
WITH country_portuguese AS (
  SELECT CountryCode FROM countrylanguage
  WHERE Language='Portuguese' AND IsOfficial='T'
)
SELECT
  *
FROM country_portuguese col_pt
JOIN countrylanguage col
  ON col_pt.CountryCode=col.CountryCode;
```

This produces the following output:

```
mysql> WITH country_portuguese AS (
    ->    SELECT CountryCode FROM countrylanguage
    ->    WHERE Language='Portuguese' AND IsOfficial='T'
    -> )
    -> SELECT
    ->    *
    -> FROM country_portuguese col_pt
    -> JOIN countrylanguage col
    ->    ON col_pt.CountryCode=col.CountryCode;
+-------------+-------------+------------------+------------+------------+
| CountryCode | CountryCode | Language         | IsOfficial | Percentage |
+-------------+-------------+------------------+------------+------------+
| BRA         | BRA         | German           | F          |        0.5 |
| BRA         | BRA         | Indian Languages | F          |        0.2 |
| BRA         | BRA         | Italian          | F          |        0.4 |
| BRA         | BRA         | Japanese         | F          |        0.4 |
| BRA         | BRA         | Portuguese       | T          |       97.5 |
| CPV         | CPV         | Crioulo          | F          |      100.0 |
| CPV         | CPV         | Portuguese       | T          |        0.0 |
| GNB         | GNB         | Balante          | F          |       14.6 |
| GNB         | GNB         | Crioulo          | F          |       36.4 |
| GNB         | GNB         | Ful              | F          |       16.6 |
| GNB         | GNB         | Malinke          | F          |        6.9 |
| GNB         | GNB         | Mandyako         | F          |        4.9 |
| GNB         | GNB         | Portuguese       | T          |        8.1 |
| MAC         | MAC         | Canton Chinese   | F          |       85.6 |
| MAC         | MAC         | English          | F          |        0.5 |
| MAC         | MAC         | Mandarin Chinese | F          |        1.2 |
| MAC         | MAC         | Portuguese       | T          |        2.3 |
| PRT         | PRT         | Portuguese       | T          |       99.0 |
| TMP         | TMP         | Portuguese       | T          |        0.0 |
| TMP         | TMP         | Sunda            | F          |        0.0 |
+-------------+-------------+------------------+------------+------------+
20 rows in set (0.00 sec)
```

Figure 5.28 – SELECT output with CTE and join

Here, you joined the CTE (`country_portuguese`) you created in the previous step with the `countrylanguage` table. This results in rows that have a match in both tables. You can already see `CountryCode` and `Language`. The part that is missing is the name of the countries.

5. Join with the `country` table to get the name of each country:

```
WITH country_portuguese AS (
    SELECT CountryCode FROM countrylanguage
    WHERE Language='Portuguese' AND IsOfficial='T'
)
```

```
SELECT
    co.Name,
    GROUP_CONCAT(Language) AS Languages
FROM country_portuguese col_pt
JOIN countrylanguage col
    ON col_pt.CountryCode=col.CountryCode
JOIN country co
    ON co.Code=col.CountryCode
GROUP BY co.Name;
```

This produces the following output:

```
mysql> WITH country_portuguese AS (
    ->    SELECT CountryCode FROM countrylanguage
    ->    WHERE Language='Portuguese' AND IsOfficial='T'
    -> )
    -> SELECT
    ->    co.Name,
    ->    GROUP_CONCAT(Language) AS Languages
    -> FROM country_portuguese col_pt
    -> JOIN countrylanguage col
    ->    ON col_pt.CountryCode=col.CountryCode
    -> JOIN country co
    ->    ON co.Code=col.CountryCode
    -> GROUP BY co.Name;
+---------------+-------------------------------------------------------+
| Name          | Languages                                             |
+---------------+-------------------------------------------------------+
| Brazil        | Indian Languages,Italian,Japanese,Portuguese,German   |
| Cape Verde    | Crioulo,Portuguese                                    |
| East Timor    | Sunda,Portuguese                                      |
| Guinea-Bissau | Ful,Malinke,Mandyako,Portuguese,Crioulo,Balante       |
| Macao         | Canton Chinese,English,Mandarin Chinese,Portuguese    |
| Portugal      | Portuguese                                            |
+---------------+-------------------------------------------------------+
6 rows in set (0.00 sec)
```

Figure 5.29 – SELECT output with CTE and multiple joins

So, we have the end product. We joined against the country table to get the list of countries. We could have used a subquery as we did in the previous exercise. Both are valid ways to get the same result. We also used GROUP BY to group the rows by country name. Then, we used GROUP_CONCAT() to list all the languages in each group and then named this column Languages.

In this exercise, we have combined CTEs with `JOINS`, `GROUP BY`, and function calls, which we covered in the previous chapter. In the next section, we will explore the `EXPLAIN` keyword.

Analyzing query performance with EXPLAIN

`EXPLAIN` is a very useful tool when it comes to performance. The SQL query is used to tell the database what you want, but `EXPLAIN` asks the database how it thinks it is going to do it.

Let's use the `city` table in the `world_simple` database as an example:

```
SELECT * FROM city WHERE ID=2460;
EXPLAIN SELECT * FROM city WHERE ID=2460;
```

This produces the following output:

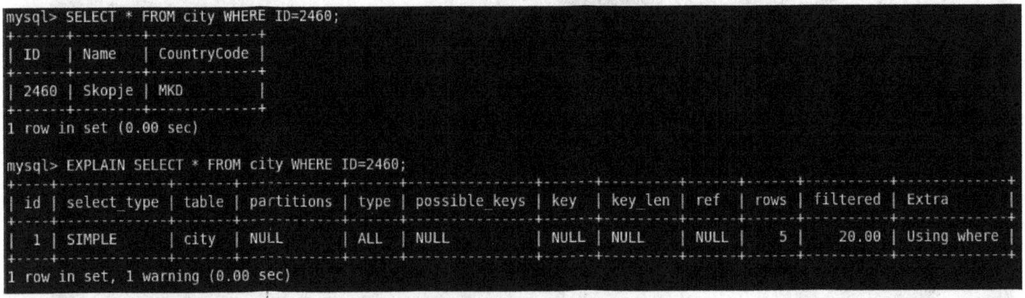

Figure 5.30 – SELECT and EXPLAIN

Note that it says `1 warning`. You can see the actual message by running `SHOW WARNINGS;`. This is expected for `EXPLAIN` as there will be a note with a rewritten version of the statement. You can ignore this for now.

We select a single city (`Skopje`) and are using an ID (`2460`) to do this lookup.

Let's go over the `EXPLAIN` output to see what each field means:

Column	Value	Meaning
`id`	`1`	The identifier for this part of the query.
`select_type`	`SIMPLE`	This means this is a simple query without the complexities of `UNION` or subqueries.
`table`	`city`	The name of the table. This shows the alias if the table is aliased.

Column	Value	Meaning
partitions	NULL	This shows which partitions are needed if the table consists of multiple partitions.
type	ALL	This is an important field. ALL means that it is scanning the whole table.
possible_keys	NULL	This is the list of indexes (or keys) being considered for this query.
key	NULL	This is the index that is used for this query.
key_len	NULL	This is the length in bytes that is being used for the query. If the index is over two 4-byte columns (meaning 8 bytes in total), then 4 bytes would indicate that only 1 column is used.
ref	NULL	This is the column that is used for comparison.
rows	5	This is the number of rows it expects to scan to get the result.
filtered	20.00	The query returns 1 row, but the database scans 5 rows. So, it filters out 20% of the rows it scans.
Extra	Using where	Additional information about how it executes this query.

Figure 5.31 – Meaning of each field

Now let's add a primary key to this table and run EXPLAIN again:

```
ALTER TABLE city ADD PRIMARY KEY (ID);
EXPLAIN SELECT * FROM city WHERE ID=2460;
```

This produces the following output:

```
mysql> ALTER TABLE city ADD PRIMARY KEY (ID);
Query OK, 0 rows affected (0.21 sec)
Records: 0  Duplicates: 0  Warnings: 0

mysql> EXPLAIN SELECT * FROM city WHERE ID=2460;
+----+-------------+-------+------------+-------+---------------+---------+---------+-------+------+----------+-------+
| id | select_type | table | partitions | type  | possible_keys | key     | key_len | ref   | rows | filtered | Extra |
+----+-------------+-------+------------+-------+---------------+---------+---------+-------+------+----------+-------+
|  1 | SIMPLE      | city  | NULL       | const | PRIMARY       | PRIMARY | 4       | const |    1 |   100.00 | NULL  |
+----+-------------+-------+------------+-------+---------------+---------+---------+-------+------+----------+-------+
1 row in set, 1 warning (0.00 sec)
```

Figure 5.32 – ALTER and EXPLAIN output showing the primary key

So, what changed? It now says `const` in the type column instead of `ALL`. This means it is doing an efficient lookup. In the `possible_keys` and `key` columns, it now says `PRIMARY`, which means it is using the primary key. The `key_len` of 4 is because an integer in MySQL is 4 bytes. It now also shows that it only scans 1 row and returns 100% of the rows it scanned. The most important columns are `type` and `rows`.

If you are requesting the whole table (for example, a query without `WHERE`), then having `ALL` in the `type` column is not a problem. But if you are using `WHERE`, then it should not show `ALL`. If it does, you probably need to add an index.

The number of rows should be roughly the same as the number of rows your query returns. Note that aggregations (`GROUP BY`) might use many more rows than it returns because of the aggregation.

For the next example, we are going to look at the `EXPLAIN` plan of a `SELECT` query with a join:

```
EXPLAIN SELECT * FROM country co LEFT JOIN city ci
ON ci.CountryCode=co.Code WHERE ci.ID=540\G
```

This produces the following output:

```
mysql> EXPLAIN SELECT * FROM country co LEFT JOIN city ci
    -> ON ci.CountryCode=co.Code WHERE ci.ID=540\G
*************************** 1. row ***************************
           id: 1
  select_type: SIMPLE
        table: ci
    partitions: NULL
         type: const
possible_keys: PRIMARY
          key: PRIMARY
      key_len: 4
          ref: const
         rows: 1
     filtered: 100.00
        Extra: NULL
*************************** 2. row ***************************
           id: 1
  select_type: SIMPLE
        table: co
    partitions: NULL
         type: ALL
possible_keys: NULL
          key: NULL
      key_len: NULL
          ref: NULL
         rows: 2
     filtered: 50.00
        Extra: Using where
2 rows in set, 1 warning (0.00 sec)
```

Figure 5.33 – EXPLAIN output for a SELECT query with a join

We use \G instead of ; to get vertical output as we did in the previous chapters. Otherwise, the output format is identical to the previous example.

Here, we see that it is using the primary key we added to the city table (here aliased as ci). However, it is still scanning the whole country table (here aliased as co).

```
ALTER TABLE country ADD PRIMARY KEY (Code);
EXPLAIN SELECT * FROM country co LEFT JOIN city ci
ON ci.CountryCode=co.Code WHERE ci.ID=540\G
```

This produces the following output:

```
mysql> ALTER TABLE country ADD PRIMARY KEY (Code);
Query OK, 0 rows affected (0.21 sec)
Records: 0  Duplicates: 0  Warnings: 0

mysql> EXPLAIN SELECT * FROM country co LEFT JOIN city ci
    -> ON ci.CountryCode=co.Code WHERE ci.ID=540\G
*************************** 1. row ***************************
           id: 1
  select_type: SIMPLE
        table: ci
    partitions: NULL
         type: const
possible_keys: PRIMARY
          key: PRIMARY
      key_len: 4
          ref: const
         rows: 1
     filtered: 100.00
        Extra: NULL
*************************** 2. row ***************************
           id: 1
  select_type: SIMPLE
        table: co
    partitions: NULL
         type: const
possible_keys: PRIMARY
          key: PRIMARY
      key_len: 3
          ref: const
         rows: 1
     filtered: 100.00
        Extra: NULL
2 rows in set, 1 warning (0.00 sec)
```

Figure 5.34 – EXPLAIN output with a join and a primary key on the Code column

You can now see that it is using the primary key on both tables. The speed difference between scanning 5 rows and 1 row is very small. But if you query a table with billions of rows, then it makes a huge difference to the runtime of the query.

Besides the format we used in these examples, which is called the `traditional` format, there are also other formats. The first of these is JSON. To set the EXPLAIN output format, use `EXPLAIN FORMAT=<format> <query>`. Consider the following example:

```
EXPLAIN FORMAT=JSON SELECT * FROM country co
LEFT JOIN city ci ON ci.CountryCode=co.Code WHERE ci.ID=540\G
```

This returns the following JSON structure:

```
{
    "query_block": {
      "select_id": 1,
      "cost_info": {
        "query_cost": "1.00"
      },
      "nested_loop": [
        {
          "table": {
            "table_name": "ci",
            "access_type": "const",
            "possible_keys": [
              "PRIMARY"
            ],
        }
        ]
    }
}
```

This is not just the same data as the traditional format, but then in a JSON format, it has more detailed information.

Another available format is called TREE. Here is an example:

```
EXPLAIN FORMAT=TREE SELECT Name, CountryCode,
(SELECT Name FROM country WHERE Code=city.CountryCode) AS
CountryName
FROM city\G
```

And it returns the following output:

```
-> Table scan on city
-> Select #2 (subquery in projection; dependent)
```

```
-> Single-row index lookup on country using PRIMARY
(Code=city.CountryCode)
```

The benefit of this format is that the indentation helps you to see the order in which the steps are executed. Note that this format only supports a subset of the possible SQL queries, and MySQL Workbench has a special feature: Visual Explain.

You get this by going to Query-Explain Current Statement, by clicking the lightning bolt icon with the magnifying glass icon, or by clicking on **Execution Plan** on the right-hand side, as shown in the following screenshot:

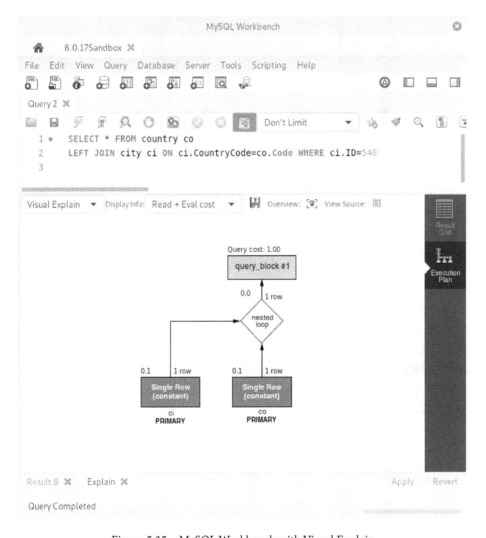

Figure 5.35 – MySQL Workbench with Visual Explain

Depending on how optimal something is, the boxes for each table will be blue, green, or red. Here is an example showing what the same query looks like if we drop the primary key of the `city` column:

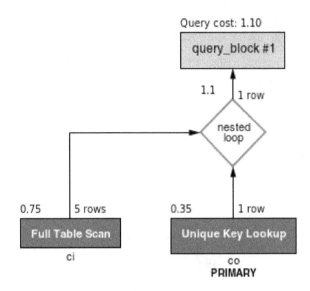

Figure 5.36 – Visual Explain from MySQL Workbench with a full scan and unique key lookup

A new feature introduced in MySQL 8.0.18 is `EXPLAIN ANALYZE`. It allows us to see what the database actually did to execute the query, as opposed to what it thinks it would do to execute your query.

Consider the following example:

```
EXPLAIN FORMAT=TREE SELECT * FROM country co
LEFT JOIN city ci ON ci.CountryCode=co.Code
WHERE ci.ID=540\G
EXPLAIN ANALYZE SELECT * FROM country co
LEFT JOIN city ci ON ci.CountryCode=co.Code
WHERE ci.ID=540\G
```

This produces the following output:

```
mysql> EXPLAIN FORMAT=TREE SELECT * FROM country co
    -> LEFT JOIN city ci ON ci.CountryCode=co.Code
    -> WHERE ci.ID=540\G
*************************** 1. row ***************************
EXPLAIN: -> Inner hash join (co.`Code` = ci.CountryCode)  (cost=1.20 rows=1)
    -> Table scan on co  (cost=0.35 rows=2)
    -> Hash
        -> Filter: (ci.ID = 540)  (cost=0.75 rows=1)
            -> Table scan on ci  (cost=0.75 rows=5)

1 row in set (0.00 sec)

mysql> EXPLAIN ANALYZE SELECT * FROM country co
    -> LEFT JOIN city ci ON ci.CountryCode=co.Code
    -> WHERE ci.ID=540\G
*************************** 1. row ***************************
EXPLAIN: -> Inner hash join (co.`Code` = ci.CountryCode)  (cost=1.20 rows=1) (actual time=0.137..0.151 rows=1 loops=1)
    -> Table scan on co  (cost=0.35 rows=2) (actual time=0.011..0.018 rows=2 loops=1)
    -> Hash
        -> Filter: (ci.ID = 540)  (cost=0.75 rows=1) (actual time=0.064..0.090 rows=1 loops=1)
            -> Table scan on ci  (cost=0.75 rows=5) (actual time=0.045..0.066 rows=5 loops=1)

1 row in set (0.00 sec)
```

Figure 5.37 – EXPLAIN and EXPLAIN ANALYZE output

As you can see in the preceding screenshot, the output looks very similar to the EXPLAIN FORMAT=TREE output, but has some more information for the actual runtimes and the actual number of rows.

The reason that the number of rows it thinks it needs to scan and the actual number of rows that it did scan may differ is because this is based on the statistics it has on the table. In the next section, you will perform an exercise using EXPLAIN.

Exercise 5.04: Using EXPLAIN

In this exercise, you will start with a query and a set of indexes you want to add. You will then execute the following query to reduce the amount of work the database has to do.

Here is the query:

```
SELECT cl.Language, cl.Percentage
FROM city ci JOIN country co ON ci.CountryCode=co.Code
JOIN countrylanguage cl ON cl.CountryCode=co.Code
WHERE
  ci.Name='San Francisco'
  AND co.Name='United States'
  AND cl.Percentage>1;
```

And these are the statements for the indexes we want to add:

```
ALTER TABLE country ADD INDEX(Name);
ALTER TABLE city ADD INDEX (Name);
```

First, run EXPLAIN on the query, then add the first index, and run EXPLAIN again. Finally, you will add the second index and run EXPLAIN again. Follow these steps to complete this exercise:

1. Connect to the MySQL client with Workbench and the appropriate user.

2. Select the world database to be used:

    ```
    USE world;
    ```

    Figure 5.38 – USE output

3. Run EXPLAIN on the query:

    ```
    EXPLAIN SELECT cl.Language, cl.Percentage
    FROM city ci JOIN country co ON ci.CountryCode=co.Code
    JOIN countrylanguage cl ON cl.CountryCode=co.Code
    WHERE
        ci.Name='San Francisco'
        AND co.Name='United States'
        AND cl.Percentage>1\G
    ```

The preceding code produces the following output:

```
mysql> EXPLAIN SELECT cl.Language, cl.Percentage
    -> FROM city ci JOIN country co ON ci.CountryCode=co.Code
    -> JOIN countrylanguage cl ON cl.CountryCode=co.Code
    -> WHERE
    ->   ci.Name='San Francisco'
    ->   AND co.Name='United States'
    ->   AND cl.Percentage>1\G
*************************** 1. row ***************************
           id: 1
  select_type: SIMPLE
        table: co
   partitions: NULL
         type: ALL
possible_keys: PRIMARY
          key: NULL
      key_len: NULL
          ref: NULL
         rows: 239
     filtered: 10.00
        Extra: Using where
*************************** 2. row ***************************
           id: 1
  select_type: SIMPLE
        table: ci
   partitions: NULL
         type: ref
possible_keys: CountryCode
          key: CountryCode
      key_len: 3
          ref: world.co.Code
         rows: 18
     filtered: 10.00
        Extra: Using where
*************************** 3. row ***************************
           id: 1
  select_type: SIMPLE
        table: cl
   partitions: NULL
         type: ref
possible_keys: PRIMARY,CountryCode
          key: PRIMARY
      key_len: 3
          ref: world.co.Code
         rows: 4
     filtered: 33.33
        Extra: Using where
3 rows in set, 1 warning (0.00 sec)
```

Figure 5.39 – EXPLAIN output for the original query

What you can see in the output is that it scans 239 rows in the country table, and then uses the CountryCode index to look up rows in the city table. Finally, it uses the primary key of the countrylanguage table to look up rows in that table. The primary key of the countrylanguage table is on (CountryCode, Language). So, it can use the code it got earlier on to do this.

4. Add the first index and run EXPLAIN again:

```
ALTER TABLE country ADD INDEX(Name);
EXPLAIN SELECT cl.Language, cl.Percentage
FROM city ci JOIN country co ON ci.CountryCode=co.Code
```

```
JOIN countrylanguage cl ON cl.CountryCode=co.Code
WHERE
   ci.Name='San Francisco'
   AND co.Name='United States'
   AND cl.Percentage>1\G
```

The preceding code produces the following output:

```
mysql> ALTER TABLE country ADD INDEX(Name);
Query OK, 0 rows affected (0.07 sec)
Records: 0  Duplicates: 0  Warnings: 0

mysql> EXPLAIN SELECT cl.Language, cl.Percentage
    -> FROM city ci JOIN country co ON ci.CountryCode=co.Code
    -> JOIN countrylanguage cl ON cl.CountryCode=co.Code
    -> WHERE
    ->    ci.Name='San Francisco'
    ->    AND co.Name='United States'
    ->    AND cl.Percentage>1\G
*************************** 1. row ***************************
           id: 1
  select_type: SIMPLE
        table: co
   partitions: NULL
         type: ref
possible_keys: PRIMARY,Name
          key: Name
      key_len: 52
          ref: const
         rows: 1
     filtered: 100.00
        Extra: Using index
*************************** 2. row ***************************
           id: 1
  select_type: SIMPLE
        table: ci
   partitions: NULL
         type: ref
possible_keys: CountryCode
          key: CountryCode
      key_len: 3
          ref: world.co.Code
         rows: 18
     filtered: 10.00
        Extra: Using where
*************************** 3. row ***************************
           id: 1
  select_type: SIMPLE
        table: cl
   partitions: NULL
         type: ref
possible_keys: PRIMARY,CountryCode
          key: PRIMARY
      key_len: 3
          ref: world.co.Code
         rows: 4
     filtered: 33.33
        Extra: Using where
3 rows in set, 1 warning (0.00 sec)
```

Figure 5.40 – EXPLAIN output after adding the first index

Here, things have changed. It now starts with the `country` table and uses the newly added index to find entries in the `country` table that match `Name='United States'`. Then, from there, it uses `CountryCode` to look up entries in the other two tables. This is a lot better.

5. Now, add the second index and run EXPLAIN again:

```
ALTER TABLE city ADD INDEX (Name);
EXPLAIN SELECT cl.Language, cl.Percentage
FROM city ci JOIN country co ON ci.CountryCode=co.Code
JOIN countrylanguage cl ON cl.CountryCode=co.Code
WHERE
    ci.Name='San Francisco'
    AND co.Name='United States'
    AND cl.Percentage>1\G
```

The preceding code produces the following output:

```
mysql> ALTER TABLE city ADD INDEX (Name);
Query OK, 0 rows affected (0.09 sec)
Records: 0  Duplicates: 0  Warnings: 0

mysql> EXPLAIN SELECT cl.Language, cl.Percentage
    -> FROM city ci JOIN country co ON ci.CountryCode=co.Code
    -> JOIN countrylanguage cl ON cl.CountryCode=co.Code
    -> WHERE
    ->    ci.Name='San Francisco'
    ->    AND co.Name='United States'
    ->    AND cl.Percentage>1\G
*************************** 1. row ***************************
           id: 1
  select_type: SIMPLE
        table: ci
   partitions: NULL
         type: ref
possible_keys: CountryCode,Name
          key: Name
      key_len: 35
          ref: const
         rows: 1
     filtered: 100.00
        Extra: NULL
*************************** 2. row ***************************
           id: 1
  select_type: SIMPLE
        table: co
   partitions: NULL
         type: eq_ref
possible_keys: PRIMARY,Name
          key: PRIMARY
      key_len: 3
          ref: world.ci.CountryCode
         rows: 1
     filtered: 5.00
        Extra: Using where
*************************** 3. row ***************************
           id: 1
  select_type: SIMPLE
        table: cl
   partitions: NULL
         type: ref
possible_keys: PRIMARY,CountryCode
          key: PRIMARY
      key_len: 3
          ref: world.ci.CountryCode
         rows: 4
     filtered: 33.33
        Extra: Using where
3 rows in set, 1 warning (0.00 sec)
```

Figure 5.41 – EXPLAIN output after adding the second index

It starts again with the `city` table and filters out `San Francisco`. Then it uses `CountryCode` to do a lookup into the other two tables.

6. To improve this even more, get rid of the `country` table as you don't strictly need this in this query. And also, you know there is only one `San Francisco`:

```
EXPLAIN SELECT cl.Language, cl.Percentage
FROM city ci
JOIN countrylanguage cl ON cl.CountryCode=ci.CountryCode
WHERE
    ci.Name='San Francisco'
    AND cl.Percentage>1\G
```

The preceding code produces the following output:

```
mysql> EXPLAIN SELECT cl.Language, cl.Percentage
    -> FROM city ci
    -> JOIN countrylanguage cl ON cl.CountryCode=ci.CountryCode
    -> WHERE
    ->    ci.Name='San Francisco'
    ->    AND cl.Percentage>1\G
*************************** 1. row ***************************
           id: 1
  select_type: SIMPLE
        table: ci
   partitions: NULL
         type: ref
possible_keys: CountryCode,Name
          key: Name
      key_len: 35
          ref: const
         rows: 1
     filtered: 100.00
        Extra: NULL
*************************** 2. row ***************************
           id: 1
  select_type: SIMPLE
        table: cl
   partitions: NULL
         type: ref
possible_keys: PRIMARY,CountryCode
          key: PRIMARY
      key_len: 3
          ref: world.ci.CountryCode
         rows: 4
     filtered: 33.33
        Extra: Using where
2 rows in set, 1 warning (0.01 sec)
```

Figure 5.42 – EXPLAIN output with the country table removed

Note that the database will never do this as it doesn't know whether it's safe to do. In this case, it is safe, but for many other cities, it is not.

Activity 5.01: The Sakila video store

You are the database administrator of the Sakila video store. As there is a lot of competition, the manager wants to do some marketing and reduce costs. The manager asks your help in obtaining the following information from the database:

- Finding the total number of films the store has with a PG rating. This is needed for advertisements.

- Finding films in which Emily Dee performed as an actor. This is also needed for advertisements.

- Finding the customers who rented the most items. The manager needs this to see what the impact of a loyalty program would be.

- Finding the film that resulted in the biggest income. This is so the manager knows what films should be bought next year.

- Finding the email address of the customer living in Turkmenistan. This is so they can be sent a questionnaire to get some feedback from a customer who doesn't live close to the store. The manager is thinking of increasing shipping costs.

> **Note**
>
> The Sakila database can be downloaded from `https://downloads.mysql.com/docs/sakila-db.zip`.

Follow these steps to complete this activity:

1. For each question, find the tables you need and join them.

2. Aggregate the data if needed.

3. Select the fields you need to answer the questions.

The following diagram created using MySQL Workbench will help you understand how the tables are related:

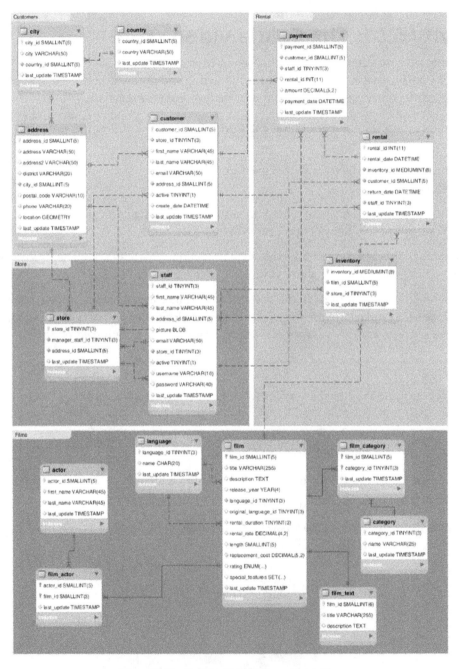

Figure 5.43 – Relationship between the tables

> **Note**
> The solution to this activity can be found in the *Appendix*.

In this exercise, you were able to help the manager with the data he needed for promotional campaigns, possible cost savings, and so on. Even if this data is not stored in a single table, you can answer the questions by combining multiple tables and then filtering and aggregating the results. In the upcoming activity, you will generate a list of years, which will display the total videos rented per year. This report will allow the manager to decide what videos to purchase.

Activity 5.02: Generating a list of years

The manager of the Sakila video store wants to buy some new videos to rent out. He wants a weekly report that shows how many videos per year of release there are in the database. This helps him to decide what videos to buy. For this activity, you will again use the Sakila database. You want a list of the number of films per year of release for the period between 2005 and 2010. Follow these steps to implement this activity:

1. Create a CTE to generate a range of years.

2. Join against the list of years we have generated.

After following these steps, the expected output should look like the following:

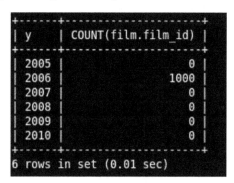

Figure 5.44 – SELECT output with film release dates between 2005 and 2010

> **Note**
> The solution to this activity can be found in the *Appendix*.

Summary

In this chapter, you learned how to combine the information in multiple tables to get the results you want. In addition to that, you learned how to use the `WITH` statement to create virtual tables that are only valid for the duration of the query, but can make the queries easier to read. And by using `WITH RECURSIVE`, you now know how to generate ranges of data that can be used for joining or for generating data for testing. With `EXPLAIN`, we can now start to understand what the database needs to do to get our results and how indexes can help to improve that.

In the next chapter, we will cover making changes to the data stored in tables and/or collections and how to remove data from tables and collections that are no longer needed. For this, we will use `UPDATE` and `DELETE` statements to work with tables and `modify()` and `remove()` to work with JSON documents inside collections.

6
Stored Procedures and Other Objects

In this chapter, we will continue exploring SQL coding and working with our database. We will create objects that can be reused and flexible enough to accept parameters. By doing this, you will learn how to create views, functions, stored procedures, and transactions to allow users to interact with a MySQL database easily.

This chapter covers the following topics:

- Introduction to database objects
- Exploring various database objects
- Working with views
- Activity 6.01 – updating the data in a view
- Working with user-defined functions
- Working with stored procedures
- Working with IN, OUT, and INOUT
- Exploring triggers
- Using transactions

Introduction to database objects

In the previous chapter, you learned how to back up and restore a database, create a database and tables with SQL commands, and set their properties. You also learned how to add, read, write, modify, and delete records using SQL commands before learning about foreign keys and indexes, and why they are essential. Finally, you learned about multi-table queries and various table joins. You will be using your knowledge of these subjects in this chapter to work with stored procedures, views, and functions.

Views are database objects that allow you to save a particular query as a table. This allows you to save results so that they can be used later. Views allow people with little SQL experience to access complex datasets that have been constructed from SQL queries. Functions can be used to create custom programming logic for your database. This is helpful in situations where you have code repeated in multiple areas and you want to avoid copying code multiple times.

Stored procedures allow you to store a set of SQL queries, to be executed when required. This is useful for completing tasks such as loading data into a database. Typically, you should use stored procedures when a query or set of queries must be repeated regularly. Triggers allow you to complete a query when another query or event occurs. For example, you can create a trigger that runs when a table is updated or when a new record is added to the table.

Exploring various database objects

There are several database objects you will work with consistently as your database portfolio expands during your career of creating and working with databases. These objects are as follows:

- **Tables**: Tables are the base objects in databases and are used to store static data. Tables contain records, which have one or more fields that display properties of the data. These tables should be designed around the Third Normal Form to ensure efficient data storage. All foreign keys, along with their constraints and indexes, should be created to ensure data integrity and speed of use.

- **Views**: Views are SQL queries that are stored in a permanent state in the database and can be used by other objects or external applications. They can consist of one or more tables with criteria filtering; however, they do not accept parameters. In certain conditions, they can be updated, though usually, they are read-only.

Now, let's learn how to work with views.

Working with views

Views are queries that are saved in a database. They are mostly used in read-only format; only under some circumstances can they be used to update data in a table. Once a view has been created, it can be used in MySQL as if it were a table or linked to an external application, such as MS Access, as a table.

Views have multiple uses. Typically, you use a view when a query may be accessed more than once. For example, let's say we had a database of customers and their orders. The sales team may want to create a query that shows the total sales for each customer for a given year. We can save this query as a view to allow the sales team to access it whenever they need to. This also allows users who are not experienced with SQL to access data that is created using SQL queries, which ensures that the databases are as simple as possible for all users.

A view can be created using the following query:

```
CREATE VIEW `<View Name>` AS
<Your query SQL here>
```

The structure is simple; just enter a name for the view and enter the respective SQL statement after AS. For example, if you wanted to create a view that contains all the data from a table named customers, the following query would work:

```
CREATE VIEW `customerData` AS
SELECT * FROM customers
```

In the next exercise, you will create a view from a single table.

Exercise 6.01 – creating a mailing list with a view

The event organizer of the Automobile Club needs you to create a list of active club members and include their names and address details. The members table contains all of the members' names, while the memberaddress table contains the members' addresses. This list will be used for the clubs mailing list. You are required to create a view to extract this information. Follow these steps:

1. Open a new SQL tab.

2. Create a SQL statement that will extract the data for the mailing list. Enter the following text in the SQL tab:

```
SELECT
  members.Surname,
```

```
      members.FirstName,
      memberaddress.StreetAddress1,
      memberaddress.StreetAddress2,
      memberaddress.Town,
      memberaddress.Postcode,
      states.State
FROM
      members
      INNER JOIN memberaddress ON memberaddress.MemberID =
      members.ID
      INNER JOIN states ON memberaddress.State = states.ID
WHERE
      members.Active <> 0
ORDER BY members.Surname, members.FirstName
```

This query joins the members and memberaddress tables to display the members' names, as well as their addresses.

3. Execute the query and examine the results, as follows:

Surname	FirstName	StreetAddress1	StreetAddress2	Town	Postcode	State
Bailey	Summer	851 Bins Spring	NULL	Cheyenneshire	7700	NSW
Balistreri	Hugh	8633 Vandervort Common	NULL	Rileyfort	3296	NSW
Baumbach	Hunter	497 Leffler Cliff	NULL	Kunzeside	7325	NSW
Baumbach	Jeremie	802 Rodriguez Trail	NULL	Kreigerhaven	7485	NSW
Bechtelar	Annette	458 Bashirian Rest	NULL	Port Earnestview	9450	NT
Bechtelar	Dayana	576 Cordelia Dale	NULL	North Malachichester	2681	NSW
Bednar	Adelle	93107 Arielle Walk	NULL	West Obieview	5590	NSW
Bergnaum	Felix	74139 Hirthe Roads	NULL	Eddieview	5177	SA

Figure 6.1 – Members mailing list

These results show a list of all the members in the autoclub database, along with their addresses.

4. Turn the SQL statement into a `Create View` statement by including the following line at the top of the preceding code snippet:

```
CREATE VIEW `vw_MembersMailingList_Active` AS
```

The SQL statement should now look as follows:

```
CREATE VIEW `vw_MembersMailingList_Active` AS
SELECT
  members.Surname,
  members.FirstName,
  memberaddress.StreetAddress1,
  memberaddress.StreetAddress2,
  memberaddress.Town,
  memberaddress.Postcode,
  states.State
FROM
  members
  INNER JOIN memberaddress ON memberaddress.MemberID = members.ID
  INNER JOIN states ON memberaddress.State = states.ID
WHERE
  members.Active <> 0
ORDER BY members.Surname, members.FirstName
```

Figure 6.2 – The CREATE VIEW line inserted at the top of SQL

5. Create the view by running the SQL statement (click the lightning bolt icon). The new view will appear in the **Views** list:

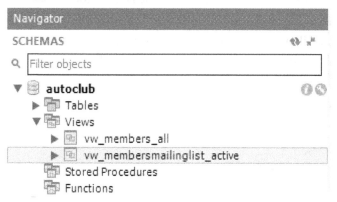

Figure 6.3 – The new vw_membersmailinglist_active view in the list

6. Test the view by right-clicking the view and choosing **Select Rows**. You should get the following result:

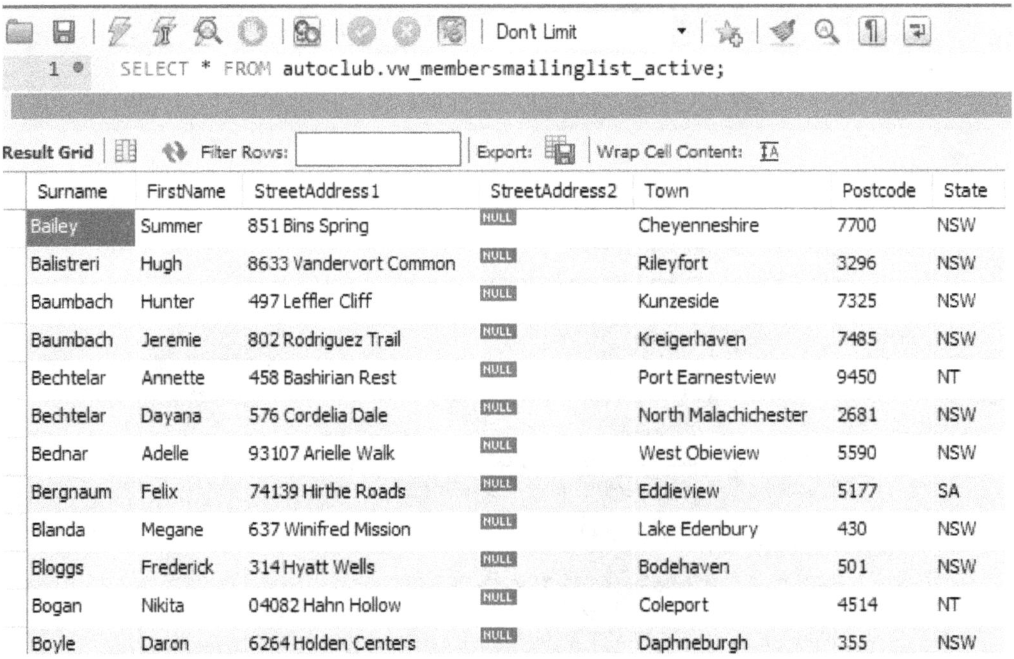

Surname	FirstName	StreetAddress1	StreetAddress2	Town	Postcode	State
Bailey	Summer	851 Bins Spring	NULL	Cheyenneshire	7700	NSW
Balistreri	Hugh	8633 Vandervort Common	NULL	Rileyfort	3296	NSW
Baumbach	Hunter	497 Leffler Cliff	NULL	Kunzeside	7325	NSW
Baumbach	Jeremie	802 Rodriguez Trail	NULL	Kreigerhaven	7485	NSW
Bechtelar	Annette	458 Bashirian Rest	NULL	Port Earnestview	9450	NT
Bechtelar	Dayana	576 Cordelia Dale	NULL	North Malachichester	2681	NSW
Bednar	Adelle	93107 Arielle Walk	NULL	West Obieview	5590	NSW
Bergnaum	Felix	74139 Hirthe Roads	NULL	Eddieview	5177	SA
Blanda	Megane	637 Winifred Mission	NULL	Lake Edenbury	430	NSW
Bloggs	Frederick	314 Hyatt Wells	NULL	Bodehaven	501	NSW
Bogan	Nikita	04082 Hahn Hollow	NULL	Coleport	4514	NT
Boyle	Daron	6264 Holden Centers	NULL	Daphneburgh	355	NSW

Figure 6.4 – The result of the vw_membersmailinglist_active view

With that, you are done. Since you have included a filter for active members only, the list will change as new members are added and existing members are made inactive.

To create a view, you will need to create the SQL for your requirements, add a single `CREATE VIEW `ViewName` AS` line as the first line in the script, and run it.

In the next section, we will look at more complex views that can be updated or changed based on the base query's content.

Updatable views

Views in MySQL are queryable, which means you can include them in another query, much like a table in MySQL. Views can also be updated as you can `INSERT`, `UPDATE`, and `DELETE` rows in the underlying table. There are specific circumstances where a view can and cannot be updatable. For a view to be updatable, the `SELECT` statement that's defining the view cannot contain the following:

- Any of the aggregate functions, including `MIN`, `MAX`, `SUM`, `AVG`, and `COUNT`
- The `DISTINCT` clause

- The GROUP BY clause

- The HAVING clause

- A UNION or UNION ALL clause

- Left joins or outer joins

- A subquery in the SELECT clause or the WHERE clause of the main query that refers to the table appearing in the FROM clause in the main query

- A reference to a view that is non-updatable in the FROM clause

- A reference to only literal values

- Multiple references to any column in the base table

So, if the select statement does not contain any of these elements, you can update the query and treat it just like a table. Fortunately, there is an easier way to determine if a view can be updated – by querying the information schema; that is, information_schema. views. This table contains columns such as the name of the view and whether it can be updated in the field. The following query shows how to see information about views via information_schema:

```
SELECT
  table_name,
  is_updatable
FROM
  information_schema.views
ORDER BY table_name;
```

The following output shows an example of what this query looks like when it's run with updatable views:

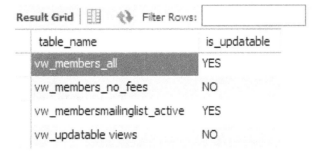

Figure 6.5 – The database views are listed with is_updatable statuses

As you can see, the `vw_members_all` and `vw_membersmailinglist_active` views can be updated. You can use these views in update queries to modify the data and insert or delete records. Any changes you make to these views will be relayed to the base table, `members`. These two views can't be updated and can only be used to read data.

In the next activity, you will confirm that you know how to update data in a view.

Activity 6.01 – updating the data in a view

One of the Automobile Clubs members, **Darby Mariella Collins** (member **ID 7**), has noticed that his **DOB** is incorrect in the system and has asked for it to be adjusted; it should be **January 11, 1990**. Since we have a view that shows information about every member, the best way to update this is by using the `vw_members_all view` function.

In this activity, you will adjust Darby Mariella Collins' date of birth in the system by doing the following:

- Confirming the date is incorrect by examining the DOB for member **ID 7** in the **members** table directly. Remember that MySQL stores dates in the **YYYY-MM-DD** format.

- Creating an update query to adjust the date while using the `vw_members_all` view as the base record source for your query.

- Confirming the date was adjusted by examining the record again in the **members** table.

Views are a simple way of saving commonly used SQL queries and are reusable, resulting in less code redundancy and quicker development time. This is because the results can be saved and accessed without the query having to be rewritten to generate them. Views are usually used as read-only record sets; however, they can be updated too. If you find that you are creating the same SQL several times in your application, it is a prime candidate to be turned into a view. If the view needs to be changed, all the code or objects using that view will pick up the changes. However, be careful when you change established views or any other object that is referenced by other objects or applications. These changes could have undesirable effects, so it is better to test them in a test environment thoroughly first.

In the next section, we will learn how to create functions and explore how they can be used in the **autoclub** database.

> **Note**
> The solution to this activity can be found in the Appendix.

Working with user-defined functions

MySQL has many built-in functions that you can call to return values or perform tasks on data, including CURRENT_DATE(), AVG(), SUM(), ABS(), and CONCAT(). These functions can be used in SQL statements, views, and stored procedures. MySQL also has another type of function, known as the **user-defined function** (**UDF**), that you can create to add new functionality to the database that is not already provided by MySQL. For example, you may want a function that can calculate and return the GST or sales tax or maybe calculate the income tax for your weekly earnings. A UDF is active when it is loaded into the database with CREATE FUNCTION and hasn't been removed with DROP FUNCTION. A function can be used while it is active.

The basic syntax for creating a UDF is as follows:

```
USE database_name;
DROP FUNCTION IF EXISTS function_name;
CREATE FUNCTION function_name ([parameter(s)])
 RETURNS data type
 DETERMINISTIC
 STATEMENTS
```

Let's look at each of the components of a basic UDF:

- USE database_name;: This ensures that the function is created in the correct database.

- DROP FUNCTION IF EXISTS function_name;: This will drop the function if it already exists to avoid an error stating that it already exists. Note that you may need to recreate the function several times during development.

- DELIMITER $$: The default delimiter is the semi-colon, ;. However, when you're defining functions, stored procedures, and triggers, you will often run multiple statements. Defining a different delimiter allows you to run all the statements as a single unit rather than individually.

- CREATE FUNCTION function_name ([parameter(s)]): This is mandatory and tells MySQL server to create a function named function_name. Parameters are optional and are defined in round brackets; multiple parameters can be separated by commas. Each parameter is declared with its name and data type.

- RETURNS data type: This is also mandatory and specifies the data type of the data that the function returns.

Several **informative statements** tell MySQL what the function does. By default, at least one of the following must be included:

- DETERMINISTIC: The function will return the same values if the same arguments are supplied to it, meaning that you always know the output, given the input.

- READS SQL DATA: This specifies if the function will read data from the database but does not modify data.

- MODIFIES SQL DATA: This specifies if the function will modify data in the database.

- CONTAINS SQL: This specifies if the function will have SQL statements but they do not read or modify data, such as SELECT CURRENT_DATE().

- <STATEMENTS>: This is the SQL code you create for the function to execute.

We will go through several exercises to demonstrate various **UDFs**. In the next exercise, you will create a function to look up a value from the lookups table.

Exercise 6.02 – creating a function

You have included a lookups table in your database to store values you will need in your database or the application that will be using it. You realize that you will be using this table a lot, so rather than creating and executing a query each time, you have decided to create a function to look up values, thus reducing your coding.

Follow these steps to create the function:

1. Open a new SQL tab and type in the following statements:

```sql
USE autoclub;

DROP FUNCTION IF EXISTS fn_Lookup;
DELIMITER $$
CREATE FUNCTION fn_Lookup(LookupKey VARCHAR(50)) RETURNS
VARCHAR (200)
READS SQL DATA

BEGIN

DECLARE TheValue VARCHAR(200);
SET TheValue = (SELECT `Value` FROM `lookups` WHERE `Key`
= LookupKey);
```

```
RETURN (RTRIM(LTRIM(TheValue)));

END $$
DELIMITER ;
```

With our function defined, let's break down each component to understand what their role is in the function's definition:

- `USE autoclub` will instruct the server to use the `autoclub` database for all the following commands. This ensures that the function is created in the correct database.

- `DROP FUNCTION IF EXISTS fn_Lookup` will remove the function if it already exists; otherwise, you will get an error when you try to create it.

- `DELIMITER $$` sets the delimiter to `$$`. This enables us to run all the statements to the point that we can reset them as a single block of statements.

- `CREATE FUNCTION fn_Lookup(LookupKey VARCHAR(50))` instructs the server to create a function named `fn_Lookup` with a single parameter named `LookupKey`. This will be a `VARCHAR` data type that's up to 50 characters in length.

- `RETURNS VARCHAR (200)` indicates that the function will return a `VARCHAR` data type that's up to 200 characters in length.

- `READS SQL DATA` is an instruction that tells MySQL that the function will read data from the database but not modify it. The function will look up a value in the database using a SQL statement, but it will not modify the data.

- `BEGIN` specifies when the code that defines the function will begin after this statement; it will end when we get to `END`.

- `DECLARE TheValue VARCHAR(200)` declares a variable named `TheValue` to use in the function. The type and size match the `RETURNS` declaration. The value that's returned from the SQL statement will be stored in this variable and then returned from the function.

- `SET TheValue = (SELECT `Value` FROM `lookups` WHERE `Key` = LookupKey` runs the SQL statement in brackets and passes the result to the `TheValue` variable. The `WHERE` clause of the SQL statement uses the passed-in parameter's `LookupKey` to filter the SQL. Since we are using a parameter, we do not need to include it in quotes like we would for a fixed string filter.

- `RETURN (RTRIM(LTRIM(TheValue)))` returns `TheValue`. The `RTRIM` and `LTRIM` functions will trim any leading and trailing spaces from `TheValue` if there are any. This is a precautionary measure to ensure clean data is returned from the function.

- `END $$` signifies the end of the function's definition code. `$$` is the custom delimiter and signifies the end of the code block as a unit.

- `DELIMITER` resets the delimiter back to the default semi-colon before exiting.

2. Execute the SQL query with the lightning bolt icon. The new function will appear in the **Functions** list:

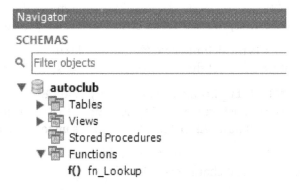

Figure 6.6 – The new fn_Lookup function

3. Test the function by executing the following query in another SQL tab:

```
SELECT fn_Lookup("autoclub");
```

As a result, you will see the root directory for the image repository:

Figure 6.7 – The result of using fn_Lookup()

You can use this function to pass in the key you want to look up; the function will return the corresponding value. If an incorrect key is passed in the function, it will return `NULL`. The function can now be used in any of your SQL code in all the objects in the database and can also be called from external applications.

Setting up a UDF can be complicated, but there are many advantages of reducing code. Let's say that the name of the `lookups` table has changed. You would need to locate all the references to it for all the lookups in your database objects, as well as in all the applications using the database, to change them. However, if they all used the `fn_Lookup` function, you would only need to change the function; the rest of the code/applications would still get their values.

In the next exercise, you will create a function that will accept two parameters, read a value from the database, and call another function.

Now that we have learned about UDFs, let's learn about stored procedures.

Working with stored procedures

Stored procedures are the workhorses of your MySQL database. Similar to UDFs, they can run multiple SQL statements, contain the logic flow, and return the results. Stored procedures are used for situations where you want to store queries that will need to be run multiple times. For example, if a set of queries need to be run daily, they can be created as stored procedures. Where UDFs return a single result, stored procedures can return a single result, or they can return entire record sets. They are ideal for moving extensive processing tasks to a MySQL server. Imagine that you are working in a sales application that's connected to a MySQL database and you need to record a sale. Your application would record a sale by doing the following:

- Determining the total payment amount to confirm payment:

  - Calculating the sale value (sales cost * item value)

  - Calculating the sales tax that applies to the sale

- Subtracting the item from the inventory table

- Checking the item's minimum stock value

- Generating a receipt

That is a lot of work for the application to do. All these tasks can be placed in a stored procedure, or a function with a single stored procedure that coordinates the logic involved, to update all of the relevant tables to complete the sale. This would result in the application only needing to call the stored procedure and pass in the relevant details.

To create a stored procedure, you can use the following syntax:

```
CREATE PROCEDURE `procedure_name` ()
BEGIN
SQL code for procedure goes here
END$$
```

In the next exercise, we'll create a simple stored procedure that returns some records from the **autoclub** database.

Exercise 6.03 – creating a stored procedure

The Automobile Club wants to be able to list all the members in their club using a stored procedure. This will allow them to easily run a query to display the data as required. To do this, you have been asked to create a stored procedure in the autoclub database. Follow these steps:

1. Open a new SQL tab and enter the following SQL statements:

```
USE `autoclub`;
DROP procedure IF EXISTS `sp_ListMembers`;
DELIMITER $$

CREATE PROCEDURE `sp_ListMembers` ()

BEGIN

SELECT * FROM members;

END$$

DELIMITER ;
```

This query will start by dropping any procedures named **sp_ListMembers** since we can't have two procedures with the same name. After this, the query creates a new stored procedure, which selects all of the data from the **members** table.

2. Run the SQL query. The new procedure, **sp_ListMembers**, will appear in the **Stored Procedures** list:

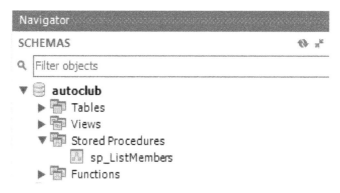

Figure 6.8 – sp_ListMembers in the Stored Procedures list

3. Open a new SQL tab and run the following command to test the stored procedure:

```
call sp_ListMembers
```

The stored procedure will run and the following output will appear in the **Result Grid** area:

ID	Surname	FirstName	MiddleNames	DOB	Signature	Photo	PhotoPath	SigPath	Active	JoinDate	InactiveDate	WhenAdded	LastMod
1	Bloggs	Frederick	NULL	1990-06-16	BLOB	BLOB	Members\Photos\MemberPhoto_1.jpg	Members\Signatures\MemberSignature_1.jpg	1	2020-01-15	NULL	2020-01-21 20:02:30	2020-01-
2	Pettit	Thomas	William	1960-10-15	NULL	NULL	Members\Photos\MemberPhoto_2.jpg	NULL	1	2020-01-20	NULL	2020-01-21 21:01:56	2020-01-
5	West	Anais	Avery	1984-08-06	NULL	NULL	NULL	NULL	1	2009-07-11	NULL	2020-01-24 17:48:10	2020-01-
6	Smaniawski	Waylon	Rita	2007-10-08	NULL	NULL	NULL	NULL	1	2020-01-24	NULL	2020-01-24 17:48:10	2020-01-
7	Collins	Darby	Marielle	1990-01-11	NULL	NULL	NULL	NULL	1	2015-10-12	NULL	2020-01-24 17:48:10	2020-01-
8	Schamberger	Dexter	D'angelo	1999-10-07	NULL	NULL	NULL	NULL	1	2020-06-01	NULL	2020-01-24 17:48:10	2020-01-
9	Wintheiser	Tania	Toy	2020-01-24	NULL	NULL	NULL	NULL	1	1993-10-01	NULL	2020-01-24 17:48:10	2020-01-
10	GuÃªann	Markus	Letha	2020-01-24	NULL	NULL	NULL	NULL	1	2020-01-24	NULL	2020-01-24 17:48:10	2020-01-
11	Welch	Kenya	Shanel	2020-01-24	NULL	NULL	NULL	NULL	1	2020-01-24	NULL	2020-01-24 17:48:10	2020-01-

Figure 6.9 – The output of running the stored procedure

Creating a basic stored procedure is very similar to creating a UDF. We did not include any parameters or results so, by default, the stored procedure returns the results of the SQL statement. This is useful for getting a list or a dataset. However, note that the dataset will be read-only, so you cannot update the record set.

In the next exercise, you will learn how to pass parameters in a stored procedure.

Exercise 6.04 – stored procedures and parameters

The Automobile Club now wants to be able to list all the data in a specific table using a stored procedure. To achieve this, you will need to create a stored procedure that takes in a table's name and outputs the data for that table. Follow these steps:

1. Open a new SQL tab and enter the following script:

```
USE `autoclub`;
DROP procedure IF EXISTS `sp_ListTableData`;
```

```
DELIMITER $$

CREATE PROCEDURE `sp_ListTableData` (IN TableName
VARCHAR(100))
BEGIN

SET @sql =CONCAT('SELECT * FROM ',TableName);
    PREPARE statement FROM @sql;
    EXECUTE statement;
DEALLOCATE PREPARE statement;

END$$

DELIMITER ;
```

This query starts by creating the sp_ListTableData procedure, which takes in a single argument named TableName. This argument is text, which specifies the table where the data will be selected. Next, a variable called @sql is created that stores the SELECT query for the provided table. Finally, the query is executed, and the results as displayed on the screen.

2. Run the preceding SQL query to build a stored procedure that will appear in the **Stored Procedures** list:

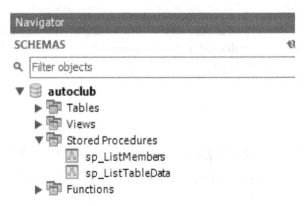

Figure 6.10 – sp_ListTableData in the Stored Procedures list

3. Open a new SQL tab. Enter the following test script and run it:

```
call sp_ListTableData("members");
```

This results in the following output:

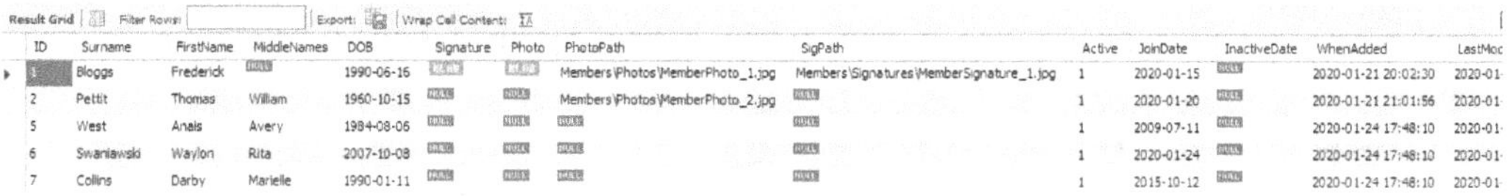

ID	Surname	FirstName	MiddleNames	DOB	Signature	Photo	PhotoPath	SigPath	Active	JoinDate	InactiveDate	WhenAdded	LastMoc
1	Bloggs	Frederick	NULL	1990-06-16	BLOB	BLOB	Members\Photos\MemberPhoto_1.jpg	Members\Signatures\MemberSignature_1.jpg	1	2020-01-15	NULL	2020-01-21 20:02:30	2020-01
2	Pettit	Thomas	William	1960-10-15	NULL	NULL	Members\Photos\MemberPhoto_2.jpg	NULL	1	2020-01-20	NULL	2020-01-21 21:01:56	2020-01
5	West	Anais	Avery	1984-08-06	NULL	NULL	NULL	NULL	1	2009-07-11	NULL	2020-01-24 17:48:10	2020-01
6	Swaniawski	Waylon	Rita	2007-10-08	NULL	NULL	NULL	NULL	1	2020-01-24	NULL	2020-01-24 17:48:10	2020-01
7	Collins	Darby	Marielle	1990-01-11	NULL	NULL	NULL	NULL	1	2015-10-12	NULL	2020-01-24 17:48:10	2020-01

Figure 6.11 – Members table data

4. Open a new SQL tab. Enter the following test script and run it:

```
call sp_ListTableData("memberaddress");
```

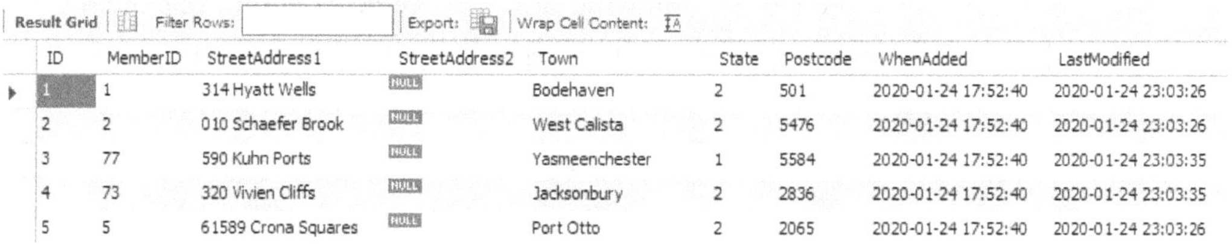

ID	MemberID	StreetAddress1	StreetAddress2	Town	State	Postcode	WhenAdded	LastModified
1	1	314 Hyatt Wells	NULL	Bodehaven	2	501	2020-01-24 17:52:40	2020-01-24 23:03:26
2	2	010 Schaefer Brook	NULL	West Calista	2	5476	2020-01-24 17:52:40	2020-01-24 23:03:26
3	77	590 Kuhn Ports	NULL	Yasmeenchester	1	5584	2020-01-24 17:52:40	2020-01-24 23:03:35
4	73	320 Vivien Cliffs	NULL	Jacksonbury	2	2836	2020-01-24 17:52:40	2020-01-24 23:03:35
5	5	61589 Crona Squares	NULL	Port Otto	2	2065	2020-01-24 17:52:40	2020-01-24 23:03:26

Figure 6.12 – MembersAddress table data

Here, you can see that the table outputted matches that the variable provided to the procedure.

In this exercise, you learned how to pass in a string value and use it as part of a SQL statement, as well as how to use the `PREPARE` statement to create some SQL from it and execute it. You also learned how to pass in table names and get the stored procedure to create a query using the table name to list its records.

In the next section, you will learn about the `IN`, `OUT`, and `INOUT` keywords and how you can customize your function parameter's functionality.

Working with IN, OUT, and INOUT

MySQL stored procedures have three directions that a parameter can be defined in. This is mandatory, which means that a parameter must be set to one of the following:

- `IN`: The value is only being passed to the stored procedure. It is used within the procedure. This is the same as providing input to the stored procedure.
- `OUT`: The value is only passed out of the stored procedure; any external variables that have been assigned to this position will take on the value that's passed out. This is similar to returning values from a stored procedure.
- `INOUT`, A variable and its value (`ExtVal`) are passed to the stored procedure (`IntVal`) and can be modified within it. When the stored procedure is completed, the external value (`ExtVal`) will equal the modified value (`IntVal`).

In the next exercise, you will learn how to use the IN and INOUT parameters in a MySQL stored procedure.

Exercise 6.05 – IN and INOUT

The Publicity Department of the Automobile Club wants to know how many Holden, Ford, Mazda, and Toyota cars belong to the club's members. They would like the stored procedure to be able to display the total number of vehicles that are of these makes. Follow these steps:

1. Open a new SQL tab and enter the following stored procedure definition. Note that there are lots of comments, so ensure that you take them into account:

```sql
USE `autoclub`;
DROP procedure IF EXISTS `sp_CountCars_MembersMakes`;


BEGIN


    SELECT
        Count(vehicle.Make) INTO @TotalInMake
    FROM
        vehicle
        INNER JOIN members ON vehicle.MemberID = members.
ID
        INNER JOIN make ON vehicle.Make = make.ID
        INNER JOIN vehiclemodel ON vehicle.Model =
vehiclemodel.ID
      WHERE
        members.Active <> 0 AND
        make.Make = CarMake;
  SET CarString = CONCAT(CarString,CarMake,"=",@
TotalInMake, "   ");


END$$


DELIMITER ;
```

2. Run the preceding SQL query to create the stored procedure. It should appear in the `Stored Procedures` list:

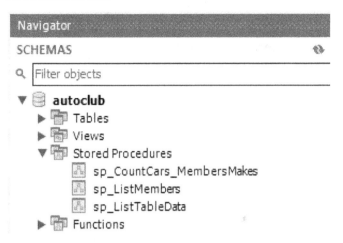

Figure 6.13 – sp_CountCars_MembersMakes in the Stored Procedures list

3. Open a new SQL tab and run the following test script:

```sql
-- Declare the variables
SET @TotalCars = 0;
SET @MakeString = "Car Make/Count :- ";

@MakeString
call sp_CountCars_MembersMakes("Holden",@TotalCars,@
MakeString);
call sp_CountCars_MembersMakes("Ford",@TotalCars,@
MakeString);
call sp_CountCars_MembersMakes("Mazda",@TotalCars,@
MakeString);
call sp_CountCars_MembersMakes("Toyota",@TotalCars,@
MakeString);

SELECT @MakeString, @TotalCars
```

Your output should look as follows:

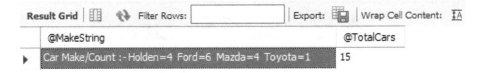

Figure 6.14 – Output of the test script

Let's look at what happened here:

- **The stored procedure**: The comments in the stored procedure explain what each line is doing. The main point is that three parameters have been passed in – one IN and two INOUT. The IN parameter is a string that will be used to filter the internal query. The two INOUT parameters will pass in values – a numeric value and a string value. Both of these values will be modified and added to each time sp_CountCars_MembersMakes is called. The external variables that are passed into the INOUT parameters will take on these new values.

- **The test script**:

```
SET @TotalCars = 0;   -- Declare and initialise @TotalCars
SET @MakeString = "Car Make/Count :- ";   -- Declare and
initialise @MakeString
```

The two SET lines at the start initialize the variables. These variables will be passed into the two INOUT parameters. Once they have been passed through the stored procedure, they will be modified and the new values will be passed into the next call of the stored procedure:

```
call sp_CountCars_MembersMakes("Holden",@TotalCars,@
MakeString);
call sp_CountCars_MembersMakes("Ford",@TotalCars,@
MakeString);
call sp_CountCars_MembersMakes("Mazda",@TotalCars,@
MakeString);
call sp_CountCars_MembersMakes("Toyota",@TotalCars,@
MakeString);
```

The stored procedure is called four times, each time with a different car make. The procedure will then count the car make and update both the INOUT variables with details of the current car make.

The following table shows the value once it's been passed through the procedure:

Stage	@MakeString	@TotalCars
Initialized values	Car Make/Count :-	0
Holden	Car Make/Count :- Holden=4	4
Ford	Car Make/Count :- Holden=4 Ford=6	10
Mazda	Car Make/Count :- Holden=4 Ford=6 Mazda=4	14
Toyota	Car Make/Count :- Holden=4 Ford=6 Mazda=4 Toyota=1	15

Figure 6.15 – The result of each of the values after each pass through the procedure

```
SELECT @MakeString, @TotalCars
```

Finally, we get to the last line, which selects the two variables, causing them to be passed as the final output of the script. The results are shown in the preceding table:

- **The stored procedure**:

Now, let's break down the stored procedure:

```
USE `autoclub`;
DROP procedure IF EXISTS `sp_CountCars_MembersMakes`;
DELIMITER $$
```

These are the standard USE, DROP, and user-defined DELIMITER functions we have been using. This is standard for all our object definitions – only the database and object name will change as required:

```
CREATE PROCEDURE `sp_CountCars_MembersMakes` (IN CarMake
VARCHAR(20), INOUT TotalCars INT, INOUT CarString
VARCHAR(255))
```

Here, we created the stored procedure by providing its name. Three parameters have been defined – IN, a string value up to 20 characters in length that accepts the car's make; our first INOUT parameter, which is an integer value that specifies the total car count that will be added by the procedure; and our second INOUT parameter, which is a string that can be up to 255 characters in length. This displays the car's make, which will also be added to the procedure:

```
BEGIN
```

The beginning of the procedure's code is as follows:

```
SELECT
    Count(vehicle.Make) INTO @TotalInMake
FROM
    vehicle
    INNER JOIN members ON vehicle.MemberID = members.ID
    INNER JOIN make ON vehicle.Make = make.ID
    INNER JOIN vehiclemodel ON vehicle.Model =
vehiclemodel.ID
WHERE
    members.Active <> 0 AND
    make.Make = CarMake;
```

Let's prepare a standard `select` statement to count how many makes of a vehicle have been passed into the database. The value of the count will be placed in the `@TotalInMake` variable, while the filters will be placed in the `WHERE` clause. Note that `CarMake` is not surrounded by quotes; it is a variable, so this isn't necessary. If we were to add quotes, then it would look for a make named `CarMake` and return zero records:

```
    SET TotalCars = TotalCars + @TotalInMake;
  SET CarString = CONCAT(CarString,CarMake,"=",@
  TotalInMake, "  ");
```

These two lines add the new values to the existing values of the `INOUT` variables that have been passed in. The first will add the current `@TotalInMake` to the passed-in `TotalCars` and the new value will be the sum of both. The second appends the current vehicle's `CarMake` (this is the passed-in `IN` parameter) and the value in `@TotalMake` to the passed-in string, thereby building on the string. At the end of the procedure, these two values are passed back through the `INOUT` parameters. Here, the external values will change to the values that have been set in the procedure:

```
END$$
DELIMITER ;
```

At this point, we can reset `DELIMITER` to its default state.

In this exercise, you learned how to use the `IN` and `INOUT` parameters in a stored procedure.

Stored procedures offer a lot of flexibility when it comes to using parameters and you can design them to perform otherwise tedious tasks with a simple call. A stored procedure can return a complete record set, but it cannot be updated. If you need an updatable record set, you will need to use a view. However, stored procedures can be programmed to add, modify, and delete records and data using parameters that have been passed in to filter the database and ensure the correct output.

In the next section, we will learn about triggers.

Exploring triggers

A trigger runs automatically when a predefined action is performed on the table. You should use triggers when data has changed in a database and you want to take action. There are two types of triggers in MySQL. The first is called a row-level trigger, which executes once for each row in the transaction. The second is called a statement-level trigger, which executes only once for each transaction.

There are three possible EVENTS a trigger can be assigned to – INSERT, UPDATE, and DELETE. A trigger can be run at a specific time concerning the event. The time can be either before or after the event occurs. A trigger can be used to validate data, log the old and new values in an audit trail, or ensure business rules are adhered to.

You can create a trigger using the following syntax:

```
CREATE TRIGGER trigger_name
 (AFTER|BEFORE) (INSERT|UPDATE|DELETE)
 ON table_name FOR EACH ROW BEGIN
 SQL to execute
 END
```

Let's look at various aspects of triggers.

Advantages of triggers

Triggers assist with data integrity, catching errors, and tasks that can be run automatically when the trigger fires rather than being scheduled. They are good for auditing data changes, logging events, and assisting in preventing invalid transactions.

Disadvantages of triggers

Triggers cannot replace all data validation procedures. They can only provide additional data validation. They cannot be seen by the client applications and their actions can be confusing to developers as they cannot see what is happening in the database layer. They use a large number of resources on the database server and do not provide any benefits when there are a lot of data events per second. This is because the triggers will be firing all the time and drain the database server's resources.

Restrictions with triggers

Triggers come with their own set of restrictions. For example, there can only be one trigger per event; you can only have one `BEFORE UPDATE` trigger on any given table but you can run several statements in them. Triggers do not return values. They cannot use the `CALL` statement and they cannot create temporary tables or views.

Often, we may want to use programming logic to enforce conditions with triggers. To achieve this, we can use several different statements. The first is called `FOR EACH`, which can be used to iterate the rows in a table. Specifically, `FOR EACH ROW` can be used to iterate through every row of a given table. The second statement that can be used is an `IF` statement, which will execute a command when a specific condition is true.

In the next exercise, you will create a trigger to enforce a business rule.

> **Note**
>
> Triggers can result in inconsistent results based on several factors, including the type of database engine that's being used with the table the trigger has been assigned to. In the following exercises, we will be setting up some triggers and also testing them with the **InnoDB** and **MyISAM** engines. Various issues will be pointed out and tested. The results will be compiled at the end so that you can decide if you wish to go down the trigger path with future databases.

Exercise 6.06 – triggers to enforce business rules

The Automobile Club wants to establish a minimum age requirement for its members. To do this, they want to ensure that every record that's inserted into the `members` table has an age field of over 18. To achieve this, you must write a trigger that runs when a new record is inserted and verify the specified age.

Follow these steps to complete this exercise:

1. Add a rule that sets the minimum age of a member to 18 as a new entry in the
    Lookups table. Open a new SQL query tab and run the following command:

    ```
    INSERT INTO `autoclub`.`lookups` (`Key`, `Value`,
    `Descriptions`) VALUES ('MinMemberAge','18','Minimum age
    in years for members');
    ```

    You should get the following result:

Key	Value	Descriptions
GSTRate	.1	GST Rate, currently 10%
ImageRepository	D:\FileRepository\	AutomobileClubimages
MinMemberAge	18	Minimum age in years for members
NULL	NULL	NULL

Figure 6.16 – The new NewMemberAge business rule

2. Check the current age of the test member so that we know what our value for
    comparison is. To do this, open a new SQL query tab, enter the following script,
    and run it. Keep this tab at hand so that you can check it later:

    ```
    SELECT Firstname, Surname, DOB FROM members WHERE ID=2
    ```

    You should get an output similar to the following:

Firstname	Surname	DOB
Thomas	Pettit	1960-10-15

Figure 6.17 – The test member's current DOB value

3. Now, open a new SQL query tab and enter the following script:

    ```
    DELIMITER $$

    DROP TRIGGER IF EXISTS autoclub.CheckMemberAge$$
    USE `autoclub`$$
    CREATE  TRIGGER `CheckMemberAge` BEFORE UPDATE ON
    `members` FOR EACH ROW BEGIN
    ```

```
    declare msg varchar(128);

    SET @MinAge = (SELECT `Value` FROM LOOKUPS WHERE
`KEY`='MinMemberAge');

    if NEW.dob > (SELECT DATE_SUB(curdate(), interval @
MinAge year)) THEN
    set msg = concat('MyTriggerError: Minimum member age
is: ', @MinAge);
    signal sqlstate '45000' set message_text = msg;
    end if;

END$$
DELIMITER ;
```

The first time you run the script, line 1 may be different because the trigger does not exist yet, but lines 2 and 3 should be the same. Any subsequent runs will be as shown:

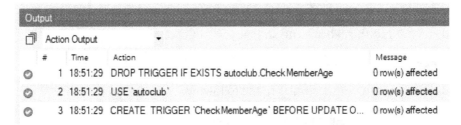

Figure 6.18 – Trigger creation messages after executing the script

4. Open a new SQL query tab and enter the following script to update the date of birth to October 15, 2006:

```
update `autoclub`.`members` SET `DOB` = '2006-10-15'
WHERE ID = 2;
```

You should get the following output:

#	Time	Action	Message
1	17:39:58	update `autoclub`.`members` SET `DOB` = '2006-10-15' WHERE ID = 2	Error Code: 1644. MyTriggerError: Minimum member age is: 18

Figure 6.19 – The script failed to execute because the member is under 18

Error Code: 1644 is generated when there is an unhandled user-defined exception condition.

5. Run the script again to check the current **DOB**:

Figure 6.20 – The test member's DOB has not changed

Note that the DOB has not changed because the trigger detected an invalid age.

6. Now, open a new SQL query tab and execute the following script:

```
update `autoclub`.`members` SET `DOB` = '2000-10-15'
WHERE ID = 2;
```

The following will be displayed in the output window, indicating that the script was successful:

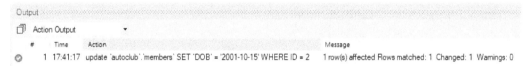

Figure 6.21 – The update script was successful in changing the test member's DOB

Quickly checking the `members` table confirms that the DOB was changed since the business rule was satisfied and the update was allowed to proceed:

Figure 6.22 – The test member's DOB has indeed been changed

First, we added a business rule to the lookup table. We did this because rules change. So, if the membership age is ever lowered to 18 or raised, it only needs to be changed in the lookup table. Then, any code, SQL, triggers, or applications that use this age limit will always pick up the current age limit.

Then, we created the rule. Let's break the following SQL down:

```
DELIMITER $$
DROP TRIGGER IF EXISTS autoclub.CheckMemberAge$$
USE `autoclub`$$
```

These are the standard user-defined `DELIMITER` and `DROP` objects we have been using. We are also adding the `USE `autoclub`$$` command to ensure we are putting the trigger in the right database. Notice that the `DROP` and `USE` lines are terminated with the `$$` delimiter instead of the usual `;`. This is because we changed it so that these lines will each run their commands. Note that the `CREATE TRIGGER` code does not use `$$` until `END$$`, which means that this entire block will run as a single command.

```
CREATE  TRIGGER `CheckMemberAge` BEFORE UPDATE ON
`members` FOR EACH ROW BEGIN
```

A few things are going on here:

- `CREATE TRIGGER 'CheckMemberAge`` is creating the trigger. Notice that the name of the trigger is enclosed in backticks. These can be removed, so long as there are no spaces in the. Note that you shouldn't use spaces as these will annoy you later.

- `BEFORE UPDATE` states that we want this trigger to run before the record is updated. Run validation triggers on the `BEFORE` event so that you can cancel them if they fail the validation procedure.

- `ON `members`` is telling us to create the trigger for the `members` table.

- `FOR EACH ROW` is saying that the trigger should run for each row or record that's been updated (not the entire table). So, if you updated 100 records in bulk, the trigger would run for each record update. If one of those 100 records were underaged, then only that one would be rejected; the others would be updated. MySQL only supports `FOR EACH ROW`; it does not support `FOR EACH STATEMENT`.

- `BEGIN` signifies that the statements are in the body of the trigger.

7. In the trigger logic, we start by declaring a new variable:

```
declare msg varchar(128);
```

8. The following statement declares a string variable that will store the error message we may need to return if the age test fails. It can hold a string that's up to 128 characters in length:

```
    SET @MinAge = (SELECT `Value` FROM LOOKUPS WHERE
`KEY`='MinMemberAge');
```

9. The following statement looks up the `MinMembersAge` value in the `Lookups` table and stores it in a variable named `MinAge` so that it can be used later in the script:

```
    if NEW.dob > (SELECT DATE_SUB(curdate(), interval @
MinAge year)) THEN
```

This is where the `NEW` DOB value is tested against the `@MinAge` value. We use the `NEW` command to reference the value we are trying to insert. The existing value is referred to as `OLD`. We will use this later in this chapter. `SELECT DATE_SUB(curdate(), interval @MinAge year)` is a separate SQL statement, so it is enclosed in brackets, `()`. It subtracts 18 years from the current data to find the comparison date. If `NEW.DOB` is greater, then the prospective member is too young and the command in the **IF-THEN** block will execute.

10. The `DATE_SUB(date, INTERVAL value interval)` function subtracts a time or date interval from a date and then returns the date:

```
    set msg = concat('MyTriggerError: Minimum member age is:
', @MinAge);
    signal sqlstate '45000' set message_text = msg;
```

If the age test results in the member being under 18, then the following two lines will be executed:

- `set msg` sets the message to be returned. Here, we are concatenating a text message and the minimum age.

- `signal sqlstate` returns an error state. A large list of error values can be used that can be located on the internet. Here, **45000** means **unhandled user-defined exception**. When it's returned, MySQL specifies **ErrorCode 1644**, which means **unhandled user-defined exception**.

- `set message_text` assigns the error message that was defined in the previous line to be returned.

- `signal sqlstate` effectively cancels the update attempt, returns the error code and message, and then drops out of the trigger; the data is not updated:

```
        end if;
```

11. The following code shows the end of the **IF END IF** block. If the age test results in an age of 18 or over, then the code within the block will not be executed and an update will occur:

```
END$$
DELIMITER ;
```

These lines end the CREATE TRIGGER block and reset the DELIMITER back to its default – that is, ;.

In short, when a record is updated, this trigger will check the age that's been provided. If it's under 18, then code will be run to cancel the update; otherwise, it will allow the update to occur.

In the next section, we will learn about transactions.

Using transactions

Depending on the application you use to connect to MySQL, you may have to execute a COMMIT statement to save the data. By default, the MySQL client is set to use autocommit, so you don't have to do this. If you want to have the option to undo the INSERT statement, then you need to use a transaction. This can be done either with a BEGIN statement or a START TRANSACTION statement. Once you have run one or more statements to modify the data, you need to use COMMIT or ROLLBACK.

The following code shows how to use a transaction:

```
BEGIN;   -- This indicates the begin of the transaction
INSERT INTO mytable VALUES (1, 'foo', 'bar', 'baz');
SELECT * FROM mytable;
COMMIT;   -- Use ROLLBACK instead of COMMIT if you don't want to
save your work
```

> **Note**
>
> In the preceding statements, the two dashes followed by a space (--) indicates a comment in MySQL. These comments can also use C style comments; for example, /* some comment */.

If you get disconnected from the MySQL server, then the database will automatically roll back your transaction. If you insert, delete, and/or update multiple rows in the same transaction, then either all the changes will be applied to the database or none at all. Only after running the COMMIT statement will the other users of the database be able to view your changes.

> **Note**
> You can ask the database to show data from other sessions that haven't been committed yet, but that's not common.

Another very useful statement is TRUNCATE, which allows you to remove all the data from a table. This is a very powerful command and should be used with care. The statement looks as follows:

```
TRUNCATE <table name>;
```

In the next exercise, you will implement a transaction.

Exercise 6.07 – implementing a transaction

In this exercise, you will use a transaction to undo changes you made to the database in the previous exercises. Follow these steps:

1. Connect to the MySQL client with Workbench and the appropriate user.

2. Create the test database:

   ```
   CREATE DATABASE test;
   ```

3. Select the test database:

   ```
   USE test;
   ```

4. Create the animals table:

   ```
   CREATE TABLE animals (id int primary key, name
   varchar(255));
   ```

5. Use the DESCRIBE command to remind yourself of the layout of the animals table:

   ```
   DESCRIBE animals;
   ```

This will produce the following output:

```
+--------+--------------+------+-----+---------+-------+
| Field  | Type         | Null | Key | Default | Extra |
+--------+--------------+------+-----+---------+-------+
| id     | int          | NO   | PRI | NULL    |       |
| name   | varchar(255) | YES  |     | NULL    |       |
+--------+--------------+------+-----+---------+-------+
2 rows in set (0.23 sec)
```

Figure 6.23 – The DESCRIBE command's output

6. Empty the `animals` table by using the `TRUNCATE` command, as follows:

```
TRUNCATE TABLE animals;
```

7. Use the `BEGIN` statement to start a transaction, as follows:

```
BEGIN;
```

8. Now, add a record to the `animals` table using `INSERT`:

```
INSERT INTO animals VALUES(1, 'dolphin');
```

9. Check that the record has been added to the table using the `SELECT` command:

```
SELECT * FROM animals;
```

This will produce the following output:

```
+----+---------+
| id | name    |
+----+---------+
|  1 | dolphin |
+----+---------+
1 row in set (0.00 sec)
```

Figure 6.24 – The SELECT command's output

10. Use `ROLLBACK` to undo all the changes to the point where you started the transaction:

```
ROLLBACK;
```

11. Check the contents of the table using a `SELECT` query:

```
SELECT * FROM animals;
```

This will produce the following output:

```
Empty set (0.00 sec)
```

Figure 6.25 – The SELECT command's output after ROLLBACK

Now, the table is back in its original state. This approach not only works for adding records but also for undoing the changes that have been made to existing records or stopping records from being deleted.

Summary

In this chapter, we covered a lot of information and learned many new skills, as well as about views, stored procedures, functions, and triggers. You learned how to create views and how to determine which views are updatable and which are read-only, and why. You also learned how to create and use functions. Finally, you learned how to create stored procedures to perform some pretty amazing tricks while using the `INOUT` parameters before learning about the good, the bad, and the ugly of triggers.

In the next chapter, we will start applying MySQL queries to web applications through Node.js. This will allow you to develop a dynamic application using data from MySQL databases.

7

Creating Database Clients in Node.js

In this chapter, you will learn how to set up your development environment to make development easier, as well as to protect your production database, by creating a development database for you to work on. You will also learn about best practices for developing client applications that work with MySQL databases.

After that, you will learn how to install Node.js modules, generate scripts to output to the console, and connect to the database to create a simple web application. You will also create a table in the database with Node.js.

This chapter covers the following topics:

- Introduction to database management with Node.js
- Best practices for SQL client development
- JavaScript using Node.js
- Connecting to MySQL
- Activity 5.01 – building a database application with Node.js

Introduction to database management with Node.js

One of the goals of databases is to provide users with a convenient way to serve data to clients and consumers. Let's say that your company creates a database containing customers. It would be valuable for employees to access this database to view data relevant to the customers they work with. For instance, they may wish to provide their customer service team with a list of products that a customer owns.

To achieve this, you need to be able to provide the customer service team with an interface that accesses your database. These clients can be developed in many ways. In this chapter, you will learn how to interface with databases through Node.js, a popular JavaScript-based service.

When you develop applications for databases, they will often retrieve, modify, and delete data from the database tables. Due to the possible data changes, you must learn how to set up a proper development environment for your application. This will allow you to test applications that potentially modify data without worrying about modifying the data that's currently in use by other users.

Once you have an appropriate development environment in place, you can start working with Node.js to create applications that interface with MySQL databases. You will start by understanding the basics of Node.js, including how to set up an application and how to output data through the terminal, browser, and user filesystem. Once you have working outputs, you must learn how Node.js interfaces with MySQL. Specifically, you must learn how to set up connections with the database, create databases and tables, and select data from a database.

By the end of this chapter, you will be able to write basic Node.js applications and work with MySQL databases within Node.js. These skills will help you develop dynamic applications that act as clients for databases. Although you will be working primarily with Node.js, many of these skills can be transitioned to other programming languages and technologies. MySQL modules are implemented in a fairly consistent way, so once you've learned the basics of querying and connecting to databases, you will be able to apply this knowledge anywhere.

Now, let's look at some of the best practices for SQL client development, including development databases and backing up data. After that, we will start working on our database.

Best practices for SQL client development

Suppose your client has asked you to alter an existing table and change the format of the field that tracks the age of the client from an integer to a float. After making this change, you find that the reports and programs that use the data are now producing errors. It turns out that many other dependencies were relying on the data to be formatted in a specific way, and now it has been changed.

To avoid these types of issues, you should follow several best practices while developing robust SQL databases. First, you should install a development MySQL server, which allows you to change and test your data without it negatively influencing the clients who use the data.

Installing a development MySQL server

When you develop an application that interacts with a database, you separate your instances into two separate databases – a production database and a development database. A **production database** is a database that contains live data that is accessed by clients and users, while a **development database** is a database where developers can change and test data, without impacting any data that is currently being used by other database users. In the real world, the production server and production database should not be used for development purposes. Development, in this context, refers to any changes that are made to the data or the format of the data in the database.

Code can go wrong while you are testing and experimenting. For instance, suppose a report expects a field named `age`, which is formatted as an `integer`. Changing the field's format to a `float` can cause the report to stop working completely since the operations it completes assumes an `integer` format for the field. It is also possible to accidentally delete or modify important data that could be lost without proper backups. So, it is good practice to create and maintain a development server and database. These are usually copies of the production server and database, without any connections to clients, reports, or any other data users. This creates an environment that can be modified without the risk of data loss or functionality breaking.

When your development is complete, you can copy your development objects and make the same changes to the production database. You can also release a new frontend application to your users if required. By working through a development database, you can do all your testing separately from the production database. This means that you can verify that everything is working correctly before making changes that impact the other users. This prevents situations where you may accidentally break or delete data that is important to users of the database.

The following are a few reasons why you should not use production servers and databases for development:

- You don't directly alter the production data while working on the development database, so you avoid losing or damaging the data that is currently in use.

- Software development could slow down the production database due to an increased number of queries being sent to the server for testing and quality assurance.

- It is easier to recover the development database if your coding causes the data to be removed or modified unintentionally. Since the development database is usually a direct copy of the production database, you can transition the data without needing to find a proper daily backup.

With a development database, you don't need an active internet connection. It is possible to use a development database as a local instance; however, a production database needs to be internet-enabled so that external clients can access it.

Overall, a development database is a valuable tool for ensuring that your database can be worked on safely without it impacting other users. With this in mind, let's learn how to create a development database.

Creating a development MySQL server

Once you have installed MySQL server on your development computer, you can begin setting up the development database. If the production database already exists, then there are several things you need to think about – all of which will be covered in this section.

One of the factors to consider is whether you can make a complete copy of the production database for development purposes. This is dependent on the following factors:

- **Size of the database**: If the database is too big, you may not be able to take a complete copy of the database. However, you may succeed in getting a subset of the data – for example, 10,000 records plus the ancillary tables.

- **Sensitivity of the data**: You may need to desensitize the data by changing sensitive information such as names, addresses, phone numbers, and any other information that could identify the people or businesses in the database. In most countries, it is a legal requirement to desensitize the test data. This process typically involves partially censoring the data by replacing the last few characters with asterisks, for example.

These factors will need to be discussed with the database owners or your manager so that you know what their preferences are. You need to ensure that you get the complete set of objects from the production database – tables, stored procedures, functions, views, and triggers. If you are developing a new database, this is not an issue; you can create your own dataset for development purposes.

Another thing you need to do is set up your development database so that it mirrors the production database as much as possible. You do so by ensuring the following:

- The development database has the same name as the production database.

- The ODBC connections are named the same. However, the IP addresses will most likely be different.

- The user access and rights are the same as they are for the production database you are mirroring.

- If the production server accepts remote connections from users, you need to set up your development environment to imitate this by opening the server for remote access and setting up an ODBC connection to use with it. Then, you can test the remote user's experience. If you are developing a database and application on your system, the server may be on the same machine you are developing the database on. In this case, the data retrieval process will be lightning-fast, but this may not reflect the user's experience. In such cases, you may want to consider testing your program's database queries using the remote connection as well.

- The database should be identical in structure to the production database, which means that they must have the same tables, fields, and schemas.

Before you start making changes related to your database, you will need to have some way to preserve the previous data in case you accidentally change or delete the wrong data. To achieve this, you will have to back it up.

Backing up before making changes

Things can go wrong in development, no matter how skilled you are. So, when they do, it is better to rebuild your development environment and recover the data that's stored in the database from a recent backup.

Consider the following situation. One of your company's clients has dabbled in coding and structuring databases. An issue was reported to the database engineer regarding an application feature that was no longer working. The company's database engineer referred to older copies of the application and the database to investigate the issue. Later, the engineer realized that the client had removed a field and modified the code 6 months earlier, which the engineer was not aware of. The client had done that as they thought it was no longer required. However, the engineer was able to fix the code and recover the field because they had a backup of the application and the database.

Fortunately, MySQL provides a simple, fast, and effective way to back up and recover when things do go wrong. The preferred method to back up when developing is using the **Data Export** tool in MySQL Workbench, which allows you to back up the entire database or just part of it in a single SQL file that you can execute to recover the database if and when it is required. It is a fast, efficient, and easy method. To determine when to back up a development database, you can follow a few simple criteria.

First, before you start making changes to the database, you should make a backup of the current database's state. Then, you should create a backup when you have made a significant change that you wish to preserve in case the data is changed or deleted. Finally, once you are done making changes for the day, you should make a final backup of the current changes. In addition to daily backups, it is ideal to take a backup once a change or feature has been fully completed as this can be useful for future reference, in case an issue in the data is found.

In other words, back up often. Most of the time, you never refer to the backup files. But when you need to and they are there, you'll be thankful that your efforts were saved.

Of course, with all these backups happening, your backup folder will quickly become crowded. Ensure that you adopt a good and easy-to-follow naming convention for the files. The following are a couple of examples:

- `DatabaseName_Full_20191015a.sql`
- `DatabaseName_country_20191015b.sql`

In the preceding examples, you have the following:

- `DatabaseName`: The name of the database.
- `Full`: This indicates a full backup of all database objects. If you have a single object such as a stored procedure or a table named `country`, then the name of the object is the obvious choice to provide.

- 20191015: This specifies the date of the backup in reverse order, YYYYMMDD. This assists in sorting for quick retrieval.

- a, b: a is specified for the first version, b for the second, and so on when you're creating multiple backups on the same day.

Of course, you can adopt any naming convention you like that makes sense to you. When the backup file count starts to become too big, archive or delete some of the older ones. It is a good idea to keep a copy of a backup for specific periods for future reference.

Now that you have a better understanding of why backups are important, let's learn how to restore a backed-up database.

> **Note**
>
> For more information on how to take backups, please refer to *Chapter 3, Modifying a Database.*

Restoring a database

Often, you must restore a database when something has gone wrong during development, and you need to revert to a recent backup. For example, if your client has raised an issue that data is missing or incorrect in the database, you may need to take a backup of the production database and restore it to your development environment so that you can work on the problem away from the production server. If this happens, do the following:

1. Connect to the production server.

2. Back up the data to a single .sql file, as described in *Chapter 3, Modifying a Database.*

3. Connect to the development server.

4. Back up the development database to a single .sql file.

5. Run the production .sql file on the development server.

These steps place the current production data on your development server. After resolving this issue, you can take the necessary steps to fix the problem in production. If you have been working on the database in development before this, when you have resolved the production issue, you can restore the development database; otherwise, you can keep the production version as your current one.

One of the main reasons for restoring a database backup is in the case of accidental data deletion. If you accidentally modify or delete important data, you need some way to recreate it. One of the ways you can do this is by recovering the data from a backup.

Note

For more information on how to restore the database, please refer to *Chapter 3, Modifying a Database.*

If data were to be deleted from a production database, you must be able to recover it in some way. Now, let's learn how data can be recovered from accidental data deletion using backups.

Recovering from accidental data deletion

You are deep in development, and you have written a small delete query for the `country` table and executed it. The server diligently runs your query, and your prompt silently comes back. You look in the output window and the last row tells you that `263 rows have been affected`. You immediately get that sinking feeling as you realize that you forgot to add the filter. You check the contents of the table, and there it is – only one row with `NULL` in all fields. It is all gone. You panic for a moment and then you remember that you have a backup and, more importantly, you are working on the development server. Imagine the chaos if you had been working in the production database.

Before you learn how to recover, let's talk about some simple ways to help prevent this situation from occurring in the first place. Remember the adage, "*An ounce (gram) of prevention is better than a pound (kilogram) of cure.*"

There are four types of queries you should be familiar with when working with databases. Some queries are responsible for creating SQL objects such as databases and tables, which are referred to as **creation queries**. Other queries read data, such as **select** queries, which are referred to as **read queries**. These two types of queries are non-destructive since they do not change the data in the database; instead, they read or create new data. The other two types of queries are **alter** and **delete queries**. An alter query is used to change data, while a delete query is used to delete data. Both queries can be destructive since they change or remove data. You need to take care when implementing destructive queries; otherwise, you could accidentally alter and delete more data than expected.

A simple way to avoid the preceding scenario is to test the filters using a read query and then change the query to an action or destructive query (`DELETE` or `UPDATE`) before committing.

In the next exercise, you will learn how to safely delete records from a table.

Exercise 7.01 – safely deleting records

You have a web page that asks users to specify the country they live in. The data is stored in the country table. A user whose Country Code is AUS wishes to delete their account from the web page you have created. You are asked to delete their data from the database. To do this safely, you need to perform a non-destructive query to verify that the correct data is being targeted. To do this, you must modify the data query and then delete the required record.

First, you will have to import the database where the data is stored. Follow these steps to do this:

> **Note**
>
> The SQL script for creating the database for this exercise can be found at https://github.com/PacktWorkshops/The-MySQL-Workshop/blob/master/Chapter07/Databases/GP_PracticeDatabase.sql.

1. Open MySQL Workbench.
2. In the **Schemas** panel, right-click on the **backuppractice** database and select **Set as Default Schema**:

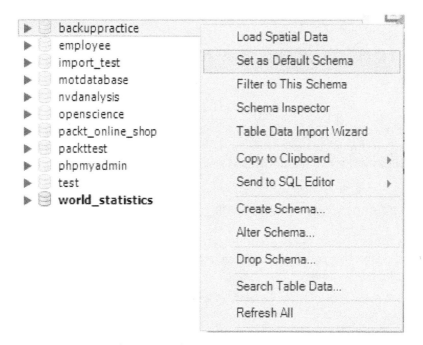

Figure 7.1 – The Set as Default Schema option for backuppractice

3. Add a new **Query** tab by clicking on the **Create new SQL tab for executing queries** icon.

4. In the new **Query** tab, type in the following SQL statement to select all the records of the `country` table:

```
SELECT * FROM backuppractice.country;
```

5. Execute the query by clicking the **Execute query** icon – that is, the lightning bolt. The data will be displayed, and the output panel will tell you that 263 rows have been returned:

SELECT * FROM backuppractice.country   263 row(s) returned

Figure 7.2 – The Output panel displaying the executed query and its results

6. Add your filter to fetch the details of the user whose county code is AUS:

```
SELECT * FROM backuppractice.country
    WHERE `Country Code`="AUS"
```

7. Rerun the query with the **Execute query** icon. This time, one row will be displayed. Check the output panel and verify that the country is Australia:

CountryID	Country Code	Country Name	ContinentID
12	AUS	Australia	5
NULL	NULL	NULL	NULL

Figure 7.3 – A single-record result from a filtered query

8. Once you are satisfied that the only records that are being returned are the targeted records, you can delete them with confidence. Replace SELECT * in your SQL query with DELETE so that it reads as follows:

```
DELETE FROM backuppractice.country
    WHERE `Country Code`="AUS";
```

9. Execute the query and check your **Output** panel. Look at the record count in the `country` table to ensure you have deleted only one record and that the final record count is 262:

DELETE FROM backuppractice.country WHERE `Country Code`="AUS"   1 row(s) affected

SELECT * FROM backuppractice.country   262 row(s) returned

Figure 7.4 – The Output panel showing the results of the DELETE query following the SELECT query

10. Check this by rerunning the following SQL query:

```
SELECT * FROM backuppractice.country
    WHERE `Country Code`="AUS";
```

You will observe that no records are returned:

```
SELECT * FROM backuppractice.country WHERE 'Country Code'="AUS"                    0 row(s) returned
```

Figure 7.5 – The results when checking whether DELETE worked

In this exercise, you learned how to safely delete records. When you are deleting or otherwise modifying data, always take the time to check what records are going to be affected by performing a SELECT query before committing to a DELETE or UPDATE query.

Now that you have a better understanding of how to prepare your database for data changes, let's learn how to create clients that allow users to read and write data to your databases. With your development databases, you will be able to safely build and deploy clients that use this data. Specifically, you will learn how Node.js can be used to interact with a database.

JavaScript using Node.js

Node.js is a JavaScript runtime environment that can be run on your computer as a standalone application. You can develop scripts to create applications that do not use the web browser as an engine to execute code. Instead, the applications are compiled and run from a server on a computer. Then, you can access and execute the applications directly on your web browser by navigating to the server that is running the code or by using your computer's command prompt. Node.js uses a runtime engine called **V8**, which is open source and written by Google. The purpose of V8 is to compile JavaScript code so that it can be run more efficiently. Traditional JavaScript is interpretive, which means that each line is translated into code that the browser can understand before it is run. However, when the program is compiled beforehand, the line-by-line translation does not need to happen, so the application can run directly with less time spent on translation. Node.js acts similar to a server in terms of functionality. This means it allows us to implement a variety of features that can be used to construct robust web applications.

A majority of Node.js's functionality will exist on the backend of a web application. The backend includes features such as **REST** and **JSON API integration**. This will allow users to send REST-based HTTP requests to the server to retrieve or alter data. This is often used in conjunction with databases to send a request to retrieve or alter data in a database. You can also use data to upload files and data to your server. An operation such as uploading a profile picture is a good example of this type of functionality.

Among other tasks, retrieving and updating data is one of the primary reasons why JavaScript is used in websites, although it offers so much more. For this reason, Node.js has been included in this book in a targeted way that shows you how to work with a MySQL server and data.

Note

This chapter explains the basics of getting started with Node.js and MySQL only. It does not teach you all the aspects of Node.js and has been included to introduce you to different methods of working with a MySQL database.

Packt Publishing offers several excellent books such as *Node.js Web Development – Fourth Edition*, which can be found at `https://www.packtpub.com/web-development/nodejs-web-development-fourth-edition`, and *Server Side Development with Node.js and Koa.js Quick Start Guide*, which can be found at `https://www.packtpub.com/application-development/server-side-development-nodejs-and-koajs-quick-start-guide`, which provide you with deeper knowledge about Node.js and web development.

Before you start creating Node.js applications, you must set up Node.js on your computer. In the next section, you will learn how to install Node.js on your system and how to create a project in Node.js.

Setting up Node.js

To start, visit `https://nodejs.org/en/` to access the most recent release of Node.js. On this page, you will get a link to a recommended and current version of your operating system. Either of these versions will be sufficient for completing the exercises in this book.

New security releases now available for 15.x, 14.x, 12.x and 10.x
release lines

Download for Windows (x64)

Other Downloads | Changelog | API Docs   Other Downloads | Changelog | API Docs

Or have a look at the Long Term Support (LTS) schedule.

Figure 7.6 – The download page for Node.js

Once you click on the respective download button, an installer will be downloaded onto your PC that you can run to install Node.js. To configure the installer, follow these steps:

1.   Once the Node.js installer launches, you will see a screen similar to the following:

Figure 7.7 – The main starting screen of the installer

Press **Next**. You will need to accept the terms and conditions mentioned on the **End-User License Agreement** page:

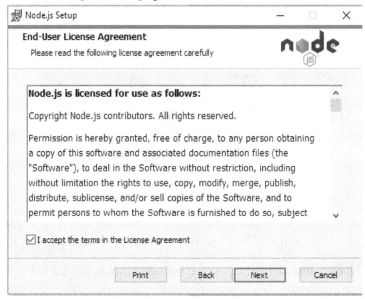

Figure 7.8 – The license agreement screen

2.   Next, the installer will ask where you want to install Node.js. You can install it at any location. For this book, leave the default location as-is – that is, in the **Program Files** directory. Click **Next**:

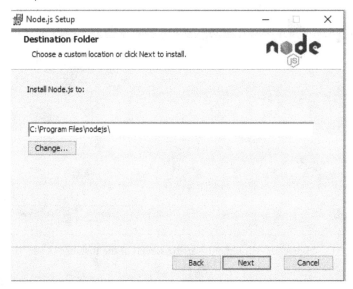

Figure 7.9 – Choosing where to install the Node.js directory

3. Node.js will provide some custom setup options. Leave everything as-is and click **Next**:

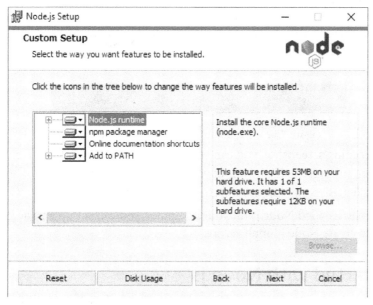

Figure 7.10 – The custom installation screen

4. Next, Node.js will give you an option for native modules. None of this is required for this book, so leave the checkbox unchecked and press **Next**:

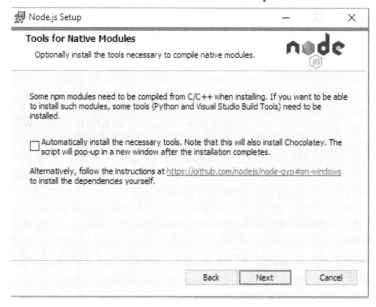

Figure 7.11 – The Native Modules screen

5.   Finally, click **Install**; the installation will complete.

Once the installer has finished running, you can verify that the installation was completed successfully by running the node -v command on your command line.

You should see various details about your version of Node.js:

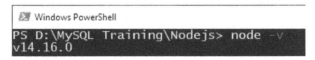

Figure 7.12 – The output of the node -v command

In the preceding screenshot, you can see that version 14.16.0 is currently installed on my system. If the output you get matches the version number you have installed, then you have successfully installed Node.js.

Now that you have installed Node.js on your computer, you can set up a basic project and learn how to create Node.js applications.

Getting started with Node.js

Before you begin, please set up Node.js. If you have not already set up Node.js and your work folder, please return to the *Setting up Node.js* section of this book and set them up; otherwise, you will not be able to work through this section.

> **Note**
>
> In this chapter, the work folder is D:\MySQL Training\Nodejs and all the references are for this folder. You can set your work folder to anything you like, but we suggest that you use the aforementioned folder drive and path for this book.

Two components come installed with Node.js by default. The first is node, which is a command-line utility that's used to run JavaScript code. This sets up and executes the program. The second is npm, a package manager that's used to install third-party modules you may need for your application. For example, you will need mysql to work with your database, so this is a component you will need to install.

When you first start a Node.js project, you must run the npm init command from the console. When you do this, Node.js will prompt you with several questions so that you can initialize your Node.js project. The following screenshot shows an example of running the npm init command:

```
Windows PowerShell
PS D:\MySQL Training\Nodejs> npm init
This utility will walk you through creating a package.json file.
It only covers the most common items, and tries to guess sensible defaults.

See `npm help init` for definitive documentation on these fields
and exactly what they do.

Use `npm install <pkg>` afterwards to install a package and
save it as a dependency in the package.json file.

Press ^C at any time to quit.
package name: (nodejs) helloworld
version: (1.0.0) 1.0.0
description: A simple hello world app
entry point: (index.js)
test command:
git repository:
keywords:
author:
license: (ISC)
About to write to D:\MySQL Training\Nodejs\package.json:

{
  "name": "helloworld",
  "version": "1.0.0",
  "description": "A simple hello world app",
  "main": "index.js",
  "scripts": {
    "test": "echo \"Error: no test specified\" && exit 1"
  },
  "author": "",
  "license": "ISC"
}


Is this OK? (yes) yes
PS D:\MySQL Training\Nodejs>
```

Figure 7.13 – The output of the npm init command

Here is a summary of the values that were inputted during initialization:

- **Package name**: This is a unique identifier for the program you are creating. In the preceding screenshot, this is denoted as `helloworld`.

- **Version**: This is the current version number of the program. In the preceding screenshot, this is denoted as `1.0.0`.

- **Description**: This is a brief description of what the program does. In the preceding screenshot, this is denoted as `A simple hello world app`.

- **Entry point**: This is the file that contains the start of your application's code. It defaults to `index.js` if nothing is inputted.

- **Test command**: This command is used to test the program. In the preceding screenshot, this is blank since there is no test command for the project.

- **Git repository**: This is the location of the GitHub repository for the program. In the preceding screenshot, this is blank since there is no GitHub repository for the project.

- **Keywords**: These are keywords that can be used to identify the program. In the preceding screenshot, this is also blank.

- **Author**: This refers to the name of the program author. In the preceding screenshot, this is also blank.

- **License**: This is the license of the program. In the preceding screenshot, it is denoted as ISC. The ISC license indicates that a project can be used in any way, so long as it is attributed to the author.

Once the init command has finished running, Node.js will create a Node project in the current directory that the command prompt window is set to. Now, you are ready to start adding dependencies to the project and write code.

Your programs will require mysql to query your database. To install mysql, you can simply use the npm install mysql command, as shown in the following screenshot:

```
PS D:\MySQL Training\Nodejs> npm install mysql
npm notice created a lockfile as package-lock.json. You should commit this file.
npm WARN helloworld@1.0.0 No repository field.

+ mysql@2.18.1
added 11 packages from 15 contributors and audited 11 packages in 2.369s
found 0 vulnerabilities
```

Figure 7.14 – The result of installing the mysql module

Once you've done this, your project folder should look as follows:

node_modules	2021-04-14 7:45 PM	File folder	
package.json	2021-04-14 7:45 PM	JSON File	1 KB
package-lock.json	2021-04-14 7:45 PM	JSON File	4 KB

Figure 7.15 – The directory of the project after installing npm init and mysql

The initialization process creates two files and a directory. The node_modules directory contains any third-party modules that have been installed for the project. The package.json file contains a copy of the project information that was supplied when the npm init command was run. The package-lock.json file stores the version numbers of all the modules that have been used in the project.

For this section's exercises and activities and to avoid repetition, take note of the following:

- When asked to create a file, you should use your text editor to create and maintain the file. All files should have a `.js` extension.

- When you're asked to run a file, you should run the file from your **command-line interface (CLI)** or console.

- When you're asked to check the results of the script's execution, you will be informed of what application to check this with – that is, Workbench, Text Editor, the browser, or Excel.

> **Note**
>
> All of this must be done in your work folder, `Drive:Path/Nodejs`, and your CLI should display the work folder path in its prompt.

Using the project you created in this section, you can now learn more about Node.js and how it can be used to create web applications. Now, let's start using Node.js.

Basics of Node.js

To learn the basic structure of Node.js applications, you must know about the different ways to output the data. In this section, you will write three programs that output text to three different locations – one to the console, one to a web browser, and one to a file. When you output text to the console, you print data to the same console window that the node command is run from. When you output the result to the web browser, you display the result in a browser window when a user navigates to the URL associated with the program. Finally, when you output the result to a file, you write the data to a file on the user's computer, which can be read with any traditional text editor.

You can run a program with node using the `node Filename.js` command, where `Filename.js` can be replaced with any file you want to run. You should use this command any time you want to run the code using Node.js.

When you work in Node.js, you may want to log data in your console. This is common for error and debugging messages, which are used to troubleshoot applications. To log data to the console, use the `console.log` method. This method takes in any text as an argument, and the text that's provided is displayed to the console when the code is run. For example, `console.log("Hello!");` will write the text `Hello!` to the console window.

One of the important features of Node.js is the ability for users to access your application through a web browser. The user must enter the respective URL to access the Node.js application, which processes their request. With this, you can do things such as display text on the browser window. To achieve this, you need to set up an HTTP server through Node.js.

To set up an HTTP server, you must import the `http` module from Node.js. You can do this using the following code:

```
var http = require('http');
```

Once you have imported the `http` module, you can use it to create a server. You can do this using the `createServer` method:

```
http.createServer(function (req, res) {

}).listen(82);
```

The preceding code creates an HTTP server that can take in requests and send responses back. The `listen` keyword tells Node.js what port the server should listen on – in this case, port `82`. To access this server, you would need to navigate to `localhost:82`, which is the local IP of your computer, through port `82`.

Your server can send responses to whoever accesses it using the `res` variable. The idea is that you construct an HTTP response so that it has a header and any data that is to be sent back. For example, the following line would write a header to your response:

```
res.writeHead(200, {'Content-Type': 'text/html'});
```

This header can be broken down into a few main pieces. `200` is the HTTP code for a successful request. It tells the user that the HTTP request was successfully received. The second portion, `'Content-Type'`, describes what type of data you are sending in the response. Many different types of content can be returned with the `Content-Type` header. The most common types are `text/html`, `json`, and `text/plain`. In this case, you are sending back some text or HTML data to the person who requested your web page. Then, you can write the actual content using `res.end`, which appends data to the end of the response. For example, using `res.end('Hello!');` would add the text `Hello!` to the HTTP response. The result of this would be the text `Hello!` being displayed on the user's screen.

When data is returned through the browser, it is expected to be of a certain type. For example, suppose you declared a `result` variable that stores the result of the sum of two numbers:

```
var result = 4+4;
```

If you want to represent this value as text, it needs to be converted into a `String` variable type. A `String` is a sequence of letters that is used to represent text in Node.js. To convert a variable into a `String`, you can use the `toString()` method, as shown here:

```
result.toString()
```

With these fundamentals in mind, let's learn how to apply them.

Exercise 7.02 – basic output in the console

Suppose that you just installed Node.js on a work computer, and you want to quickly test whether everything is working correctly. One easy way to verify this is by outputting data to the console. You are required to add two numbers (3 and 4) and display the result on the console.

Follow these steps:

1. Open a new file in your text editor.

2. Enter the following text to add two numbers (3 and 4) and display the result in the console:

    ```
    console.log(3+4);
    ```

3. Save the file as `Add-OutToConsole.js` in your working folder (`D:\MySQL Training\Nodejs`).

4. In your CLI, type the following command and press *Enter*:

    ```
    node HelloWorld-Console.js
    ```

    The output will be as follows:

```
PS D:\MySQL Training\Nodejs> node HelloWorld-Console.js
7
PS D:\MySQL Training\Nodejs>
```

Figure 7.16 – Expected output for the HelloWorld-Console.js script

The text that is outputted is what was written in the `console.log` method arguments (that is, the sum of 3 and 4). In general, you can change what is between the brackets of the `console.log` method to anything you wish. This will be printed to the console, as shown in the preceding screenshot.

> **Note**
>
> The code for this exercise can be found at `https://github.com/PacktWorkshops/The-MySQL-Workshop/blob/master/Chapter07/Exercise7.02/Add-OutToConsole.js`.

In this exercise, you created your first Node.js script. Outputting to the console is easy to implement since it only requires a single line of code. As you have learned, you can use the console to output messages that indicate the status of your programs as they execute and indicate when they have finished. You can also output messages with variable values while developing so that you can monitor what your scripts are doing, check the values of the variables, and more. This will assist you with debugging.

In the next exercise, you will use a browser to get the output of the Node.js script.

Exercise 7.03 – testing outputs in a browser

Now, you must test the output of a Node.js script in a web browser. You have been tasked with adding two numbers (4 and 4) and displaying the result to the user who accesses the web page on the web browser on port 82 of their machine.

Follow these steps:

1. Create a new file and enter the following text. Since you are outputting text to the web browser, you need to include the `http` module:

```
var http = require('http');
```

2. Create a server to monitor the request from a browser and instruct it to send a response back. Add the following lines of code after the `require` statement for `http`:

```
http.createServer(function (req, res) {
    res.writeHead(200, {'Content-Type': 'text/html'});
    var result = 4+4;
```

3. Once a request has been received from the browser, you must send a response back to be displayed on the browser. Add the following code:

```
res.end(result.toString());
```

4. Finally, tell the server what port to monitor for a request. Use port 82 for the web server for all the exercises in this chapter so that you do not clash with anything you may already have by using the standard port number 80. Add the following line:

```
}).listen(82); //The brackets close off the createServer
block
```

5. Save and name the file Add-OutToBrowser.js. The content of this file should look as follows:

```
var http = require('http');

http.createServer(function (req, res) {
  res.writeHead(200, {'Content-Type': 'text/html'});

  var result = 4+4;
  res.end(result.toString());

}).listen(82);
```

6. In your CLI, type the following command to run the code you have written and press *Enter*:

```
node HelloWorld-Browser.js
```

This time, your cursor will not come back and no output will be sent to the console. The Node.js script is running in the background and monitoring port 82 for a request from a browser.

7. To test this, open a web browser.

8. In the address bar, enter the following address and press *Enter*:

```
localhost:82
```

> **Note**
>
> `localhost` and `127.0.0.1` both refer to your computer and are interchangeable.

The server will respond, and the browser will display the sum of two numbers (4 and 4):

8

Figure 7.17 – The expected browser output for the HelloWorld-Browser.js script

To exit the script, press *Ctrl + C* (hold down the *Ctrl* key and press *C*) in the CLI window, and wait for your command prompt to return. You may need to do this a few times.

> **Note**
>
> The code for this exercise can be found at `https://github.com/` `PacktWorkshops/The-MySQL-Workshop/blob/master/` `Chapter07/Exercise7.03/Add-OutToBrowser.js`.

The techniques shown in this exercise are used often in Node.js when a dynamically built web page is required.

In the next section, you will learn how the output can be written to the files on your system.

Writing outputs to files

Node.js can be used to monitor a designated port for a web request and respond with whatever you have programmed it to. Another location you can write to is a file on the computer that the server is running from. Let's take a look at a few important details that are required to achieve this.

To work with the filesystem in Node.js, you need to import the `fs` module. This can be done in a similar way to importing the `http` module, as follows:

```
var fs = require('fs');
```

From here, you must create a new file stream that can be used to write data to a file. A file stream is a connection from Node.js to the computer's filesystem to transfer data between the Node.js application and the filesystem. To create a file stream, you can use the `createWriteStream` method. This method takes in a filename, which is the name of the file you want to write to. The following code shows how to create a file stream for a file named `Hello.txt`:

```
var stream = fs.createWriteStream("Hello.txt");
```

Once you've created a stream, you can start writing data to the file. To do so, you can use the `write` method. The `write` method takes in some data, and that data will be written into the target file. For example, the following code writes the text `Hello World` to the file:

```
stream.write("Hello\n");
```

Note that `\n` simply indicates a new line. We've added it here to indicate that anything that's added to the file after this write will be on a separate line.

Finally, you should always close the stream once you are done with it. To do this, you just need to use the `end` method:

```
stream.end();
```

With that, you have the tools you need to write data to files on your system.

In the next exercise, you will learn how file output can be implemented in an application.

Exercise 7.04 – writing to a disk file

Your manager wants you to create a `Log.txt` file that stores a log of the application that's running. You have been asked to create a Node.js file that writes `Application Started Successfully!` to the `Log.txt` file. Follow these steps:

1. Create a new file in your text editor.
2. To work with a file, the script requires a reference to the filesystem. Add the following line to tell the script that you want to use the filesystem module and assign it to a variable named `fs` for ease of reference:

   ```
   var fs = require('fs');
   ```

3. Create a new file with the filesystem variable (`fs`) for writing purposes and assign the file to a variable named `stream` using the following command:

```
var stream = fs.createWriteStream("Log.txt");
```

To write to the file, refer to the `stream` variable. Note that `\n` forces a new line in the output:

```
stream.write("Application Started Successfully!\n");
```

4. Finally, close the file:

```
stream.end();
```

5. Save the file as `Log-ToDiskFile.js`. Run the file in your CLI with the following command:

```
node HelloWorld-DiskFile.js
```

This time, your cursor will come back when the program has finished, as shown in the following screenshot:

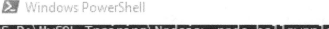

Figure 7.18 – Console output for the Helloworld-Diskfile.js script

6. Locate the `Log.txt` file in `D:\MySQL Training\Nodejs` and open it. You will see that `Application Started Successfully!` is written in the file:

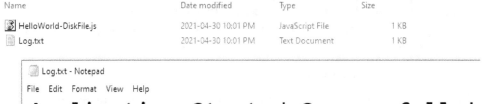

Figure 7.19 – The expected output for HelloWorld.txt in the folder and its contents

The preceding output shows that the `Log.txt` file has been successfully created and that the content that's been written through `stream.write` has been written to the file.

> **Note**
>
> The code for this exercise can be found at `https://github.com/PacktWorkshops/The-MySQL-Workshop/blob/master/Chapter07/Exercise7.04/Log-ToDiskFile.js`.

Generating files with Node.js is easy to do and is useful for reporting on data or generating log files. You have now created three separate scripts to test the three main outputs. You will use these in the upcoming exercises.

Now that you have tested your main outputs and have written your first Node.js scripts, you are ready to start writing scripts to work with a MySQL database.

Connecting to MySQL

Many web applications will dynamically generate content based on the user who is currently accessing them. Many companies desire dynamic web applications so that clients can view data specific to themselves. This is important from a usability and privacy perspective. The application is generally easier to use if it has been personalized to you. If you want to display data for a user, it should be data for that user only rather than all users.

To accomplish this, you must use databases to store the dynamic data you wish to display. Specifically, MySQL integrates well with Node.js, with specific modules written just to work with MySQL. In this section, you will learn how to connect to MySQL databases using Node.js and use these connections to query the databases. This will allow you to write dynamic applications using databases.

You should already have the required connection information from the *Prerequisites* section in the *Preface*. The information you will need includes the IP address of the MySQL server, the port number of the MySQL server, and the username and password for the MySQL user. If not, go back to the *Prerequisites* section, collect the necessary information, and come back here.

To connect to a MySQL database through Node.js, you will need to learn a few new Node.js methods. First, you must import the `mysql` module, as you did for the `http` and `fs` modules previously:

```
var mysql = require('mysql');
```

Once you have imported the `mysql` module, you need to establish a connection with your MySQL server. To do this, you will need to provide `node` with a few pieces of information. The first is the host, which is the IP address of the MySQL server. If you are working locally, this will be `localhost`. Next, you need to provide the port that MySQL is listening on, which is set to `3306` by default. Once you've done this, you must provide the username and password of the account you wish to use to connect to the MySQL server. The full code for creating a MySQL connection for Node.js looks as follows:

```
var mysqlconnection = mysql.createConnection({
host: "<Server's IP>",
port: "3306",
user: "<UserName>",
password: "<Password>"
});
```

Once you have created the connection, you can tell Node.js to attempt to connect to the specified MySQL server. This connection could succeed or fail, depending on the server's availability and the accuracy of the credentials that have been supplied. Therefore, you should ensure that you handle any errors that occur during the connection. You can do this by looking at the `err` variable when a connection attempt is made:

```
mysqlconnection.connect(function(err) {
  if (err) {
    throw err;
}
```

Once you have set up your connection, you should exit the process to ensure you do not tie up your database with connections that are no longer in use. To do this, you can simply add `process.exit();` to the end of your `connect` function:

```
mysqlconnection.connect(function(err) {
  if (err) {
    throw err;
}
process.exit();
```

Now, you can start connecting to a MySQL database using Node.js.

Exercise 7.05 – connecting to the MySQL server

Your manager has asked you to create a Node.js application that can connect to a MySQL database. The goal is to eventually use this connection to get data from the server. However, for now, you just want to get the connection working. To do so, your manager has asked you to check that the application connects to the database and prints a message when the connection is successfully made.

Follow these steps:

1. Create a file named `MySQLConnection.js` in `D:\MySQL Training\Nodejs`.

2. Since you are using the MySQL database, you will need a reference to MySQL to use it. Enter the following command:

```
var mysql = require('mysql');
```

3. Set up the connection details to make the connection to the database. Fill in the server IP and the account details to log in to the server:

```
var mysqlconnection = mysql.createConnection({
host: "<Server IP Address>",
port: "3306",
user: "<UserName>",
password: "<Password>"
});
```

4. Now, make the connection and set up error handling using the following command:

```
mysqlconnection.connect(function(err) {
```

5. Test for errors and throw an error message, including the server's error code if one occurs:

```
if (err){
    throw err;
```

`err` is used to display the error code and information about the error that was found while trying to connect to the server.

6. Enter the following code to confirm whether the connection succeeded in the console:

```
}else{
    console.log("Connected to MySQL!");
}
```

7. Now, stop the script so that the cursor comes back to the CLI by using the following command:

```
process.exit();
```

8. Close off the connection block:

```
});
//End the Connection Block
```

The complete script should look as follows:

```
var mysql = require('mysql');

var mysqlconnection = mysql.createConnection({
host: "<Servers IP>",
port: "3306",
user: "<UserName>",
password: "<Password>"
});

mysqlconnection.connect(function(err) {
    if (err) {
        throw err;
    }else{
        console.log("Connected to MySQL!");
    }
process.exit();

});
//End the Connection Block
```

9. Run the file in the CLI. A single response will be returned:

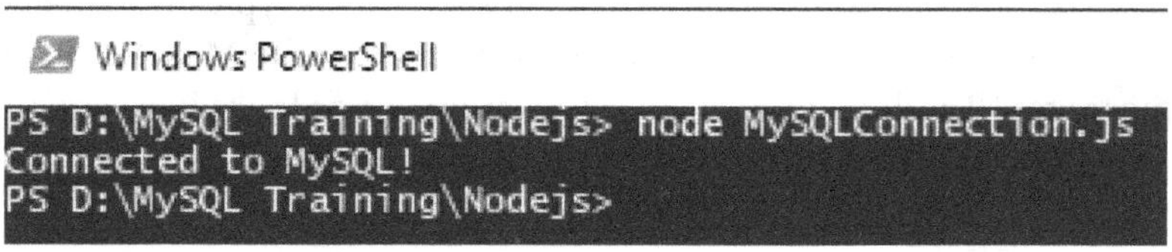

▶ Windows PowerShell

```
PS D:\MySQL Training\Nodejs> node MySQLConnection.js
Connected to MySQL!
PS D:\MySQL Training\Nodejs>
```

Figure 7.20 – Console verification that a MySQL connection has been made

As shown in the preceding screenshot, you have successfully connected to the MySQL server and verified the connection.

If an error occurs while you're executing the script, you will get an error message. If this happens, check your connection details and try again.

> **Note**
>
> The script file for this exercise can be found at `https://github.com/PacktWorkshops/The-MySQL-Workshop/blob/master/Chapter07/Exercise7.05/MySQLConnection.js`.

When you're establishing connections with database servers, you will often encounter errors if the connection has been set up incorrectly or if there are server issues. In the next section, you will learn about common errors that occur in connections and how to troubleshoot them.

Troubleshooting connection errors

In some cases, you may encounter errors while attempting to connect to a MySQL database through Node.js. It is helpful to have some knowledge of common connection problems so that you understand how to troubleshoot these issues when they occur. Let's look at two common errors that occur when setting up MySQL connections.

> **Note**
>
> To demonstrate any connection issues you may get, you must use the same `MySQLConnection.js` file you created in the previous exercise to introduce the errors. There are two resource files named `MySQLConnection 1.js` and `MySQLConnection 2.js` on GitHub at `https://github.com/PacktWorkshops/The-MySQL-Workshop/tree/master/Chapter07/Databases`. These files have already been changed to demonstrate these errors. You can use these or modify your existing `MySQLConnection.js` file.

The first type of error is called a **timeout error**. A timeout error occurs when it takes too long for Node.js to establish a connection to a target server. This typically happens for one of two reasons – either the server IP has been set incorrectly or the MySQL server is not available on the target IP address or port. To see what this error looks like, you can alter the previous exercise's code to try to connect to an IP that does not exist:

```
var mysql = require('mysql');

var mysqlconnection = mysql.createConnection({
host: "192.0.0.2",
port: "3306",
user: "root",
password: ""
});

mysqlconnection.connect(function(err) {
   if (err) {
      throw err;
   }else{
      console.log("Connected to MySQL!");
}
process.exit();

});
//End the Connection Block
```

In the preceding code, we have changed the IP address to 192.0.0.2, which does not exist. When you try to run this code, you will get an error, as shown here:

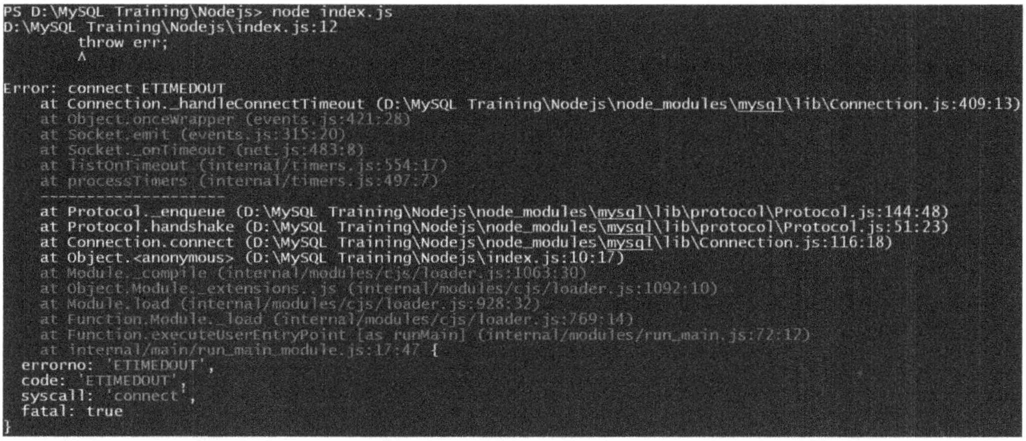

Figure 7.21 – The timeout error that's received due to an invalid IP address

Here, you can see that the error is `Error: connect ETIMEDOUT`. This error statement is used when a timeout error occurs. This tells you that either your IP address or port is incorrect. One additional thing to check with this type of error is if the MySQL instance is active and running on the target server. Resolving these three issues will typically resolve any `ETIMEDOUT` error you may encounter.

The second common type of connection error is an **access denied error**. This most commonly occurs when the username or password that's been supplied to the MySQL server is incorrect. The following code shows an example of trying to authenticate with a user who does not exist:

```
var mysql = require('mysql');

var mysqlconnection = mysql.createConnection({
host: "127.0.0.1",
port: "3306",
user: "root2",
password: ""
});

mysqlconnection.connect(function(err) {
  if (err) {
    throw err;
  }else{
    console.log("Connected to MySQL!");
```

```
    }
process.exit();


});
//End the Connection Block
```

When you attempt to run this code, you will get an error, as follows:

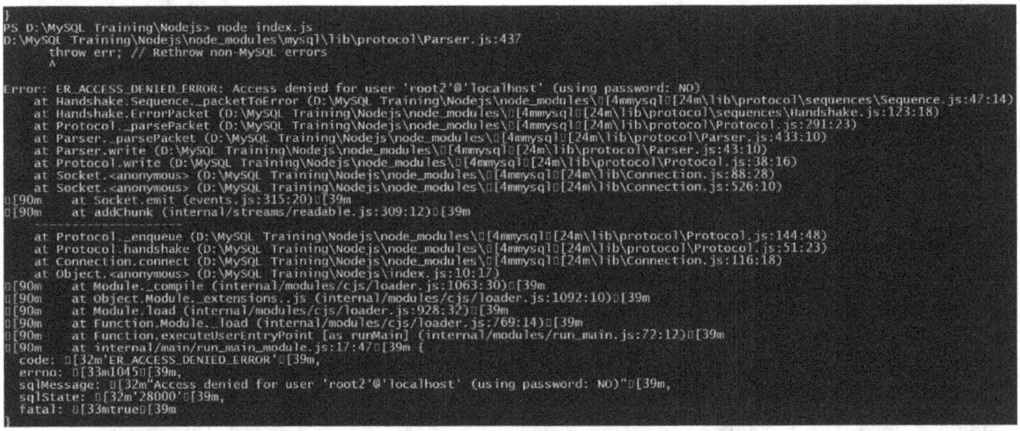

Figure 7.22 – The error that's received when an invalid username is supplied

This error is ER_ACCESS_DENIED_ERROR, which indicates that your credentials were not correct. With this type of error, you simply need to adjust the username and password so that they are valid credentials for the server.

> **Note**
>
> Make sure that you put the correct details in your script and save it for the next section.

As you work through the Node.js exercises in this book and your future development, you will create many scripts that connect to a MySQL server. You can add the connection details to each script. However, if the IP address of the server changes or the user account changes, you will have to change these details in every script. In the next section, you will learn how to tackle such issues.

Modularizing the MySQL connection

Modularizing involves placing your MySQL connection code in a separate file from your actual program logic. When you want to make a database connection, you simply import your MySQL connection module, just as you would import any other module.

To start, create a new file in D:\MySQL Training\Nodejs called MySQLConnection.js. In this file, place your current MySQL connection logic and make some minor changes:

```javascript
var mysql = require('mysql');

var mysqlconnection = mysql.createConnection({
host: "<Servers IP>",
port: "3306",
user: "<UserName>",
password: "<Password>"
});

mysqlconnection.connect(function(err) {
  if (err) {
    throw err;
  }else{
    console.log("Connected to MySQL!");
}

});

module.exports = mysqlconnection;
//End the Connection Block
```

The main changes here are that you have removed `process.exit();` and added the `modules.export = mysqlconnection;` line to the end of the file. `process.exit();` has been removed to keep the connection to the database active. This means that when this module is run, a database connection is created that can be used in other files. Once you are done with it, you can end the process in the file that uses it last. `modules.export` is used to export this file as `mysqlconnection`, which means that when you want to connect to the database, you can import the file using the `require` method, as shown in the following code snippet:

```
var mysqlconnection = require("./mysqlconnection.js");
```

On the topic of using your database connection, one idea you should discuss is how to query your database. You can do this using the `query` method of your connection. The following code shows how this can be done:

```
mysqlconnection.query("SELECT * FROM backuppractice.country;",
    function (err, SQLresult) {


}
```

When this query is run, the results are stored inside the `SQLresult` variable, and any errors will be stored in the `err` variable. The `SQLresult` variable acts as an array of objects. Each entry in the array has attributes equal to the fields of the table the data is queried from. For example, if you want to get the value of `Country` for the first record of the result, you can use `SQLresult[0].Country`. The item at index 0 is the first record of the set of results the query returns. In this query, the `Country` attribute from this record is fetched.

In the next exercise, you will learn how to use your modularized SQL connection logic to query a database.

Exercise 7.06 – modularizing the MySQL connection

You have been asked to determine the number of records that are present in the `country` table of the `backuppractice` database. In addition to this, your manager has asked you to modularize the MySQL connection so that the script can be run in multiple locations without the code having to be repeated. Follow these steps:

1. Open the `MySQLConnection.js` file that you created in *Exercise 5.05*, *Connecting to the MySQL server*, in your text editor.

2. You don't want to exit the script as you did previously, so remove the following line:

```
process.exit();
```

3. Add the following line to the end of the script to modularize the MySQL connection:

```
module.exports = mysqlconnection;
```

4. Save the script and keep the same name.

> **Note**
>
> You can find the revised MySQLConnection.js file at https://
> github.com/PacktWorkshops/The-MySQL-Workshop/blob/
> master/Chapter07/Exercise7.06/MySQLConnection.js.

5. To test the new module, create another file named TestModule.js.

6. Add a call to the MySQLConnection.js file to make the connection and assign the connection to a variable using the following command:

```
var mysqlconnection = require("./mysqlconnection.js");
```

7. Execute a query to count the records in the country table in the backuppractice database with the connection module. Include error handling and put the results in a results object named SQLresult:

```
mysqlconnection.query("SELECT Count(*) AS CountryCount \
   FROM backuppractice.country;", function (err,
SQLresult) {
```

8. Test for errors and print a message with an error code to the console if there is one:

```
if (err) throw "Problem counting Countries:- " + err.
code;
```

9. If no error occurs, print the result of the query. With a count query, there will only be one result, so you can simply access the first entry of sqlResult and get the countryCount field from it. This will return the result of the count query, which can be displayed on the screen using the console.log method:

```
console.log("Country count :- " + SQLresult[0].
CountryCount);
```

10. Exit the script:

```
process.exit();
```

11. Close off the query bracketing:

```
});
```

> **Note**
>
> The complete script can be found at `https://github.com/`
> `PacktWorkshops/The-MySQL-Workshop/blob/master/`
> `Chapter07/Exercise7.06/TestModule.js`.

12. Save and run the script using the `node TestModule.js` command. You should get the following result:

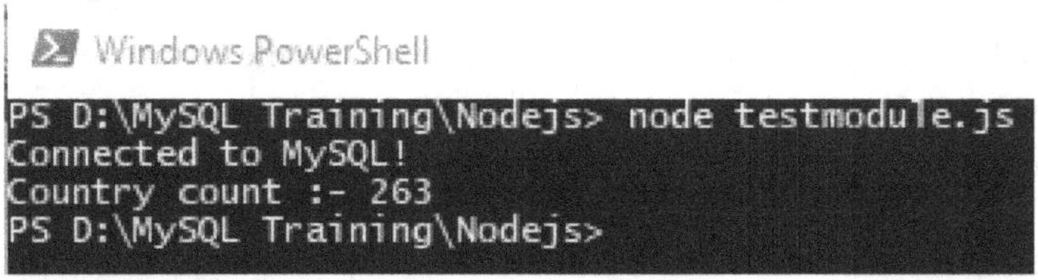

Figure 7.23 – The script's output for connection verification and country count

As you can see, the first result line (`Connected to MySQL!`) was generated in the `MySQLConnection.js` module, while the second result line (`Country count :- 263`) was generated in the `TestModule.js` script.

By modularizing the connection script, you reduce the need to enter the connection details in all your scripts to access the MySQL server. More importantly, if the details change in the future, you only need to update them in one script. You will be using this modularized script in all the upcoming exercises. You can modularize scripts to handle tasks you may perform in many of your programs, such as generating files, reports, and printing.

Now that you have simplified the connection process, let's start working with the MySQL server. The first thing you must do is create a new database in MySQL.

Creating databases in Node.js

Often, when an application is run by a user for the first time, all the databases that are required by the application need to be constructed. This initialization routine can be implemented through Node.js.

The process of creating a database is similar to the process of running a SELECT query in Node.js. First, you will need to establish a connection to the database using the modularized code you created in the previous exercise:

```
var mysqlconnection = require("./mysqlconnection.js");
```

Once you've done this, you can use the query method to run the desired queries. If you want to create a database, you can simply run a CREATE query, as shown here:

```
mysqlconnection.query("CREATE DATABASE `DATABASE_NAME`",
    function (err) {
```

One note about this query is the presence of backtick characters (`). These characters are not generally required in SQL queries unless there is a space in the name of the database. For example, if your database was named All Countries, the query would look as follows:

```
mysqlconnection.query("CREATE DATABASE `All Countries`",
    function (err) {
```

However, if you were to remove the space and use the name AllCountries, the query would not need backticks:

```
mysqlconnection.query("CREATE DATABASE AllCountries",
    function (err) {
```

Generally, most programmers will prefer to always include backticks, regardless of whether there is a space present.

Now that you know how to create a database with Node.js, let's look at a full example of an application that uses the CREATE query.

Exercise 7.07 – creating a new database

You have been asked to create an application that can track statistics about different countries. This application requires a database that can track statistics data about various countries, including the name of the country, its population, and its location. You have been asked to name the database world_statistics.

Follow these steps to create the required database:

1. Create a new file and name it `MySQLCreateDatabase.js`.

2. Add the `MySQLConnection.js` module you created in *Exercise 5.06, Modularizing the MySQL connection.*

> **Note**
>
> You can find the `MySQLConnection.js` file `https://github.com/PacktWorkshops/The-MySQL-Workshop/blob/master/Chapter07/Exercise7.06/MySQLConnection.js`.

```
var mysqlconnection = require("./mysqlconnection.js");
```

3. Use the following command to create a new database named `world_statistics` and include error handling:

```
mysqlconnection.query("CREATE DATABASE `world_
statistics`",
    function (err) {
```

4. Test for an error and generate an error message if one occurs:

```
        if (err) throw "Problem creating the database:- " +
            err.code;
```

5. Print a message to the console that indicates that the database was created:

```
        console.log("Database created");
```

6. Exit the script and close off the query bracketing:

```
    process.exit();
    });
```

> **Note**
>
> The complete script can be found at `https://github.com/PacktWorkshops/The-MySQL-Workshop/blob/master/Chapter07/Exercise7.07/MySQLCreateDatabase.js`.

7. Run the script. You should see the result of the query displayed in your command prompt:

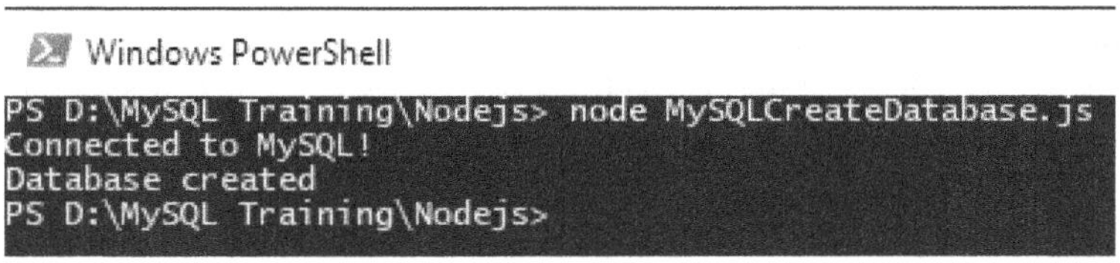

Figure 7.24 – The console's output verifying that the connection and database were created

8. In the Workbench window, refresh the schemas. The new database will appear in the list:

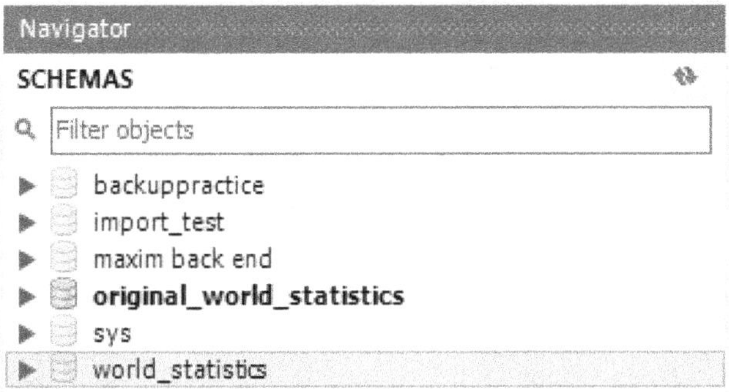

Figure 7.25 – The SCHEMAS list showing the new database

The ability to issue MySQL commands to the server using SQL from your scripts gives you a lot of control over the server and databases it contains. You can create programs to automatically set up the database for the users.

Now, let's learn how to create tables using Node.js.

Creating tables in Node.js

Now that you have created a database, you will need to add tables to it. Node.js can run queries to create tables, similar to how you created the database in the previous exercise. Start by creating a connection to the database you are working with:

```
var mysqlconnection = require("./mysqlconnection.js");
```

Now, you can define a query to create a table in your database:

```
var sql = "CREATE TABLE `world_statistics`.`test` (
  `ID` int(11) NOT NULL AUTO_INCREMENT,
  `Name` varchar(13) DEFAULT NULL,\
  PRIMARY KEY (`ID`) \
);";
```

As we mentioned previously, backticks can be added to the table name and field names if they contain spaces. They are not necessary if no spaces are present, but we will continue to include them in the queries so that we have consistent formatting. Once the query has been defined, it needs to be executed against the database:

```
mysqlconnection.query(sql, function (err) {
}
```

Now, let's create a table using a Node.js script.

Exercise 7.08 – creating a table in a database

In the previous exercise, you created a database to store statistical data. Your manager would now like you to create a table in the data that can store the name of the continents in the world. Your manager has specified that the table should be named `continents` and that it should have two fields. The first field will be a unique identifier called `ContinentID`. The second field will be the continent's name and will be called `Continent`. You must specify `ContinentID` as the primary key of your table. Follow these steps to create a table in the database:

1. Create a new file and name it `MySQLCreateTable.js`.

2. Add the `MySQLConnection.js` module:

   ```
   var mysqlconnection = require("../mysqlconnection.js");
   ```

   > **Note**
   >
   > You can find the `MySQLConnection.js` file at `https://github.com/PacktWorkshops/The-MySQL-Workshop/blob/master/Chapter07/Exercise7.06/MySQLConnection.js`.

3. Create a variable named `sql` and set the value of the variable equal to the following query, which will create a new table named `continents` and define its fields and properties:

   ```
   var sql = "CREATE TABLE `world_statistics`.`continents` ( \
       `ContinentID` int(11) NOT NULL AUTO_INCREMENT, \
       `Continent` varchar(13) DEFAULT NULL, \
       PRIMARY KEY (`ContinentID`) \
   );";
   ```

4. Run the SQL query with the included error handling:

```
mysqlconnection.query(sql, function (err) {
```

5. Test for errors and display a message and error code if there is one:

```
if (err) throw "Problem creating the table:- " + err.
code;
```

6. Display a message on the console indicating that the table was created using the following command:

```
console.log("Table created");
```

7. Exit the script and close off the query bracketing:

```
process.exit();
});
```

> **Note**
>
> The complete script can be found at https://github.com/
> PacktWorkshops/The-MySQL-Workshop/blob/master/
> Chapter07/Exercise7.08/MySQLCreateTable.js.

8. Save the file and run the script. You should see the following output:

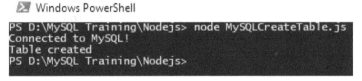

Figure 7.26 – The console output for the MySQLCreateTable.js script

Here, you can see that the table was successfully created in the database.

9. Refresh the schema in Workbench. The new table will be visible:

Figure 7.27 – The schemas list showing a new table in world_statistics

10. Right-click the new table – that is, `continents`. Select `Alter Table` and examine it. Your table should contain the fields and properties you defined in the SQL statement:

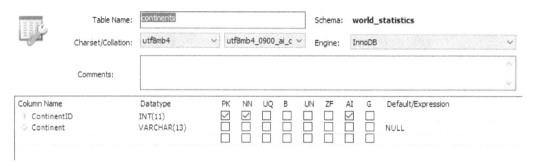

Figure 7.28 – The Alter Table view displaying table properties

When you define a table, not only can you define the field name, field type, and primary key, but also any other property a table can have, such as indexes, collation, character sets, and more in the script.

In the next section, you will put your learning to the test by employing what you have learned in this chapter.

Activity 7.01 – building a database application with Node.js

You work for a marketing company called *Marketing Our Thing*. You have been asked by Fred, the Marketing Head, to create a small database with two tables to store the details of its customers and the purchases they have made. Fred has provided you with details of what he wants in a `Requirements.txt` document. You are been asked to create two scripts that will run in Node.js – one to create the database and another to create the tables and fields. Both scripts should take advantage of the `mysqlconnection.js` file to create a data connection.

The following are the requirements:

- **Database Name:** `MOTdatabase`
- **Table Name:** `Customers`

- **Table Definition**:

Field Name	Data Type	Other
CUSTID	Int	Primary Key
CustName	VarChar(50)	NOT NULL

Figure 7.29 – The Customers table

- **Table Name**: CustomerPurchases

- **Table Definition**:

Field Name	Data Type	Other
CPID	Int	Primary Key
CustID	Int	NOT NULL
SKU	VarChar(20)	NOT NULL
SaleDateTime	VarChar(25)	NOT NULL
Quantity	Int	NOT NULL

Figure 7.30 – The CustomerPurchases table

> **Note**
>
> You can find the Requirements.txt file at https://github.com/
> PacktWorkshops/The-MySQL-Workshop/blob/master/
> Chapter05/Activity5.01/Requirements.txt.

Follow these steps to complete this activity:

1. Set up a database connection using the mysqlconnection.js file that you created in *Exercise 5.06, Modularizing the MySQL connection.*

2. Create a script called motdatabase.js and use the query method to create MOTdatabase.

3. Run the motdatabase.js script in your command prompt.

4. Refresh the database schema in MySQL Workbench. You should see `motdatabase` in the schema list:

Figure 7.31 – The schema of motdatabase in MySQL Workbench

5. Create another Node.js file named `mottables.js` to create two tables called `Customers` and `CusomterPurchases` using the `query` method, as per the requirements provided by the marketing head.

6. Run the `mottables.js` script via the console.

7. Refresh the schema to see the two new tables, `Customer` and `CustomerPurchases`, in `motdatabase`:

Figure 7.32 – The tables that are currently in motdatabase

8. Examine the fields and properties of the tables.

> **Note**
> The solution to this activity can be found in the *Appendix.*

With that, you have learned how to use Node.js to set up a database in MySQL. You now know how to create client applications through Node.js while using MySQL as a database.

Summary

You have worked your way through a lot in this chapter, so let's recap what you have learned. In the *Best practices for SQL client development* section, you learned about the importance of creating a development server, including how to duplicate the production database, the importance of creating regular backups during development and how to do so easily and quickly, and how to recover from accidental loss or damage by restoring the full database or just the tables that were lost or damaged.

You also learned how to install modules, connect to the database, and modularize the connection script so that it can be reused in other scripts using Node.js. Finally, you learned how to create a database and add tables using Node.js.

In the next chapter, you will learn how to modify the structure of tables and data within tables. You will also learn how to output the data to the console and the browser by using text and Excel files, including formatting outputs and creating dynamic outputs through Excel formulas.

8
Working with Data Using Node.js

In this chapter, you will continue working with Node.js by inserting, updating, and reading records from the database. You will output data to the web browser through Node.js and build data tables in HTML using Node.js. Additionally, you will learn about **Open Database Connectivity** (**ODBC**) connections in detail, which allow connections to be made to a database through programs such as MS Excel and MS Access.

In this chapter, we will cover the following main topics:

- Interacting with databases
- Inserting records in Node.js
- Updating the records of a table
- Displaying data in browsers
- ODBC connections

Interacting with databases

Let's suppose that your company requires an application that can interact with databases to insert, update, and display data in a user-friendly way. Currently, you have only learned how to connect and view data in Node.js. To insert, update, and display data, first, you will need to understand how Node.js handles queries that change data and outputs through the browser.

In *Chapter 5, Correlating Data Across Tables*, you were introduced to Node.js and learned how to set it up with libraries and some basic functionality. This chapter picks up where you left off and will teach you how to work with data that is present in the MySQL server using Node.js. As you work through this chapter, you will learn how to implement some particularly useful SQL queries, such as `insert` and `update`, which will allow you to work with data and Node.js. These methods will allow you to insert and update data within a database. In addition to this, you will learn how blocking queries and non-blocking queries work in Node.js to ensure queries run in the correct order.

Also, you will learn about outputting data in a way that is easy to read for any user. To accomplish this task, you will learn about HTML tables, along with other HTML tags that help structure web pages. Once you have a good understanding of the basic syntax of HTML, you will learn how to build HTML outputs in Node.js. This will include iterating database query results in order to add them into tables. The result will be HTML-formatted outputs that are generated when a web page is visited.

You will finish off this chapter by looking at ODBC connections, the various types of ODBC connections, how and when to use them, and how to create them.

By the end of this chapter, you should have full knowledge of how to work with MySQL databases through Node.js. These skills will enable you to create dynamic web applications that can read, update, and insert data into MySQL databases.

Now, let's start by investigating queries that insert and update the records within a database.

Inserting records in Node.js

When you first create a database for an application, it will not have any data contained inside it. As the user interacts with the application, often, you will want to store data from the interactions in the database, to be used later. For example, let's suppose that a company wants you to create an application where a user can input their tasks for the week. Each time they open the application, they see their current tasks. When a user adds a new task, the application needs to add that task to the database. This is so that it is saved and accessible each time the application is loaded. To achieve this, you will need to learn how to insert data into your database.

Inserting data into a database involves running queries against the database. In *Exercise 5.06 – modularizing the MySQL connection* of *Chapter 5, Correlating Data Across Tables*, you learned how to query a database for data using a SELECT query. In this section, you will use the same query method but with an insert query instead. Before you look at how INSERT queries work with Node.js, let's discuss a new query concept known as a **parameterized query**.

When you insert data into a database, you will add the data to the query method. Typically, this data is provided by the user of the product. In the preceding example, the data provided could be a new task that a user wishes to add to the application. The ideal way to add this data is by using a parameterized query. A parameterized query puts a placeholder (?) in the query for data that will be provided by the user. The following code shows an example of a parameterized query:

```
var sql = "INSERT INTO customers (customerName) VALUES ?";
```

The ? character indicates a placeholder for the values that will be inserted into the customers table. When you want to run the query, you provide the values for the query, and MySQL will replace the ? character with the provided values. The values are provided as a two-dimensional array, where the outer array represents the set of records being inserted, and the inner array represents the fields being inserted into the table. For example, if you wanted to insert a single record into the customers table, first, you would define the record:

```
var record = [[Joy]];
```

The preceding code represents a single record being inserted, with a single field, which has the value of Joy. Now, when you run the query method, you provide the query and the record being inserted. The results of the query will be written into the result variable. If any errors occur during the execution of the query, they will be written into the err variable:

```
mysqlconnection.query(sql, [record], function (err, result)
```

When this query is run, the ? character that was written into the sql variable is replaced with the record stored inside the record variable. So, the following query is what is executed against the database:

```
INSERT INTO customers (customerName) VALUES 'Joy';
```

After executing the query, you can use the `result` object to verify that the query has terminated successfully. There are two properties that are useful to verify the results. The first is `result.affectedRows`, which tells you how many rows were changed by the query. For example, if one row is inserted into the database, the `affectedRows` property would be 1. The second property is `insertID`, which is the ID of the record that was inserted by the query.

One additional note about parameterized queries is the representation of fields that contain spaces. If you wish to query a field with spaces in the name, you must use the ' character to indicate that the name contains spaces. The following example shows a parameterized query with this feature:

```
var sql = "INSERT INTO world_statistics.continents ('continents
in world') VALUES ?";
```

With this understanding, in the following exercise, let's look at an example of how to apply these concepts.

Exercise 8.01 – inserting a record into a table

> **Note**
>
> In this exercise, the database being used was created in *Exercise 7.01*. You can find the database at `https://github.com/PacktWorkshops/The-MySQL-Workshop/tree/master/Chapter07/Databases`.

The user, Roy, lives in Africa. Your manager has asked you to add Roy's continent, Africa, into the `continents` table of the `world_statistics` database. To verify that the operation runs successfully, your manager has asked you to print the results of the query and any errors to the console. To insert Roy's continent, Africa, into the `continents` table, perform the following steps:

1. Create a script file and name it `MySQLInsertOneRecord.js`.

2. Add the `mysqlconnection` module to the top of `MySQLInsertOneRecord.js`:

```
var mysqlconnection = require("MySQLConnection.js");
```

> **Note**
>
> Use the `MySQLConnection.js` file created in *Exercise 5.06 – modularizing the MySQL connection*, to take advantage of modularization. The file is located in GitHub at `https://github.com/PacktWorkshops/The-MySQL-Workshop/blob/master/Chapter07/Exercise7.06/MySQLConnection.js`.

3. Define the following SQL query to insert Roy's record into the table. Utilize the concept of a parameterized query with a placeholder for the values being inserted:

```
var sql = "INSERT INTO world_statistics.continents
(continent) VALUES ?";
```

4. Next, define a record array, with a single field contained within it. This record array will store the value you want to insert into the database, which is the continent that Roy lives in, Africa:

```
var record = [['Africa']];
```

> **Note**
>
> The quotation marks surrounding the data item (in this case, `Africa`) can be either the standard double quotes or single quotes.

5. To execute the SQL query, use the `mysqlconnection.query` method. Pass in the SQL statement and the record array and set up error handling. Store the result of the query being executed against the database in the object named `result`:

```
mysqlconnection.query(sql, [record], function (err,
result) {
```

6. Test for an error using the following command. Print a message and the error code if there is one:

```
if (err) throw "Problem inserting the data" + err.
code;
```

7. Verify that the query has been completed successfully by printing the variable result to the console using `console.log`. Print `affectedrows` and `insertId` separately to verify that the data was inserted successfully:

```
console.log(result);
console.log("Number of rows affected : " + result.
affectedRows);
console.log("New records ID : " + result.insertId);
```

8. Exit the script and close the bracketing for `mysqlconnection.query`:

```
process.exit();
});
```

Your complete script should look like the following:

```
var mysqlconnection = require("./mysqlconnection.js");

var sql = "INSERT INTO world_statistics.continents
(continent) VALUES ?";

var record = [['Africa']];

  mysqlconnection.query(sql, [record], function (err,
result){

    if (err) throw "Problem inserting the data" + err.
code;

    console.log(result);
    console.log("Number of rows affected : " + result.
affectedRows);
    console.log("New records ID : " + result.insertId);

  process.exit();

});
```

9. Save and run the script. You will get a response similar to the following screenshot on your console:

```
Windows PowerShell

PS D:\MySQL Training\Nodejs> node mysqlinsertonerecord.js
Connected to MySQL!
ResultSetHeader {
  fieldCount: 0,
  affectedRows: 1,
  insertId: 1,
  info: '',
  serverStatus: 2,
  warningStatus: 0
}
Number of rows affected : 1
New records ID : 1
PS D:\MySQL Training\Nodejs>
```

Figure 8.1: The console output showing detailed and selective results

> **Note**
>
> You can get different values for the fieldCount, affectedRows, and insertID fields. These values depend on the number of records in the continents table. The values might appear larger due to more records being present or smaller if there are fewer records present.

10. To see the newly inserted record, go to Workbench, and click on **Select Rows** in the **continents** table. You should see the following result:

Figure 8.2: The table contents after running the script

From this result, you can see that the Continent record with the value of Africa has now been added to the database table.

Inserting data into a table is not difficult and takes little coding. It is made easier with the ability to replace the values in the SQL statement with a single question mark (?) and define the data in another variable or from some other source, such as a file or an API call.

Often, you will want to insert multiple records at the same time. In the next section, we will look at how multiple records can be inserted into a MySQL database using Node.js.

Inserting multiple records

In some cases, you might need to insert multiple records at a single point in time. For example, if you had an application that kept track of a user's daily tasks, you might want to add a feature for the user to add multiple tasks at once. In this scenario, your first thought would be that you might need to run multiple insert queries against your database. However, there is a more efficient option, which takes advantage of parameterized queries.

Recall that when you wanted to insert data into a table, you needed to use a two-dimensional array to store the values to be inserted. Using this object, you can also insert multiple records with your INSERT statement. For example, in the preceding exercise, you were inserting customers inside a database table. If you had two customers to insert, you would simply add another array into the record array, as follows:

```
var record = [['Joy'],['James']];
```

This will allow you to insert two records—one with the value of Joy and the other with the value of James. Each array within the record array represents an individual record. When MySQL runs the query, it will insert each record that is present in the array.

In the next exercise, we will look at how to insert multiple records into a table.

Exercise 8.02 – inserting multiple records into a table

User James has lived on multiple continents: Asia, Europe, North America, and Oceania. Your manager has asked you to enter his continents into a new table, called userContinents, in the world_statistics database. This table will contain the user's name and the continents they have lived on. To insert James' details into the userContinents table, perform the following steps:

1. Create a script file and name it MySQLInsertMultipleRecordsContinents. js.

2. Insert the connection module using the following command:

```
var mysqlconnection = require("MySQLConnection.js");
```

3. First, set up a SQL query to create our new `userContinents` table. This table will contain a continent ID, the user's name, and the continent name:

```
var sql = "CREATE TABLE 'world_
statistics'.'userContinents' ( \
    'ContinentID' int(11) NOT NULL AUTO_INCREMENT, \
    'Continent' varchar(13) DEFAULT NULL, \
    PRIMARY KEY ('ContinentID')\
);"
```

4. Next, run the `create table` query to add the table to your database:

```
//Execute the SQL, include error checking
mysqlconnection.query(sql, function (err) {

    //Handle any errors
    if (err) throw "Problem creating the table:- " + err.
code;

    //Otherwise tell user that the table was created
    console.log("Table created");

//And leave
process.exit();

//Close off the block bracketing
});
```

5. Next, set up the SQL query to insert multiple records into the database:

```
sql = "INSERT INTO world_statistics.userContinents
(continent) VALUES ?";
```

6. Since you have been asked to insert multiple values, enter the following code to define multiple records for the `userContinents` table:

```
var record = [['Asia'],['Europe'],['North
America'],['Oceania'];
```

7. Enter the following code. It will start by executing the query using the `query` method. Once this has been completed, the result, the number of rows inserted, and the ID of the inserted rows are outputted to the console through the `console.log` method:

```
mysqlconnection.query(sql, [record], function (err,
result) {
    if (err) throw "Problem inserting the data" + err.
code;
    console.log(result);
    console.log("Number of rows affected : " + result.
affectedRows);
    console.log("New records ID : " + result.insertId);
process.exit();
});
```

The complete script should look like the following:

```
var mysqlconnection = require("MySQLConnection.js");
var sql = "CREATE TABLE 'world_
statistics'.'userContinents' ( \
  'ContinentID' int(11) NOT NULL AUTO_INCREMENT, \
  'Continent' varchar(13) DEFAULT NULL, \
  PRIMARY KEY ('ContinentID')\
);"
//Execute the SQL, include error checking
mysqlconnection.query(sql, function (err) {
  //Handle any errors
  if (err) throw "Problem creating the table:- " + err.
code;
  //Otherwise tell user that the table was created
  console.log("Table created");
//Close off the block bracketing
});
sql = "INSERT INTO world_statistics.usercontinents
(Continent) VALUES ?";
var record = [['Asia'],['Europe'],['North
America'],['Oceania']];
mysqlconnection.query(sql, [record], function (err,
result) {
```

```
    if (err) throw "Problem inserting the data" + err.
code;
    console.log(result);
    console.log("Number of rows affected : " + result.
affectedRows);
    console.log("New records ID : " + result.insertId);
process.exit();
});
```

8. Save and run the file. The results in the console will appear as follows:

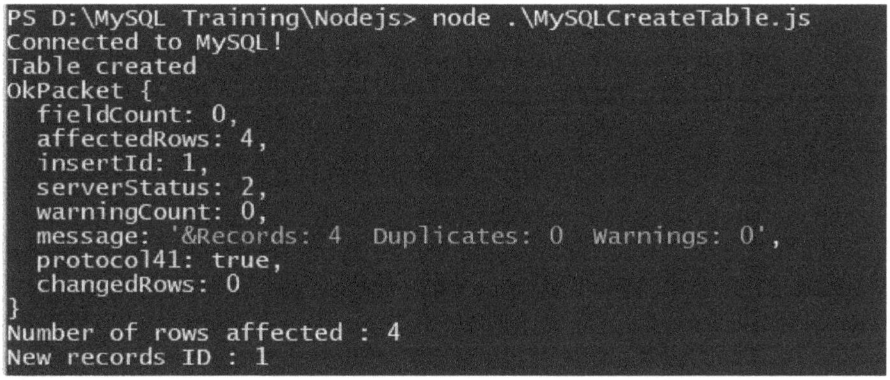

```
PS D:\MySQL Training\Nodejs> node .\MySQLCreateTable.js
Connected to MySQL!
Table created
OkPacket {
  fieldCount: 0,
  affectedRows: 4,
  insertId: 1,
  serverStatus: 2,
  warningCount: 0,
  message: '&Records: 4  Duplicates: 0  Warnings: 0',
  protocol41: true,
  changedRows: 0
}
Number of rows affected : 4
New records ID : 1
```

Figure 8.3: Detailed and selective logging of the script results

This output shows that the query was successfully executed, and four rows were inserted. Note that the number of rows affected will match the number of records provided to the parameterized query.

9. Now, view the table's contents in Workbench:

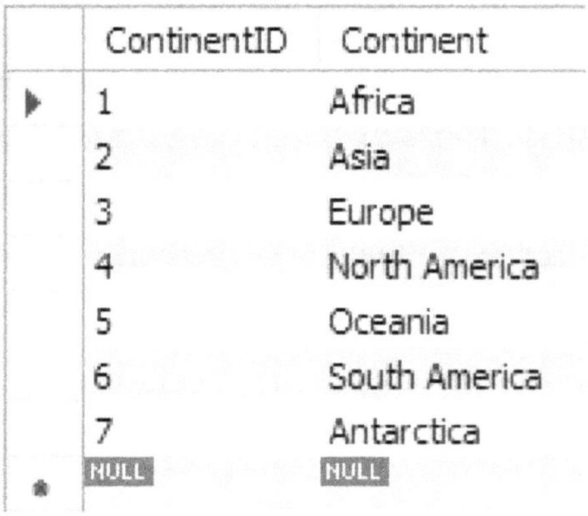

ContinentID	Continent
1	Africa
2	Asia
3	Europe
4	North America
5	Oceania
6	South America
7	Antarctica
NULL	NULL

Figure 8.4: The table's contents after the script has been executed

Note that the record count is now 7, which implies six more records have been successfully added to the table.

Inserting multiple records is no more difficult than inserting one record. The only difference is the number of records you define in the record variable and the way they are constructed.

Now, in the following section, you will extend your skills by inserting multiple field records into a table.

Inserting with multiple fields

In the previous section, you learned how to insert multiple records into a database through Node.js. You might have noticed that each of the records that you inserted only had a single field—in this case, the name of the continent. However, most of the databases you work with will have many different fields within them, which means you will eventually need to run insert queries with multiple fields.

The code changes required to accommodate multiple fields are small. You simply need to add each field into the array for the parameterized function. For example, let's suppose you are working on an application that keeps track of the date and location a user has logged in from. In this case, there are two fields that you need to insert into the database: the date of the login and the location of the user. The following code shows how a record can be set up for this scenario:

```
var record = [['02/03/2021','North
America'],['01/05/2020','Europe']]
```

Next, when you write your query, you will need to add in every field that you want to insert data into. For example, the following code shows how a query can be written to insert the record data:

```
var mysqlconnection = require("./mysqlconnection.js");
var sql = "INSERT INTO loginRecord(loginDate,loginContinent)
VALUES ?";
```

From here, you will simply need to run the query method, which is the same process as any other Node.js MySQL query:

```
mysqlconnection.query(sql, [record], function (err, result) {
}
```

Since the query is parameterized, the ? character can be replaced with any variable. In the preceding code, you are replacing the ? character with a set of records with multiple fields. It all comes down to how you format the data variable and the fields you include within the SQL query. The structure is as follows:

```
Fields in SQL ('Field name 1','Field name 2')
Data records [ ['data 1','data 2'],['data 1','data 2'],['data
1','data 2'], … ];
```

Here are a few points to remember:

- Field names with spaces must be enclosed in backticks (').
- Data points must use either single (') or double (") quotes and be separated by a single comma (,).
- Records must be enclosed in square brackets ([]) and be separated by a single comma (,).
- The entire record/data definition must be enclosed in square brackets.
- The fields that we are inserting our records into must be enclosed in round brackets.

Using these points, in the following exercise, you will insert multiple field records.

Exercise 8.03 – populating records from the existing tables

Let's suppose the travel agency wants to get a list of all of the countries with a countryID value of less than 10. You are told that this would represent a single region of countries and should be stored in a table named Region1. The table should contain the countryID and Country Code values of all the respective countries.

In this exercise, you are asked to create a table, named Region1, in the world_ statistics database and store the data of the countries with a countryID value of less than 10. To achieve the goal of this exercise, perform the following steps:

1. Create a script file and name it MySQLInsertRecordsFromAnotherTable.js.
2. Add the MySQL connection module to the top of the code file:

```
var mysqlconnection = require("MySQLConnection.js");
```

3. Next, write a query to create the new `Region1` table:

```
var sql = "CREATE TABLE world_statistics.
Region1(CountryID INT, 'Country Code' VARCHAR(45));";
```

4. Execute the query, print the number of rows affected, and use `insertID` to verify that the query was completed successfully:

```
mysqlconnection.query(sql, function (err, result) {

    if (err) throw "Problem creatings the data" + err.
code;

    console.log(result);
    console.log("Number of rows affected : " + result.
affectedRows)
    console.log("New records ID : " + result.insertId);

});
```

5. Define a variable, named `records`, to store a `select` query to display the required data for the table:

```
var records = "SELECT 'ContinentID', 'Country Code' FROM
world_statistics.countries WHERE 'CountryID' < 10 ORDER
BY 'CountryID'";
```

6. Define a variable, named `sql`, and set it equal to an `INSERT` query for the country table:

```
var sql = "INSERT INTO world_statistics.country
('CountryID','Country Code')";
```

7. Next, concatenate the SQL queries together to insert the results of the records inside the definition that was written for the `sql` variable:

```
sql = sql + " " + records;
```

The fully constructed query is now stored in the `sql` variable, and it can be executed using the `query` method:

```
mysqlconnection.query(sql, function (err, result) {
```

8. The next set of lines will be run once the INSERT query has been completed. The result of the query, the number of rows inserted, and the last ID inserted will be displayed in the console through the console.log method. Enter the following commands:

```
      if (err) throw "Problem inserting the data" + err.
    code;
        console.log(result);
        console.log("Number of rows affected : " + result.
    affectedRows);
        console.log("New records ID : " + result.insertId);
    process.exit();
    });
```

9. Save and run the script. Your results in the console should appear similar to the following:

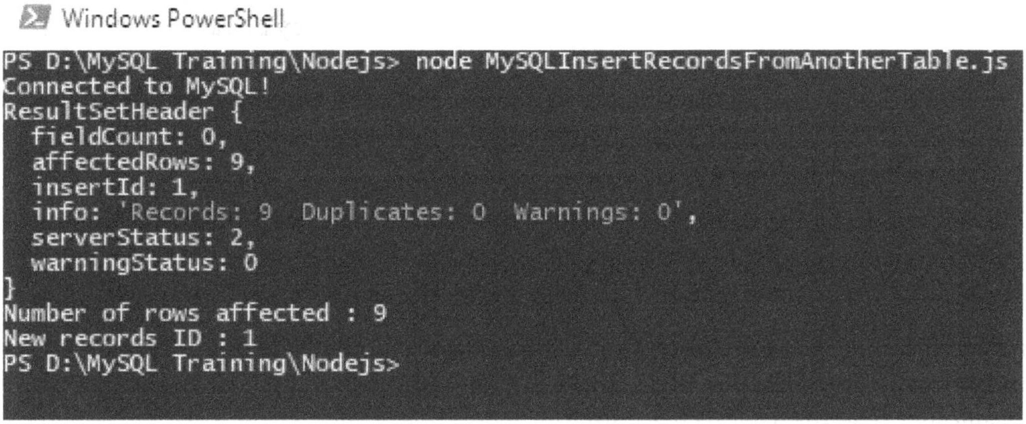

Windows PowerShell

```
PS D:\MySQL Training\Nodejs> node MySQLInsertRecordsFromAnotherTable.js
Connected to MySQL!
ResultSetHeader {
  fieldCount: 0,
  affectedRows: 9,
  insertId: 1,
  info: 'Records: 9  Duplicates: 0  Warnings: 0',
  serverStatus: 2,
  warningStatus: 0
}
Number of rows affected : 9
New records ID : 1
PS D:\MySQL Training\Nodejs>
```

Figure 8.5: The console output indicating that nine records have been inserted

The preceding output shows that nine rows were inserted into the database table. The rows that were inserted into the database table correspond to all of the countries that have a CountryID value of less than 10. To further verify this, you can query the table through MySQL Workbench to see the inserted records.

The results in Workbench should appear as follows:

CountryID	Country Code
4	ABW
2	AFG
1	AGO
3	ALB
3	AND
NULL	ARB
2	ARE
6	ARG
2	ARM

Figure 8.6: The contents of the table after the script has been executed

The preceding result shows that the values are the same as the ones expected from the country table—that is, all the countries with a CountryID value of less than 10. Note that there is also an ID that is NULL in the results. When SQL attempts to filter the table, it will always allow NULL values in the results, unless you specify the CountryID IS NOT NULL condition. Since we have not specified this option in the WHERE clause, the NULL values are displayed.

With this, you now understand how to build data from existing tables. These situations happen frequently in order to create subsets of data. These smaller sets of data are often favorable to allow for more efficient querying, due to less data being present.

In the next section, you will learn how to make updates and modifications to the data that has been inserted into the tables. This will allow you to write applications that can dynamically update data as it is being used.

Updating the records of a table

Often, you will want to update the data stored within a database table. For example, consider the task list from the *Inserting records in Node.js* section. When a user adds a task to their database, it is initially marked as incomplete. Rather than having a separate table to store completed tasks, you could, instead, have a field for the task record that keeps track of its status—that is, if the task has been completed or not. Once the task is complete, you can simply modify the record to set the completed field to yes. Often, updating an existing record is faster than inserting a new record, so this is a more efficient option.

One important concept to bear in mind before looking at UPDATE queries is the idea of blocking queries and non-blocking queries in MySQL. A **blocking query** is a query that needs to be completed in full before the next action can be executed. For example, setting the database using a USE query would be blocking, as the database needs to be set before any queries can be executed.

In Node.js, you can execute a blocking query by embedding the query directly into the connection query method. The following code runs a query to use the world_ statistics database, printing that the database is being used when the query is successful:

```
mysqlconnection.query("USE world_statistics", function (err,
result) {
    if (err) throw err;
    console.log("Using world_statistics database");
```

This query type should be used whenever you wish to wait for a query to finish before continuing to the next section of code.

Non-blocking queries can be used in situations where a query's completion is not essential to the code that follows. For example, if you wanted to insert a record in to two different tables, each insert could be done without blocking, since both queries are independent of each other. The following code shows an example of this:

```
var sql = "INSERT INTO world_statistics.country
('CountryID','Country Code') VALUES ('1','CAD')";

mysqlconneciton.query(sql, function(err,result)
{
    if (err) throw err;
    console.log("inserting country");
}
```

Now, we'll take a look at how to update records in a table through Node.js. Updating records can be done using an update query in SQL. The update query has the following format:

```
UPDATE table_name SET field1 = new-value1, field2 = new-value2
[WHERE Clause]
```

The fields in the `table_name` table can be updated to any values required. The `WHERE` clause can be used to specify updating records that only meet a specific condition.

In the next exercise, you will look at how update queries can be executed through Node.js.

Exercise 8.04 – updating a single record

Facts change and databases need to be updated to stay current and useful. In *Exercise 8.02 – inserting multiple records into a table*, you created a table that contained the names of continents. You have been informed that some countries consider Oceania as a continent, while others call this continent Australia. In the `world_statistics.continents` table, currently, you only have `Oceania`, whose value of `continent ID` is 5. The customers of your company have asked for this to be updated to `Australia/Oceania`, which is a more appropriate name for the continent. Additionally, you have noticed that the field size is too small to contain the full text of `Australia/Oceania`. To solve this problem, first, you must make the field size larger, then update the field to have a continent value of `Australia/Oceania`.

To complete this exercise, perform the following steps:

1. Create a new script file and name it `UpdateOneRecord.js`.

2. Include the `mysqlconnection` module at the top of the JavaScript file:

    ```
    var mysqlconnection = require("MySQLConnection.js");
    ```

3. Issue your first blocking query. Instruct the server to use the `world_statistics` database, include error handling, and notify the users on the console when done:

    ```
    mysqlconnection.query("USE world_statistics", function
    (err, result) {
        if (err) throw err;
        console.log("Using world_statistics database");
    ```

4. In MySQL Workbench, right-click on the `continents` table, and select the `Alter table` option. From the results, you can see that the `Continent` field is `VARCHAR(13)`, so it is 13 characters in size:

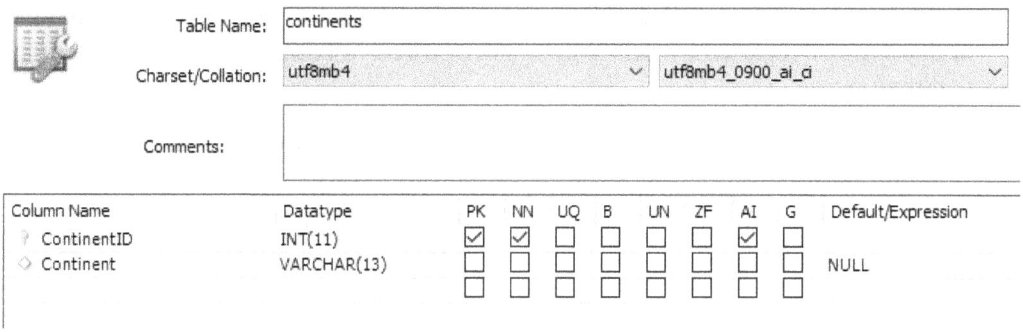

Column Name	Datatype	PK	NN	UQ	B	UN	ZF	AI	G	Default/Expression
ContinentID	INT(11)	☑	☑	☐	☐	☐	☐	☑	☐	
Continent	VARCHAR(13)	☐	☐	☐	☐	☐	☐	☐	☐	NULL
		☐	☐	☐	☐	☐	☐	☐	☐	

Figure 8.7: The field is too short to fit the data

5. Since the field is too small to contain the value of Australia/Oceania, adjust the size using an ALTER query. Update it with 17 characters to fit the complete text of Australia/Oceania. Enter the following commands to change the column size, add error checking, and log to the console when done:

```
var ChangeCol = "ALTER TABLE 'continents' "
    ChangeCol = ChangeCol + "CHANGE COLUMN 'Continent'
'Continent' VARCHAR(17 ) NULL DEFAULT NULL;"
    mysqlconnection.query(ChangeCol, function (err) {
        if (err) throw err;
            console.log("Column Continent has been
resized");
```

You will notice in ChangeCol that 'Continent' appears twice. The first occurrence tells the server what field to change, while the second is the field's new name. Since you are not changing its name, it remains as 'Continent'. Now you have changed the column size to 17.

6. Update the record and enter the following command to build the variable containing the new value and the ID of the record you want to update:

```
var updateValues = ["Australia/Oceana",5];
```

7. Build the SQL query to update the continent field. There are two ? symbols. Both of them will be replaced by the values in the updateValues record in the order that they appear:

```
var sql = "UPDATE continents SET Continent = ? WHERE
ContinentID = ? ";
```

8. Execute the SQL query to update the record. Include error handling and inform the user:

```
        mysqlconnection.query(sql, updateValues,
function (err, result) {
        if (err) throw err;
        console.log("Record has been updated");
        process.exit();
```

9. Finally, close the brackets for the three query executions that you made:

```
        });
    });
});
```

The entire script will appear as follows:

```
var mysqlconnection = require("MySQLConnection.js");

mysqlconnection.query("USE world_statistics", function
(err) {
    if (err) throw err;

    console.log("Using world_statistics database");

    var ChangeCol = "ALTER TABLE 'continents' "
    ChangeCol = ChangeCol + "CHANGE COLUMN 'Continent'
'Continent' VARCHAR(20) NULL DEFAULT NULL;"
    mysqlconnection.query(ChangeCol, function (err) {
        if (err) throw err;

        console.log("Column Continent has been
resized");
```

```
            var updateValues = ["Australia/Oceana",5];

            var sql = "UPDATE continents SET Continent = ?
WHERE ContinentID = ? ";

            mysqlconnection.query(sql, updateValues,
function (err, result) {
                if (err) throw err;

                console.log("Record has been updated");

                process.exit();

            });
        });
    });
```

10. Save and run the file.

    Once the program has been run, the command line will verify that the record has been updated:

```
PS D:\MySQL Training\Nodejs> node index.js
Connected to MySQL!
Using world_statistics database
Column Continent has been resized
Record has been updated
```

Figure 8.8: The output of the program when successfully completed

11. Navigate to MySQL Workbench, right-click on the `continents` table, and pick the `Select Rows` option. You should see that the record has now been updated to `Australia/Oceania`:

	ContinentID	Continent
▸	5	Australia/Oceania

Figure 8.9: The new continent value, Australia/Oceania, in the continents table

    Hence, by using update queries in Node.js, you were able to successfully update the incorrect entry in the `continents` table.

The next activity will challenge and bring many of the skills that you have learned so far together into a common real-life scenario working with databases.

Activity 8.01 – multiple updates

The manager of the ABC company wants to add new details to the existing country table in the world_statistics database, that is, the capital city of each country, the independence status of each country, and their currency types. You have been tasked with making these changes and updating the database. This new information has been provided through SQL scripts so that you can create and populate a temporary table and create the new countryalldetails table. To implement this activity, perform the following steps:

1. Create a database connection to the world_statistics database.

2. Create the countryalldetails table with the following columns: CountryID, ContinentID, CountryCode, CountryName, Is_independent, Currency, and Capital.

3. Load the data found at https://github.com/PacktWorkshops/ The-MySQL-Workshop/blob/master/Chapter08/Activity8.01/ CountryDetails.sql into the countryalldetails table.

4. Using the new temporary table, populate the countryalldetails table with the new values. This can be done using an UPDATE query, where the country codes in the temp table are joined to the countryalldetails table.

    After performing the preceding steps, the expected output on the console should be similar to the following:

```
Windows PowerShell

PS D:\MySQL Training\Nodejs> node Activity-MultipleUpdates.js
Connected to MySQL!
Using World_Statistics
Column Capital created
Column Is_Independent created
Column Currency created
Capital is updated
Number of rows affected : 263
Is_Independent is updated
Number of rows affected : 263
Currency is updated
Number of rows affected : 263
PS D:\MySQL Training\Nodejs>
```

Figure 8.10: Console messages indicating the progress of the script

The `Capital`, `Is_Independent`, and `Currency` columns should be visible on the Workbench GUI:

Figure 8.11: A schema displaying the country table with its new fields

The `country` table should look similar to the following screenshot:

CountryID	ContinentID	Country Code	Country Name	Capital	Is_Independent	Currency
1	4	ABW	Aruba	Oranjestad	Part of NL	AWG
2	2	AFG	Afghanistan	Kabul	Yes	AFN
3	1	AGO	Angola	Luanda	Yes	AOA
4	3	ALB	Albania	Tirana	Yes	ALL
5	3	AND	Andorra	Andorra la Vella	Yes	EUR
6	NULL	ARB	Arab World	NULL	NULL	NULL
7	2	ARE	United Arab Emirates	Abu Dhabi	Yes	AED
8	6	ARG	Argentina	Buenos Aires	Yes	ARS

Figure 8.12: The Select Rows view, showing the new fields populated with data

The `countryalldetails` table should have a structure similar to the following screenshot. From this, you can see that three new fields, `Capital`, `Is_Independent`, and `Currency` have been added:

Column Name	Datatype	PK	NN	UQ	B	UN	ZF	AI	G	Default/Expression
CountryID	INT(11)	☑	☑	☐	☐	☐	☐	☑	☐	
ContinentID	INT(11)	☐	☐	☐	☐	☐	☐	☐	☐	NULL
Country Code	VARCHAR(5)	☐	☐	☐	☐	☐	☐	☐	☐	NULL
Country Name	VARCHAR(50)	☐	☐	☐	☐	☐	☐	☐	☐	NULL
Capital	VARCHAR(50)	☐	☐	☐	☐	☐	☐	☐	☐	NULL
Is_Indepent	VARCHAR(25)	☐	☐	☐	☐	☐	☐	☐	☐	NULL
Currency	VARCHAR(5)	☐	☐	☐	☐	☐	☐	☐	☐	NULL

Table Name: countryalldetails   Schema: world_statistics

Charset/Collation: utf8mb4   utf8mb4_0900_ai_ci   Engine: InnoDB

Comments:

Figure 8.13: The new fields in the Alter Table view

> **Note**
> The solution for this activity can be found in the *Appendix* section.

In this activity, you inserted multiple details into the table. With this, you have now updated the `country` table to provide valuable information related to the capital, the independence status, and the currency of the country.

In the next section, you will learn how to format the data before viewing it on the web browser.

Displaying data in browsers

So far, you have learned how to execute various queries through Node.js. One of the common uses of Node.js is to display data to the user through a web browser. Since Node.js is built through JavaScript, it can naturally build web pages that the user of the application can view and interact with.

You have already learned how to interact with the MySQL server through Node.js, so this section will spend some time discussing how the data can be formatted and displayed. To begin, you will need to install a new Node.js module called `numeral`. To install `numeral` for a given project, you need to run the following command in your command line:

```
npm install numeral
```

The `numeral` module can be used to format numeric values. For example, let's suppose that you have a set of decimal numbers, and you wish to display them with two decimal places (so, a number such as 1.231 becomes 1.23). The `numeral` module provides a formatting method to allow you to do this, as follows:

```
numeral(Field).format('0.00')
```

The preceding code will take the value of `Field`, and format it with two decimals. This allows you to have a consistent display for any numeric fields you get from a database. There are many other formats available through the numeral module that might be useful. For instance, the `0.00a` format is commonly used to condense large numbers. When a number is formatted as `0.00a`, it will condense the number down to two decimals, and add either `m` for million or `b` for billion. For example, 1,240,000 would become 1.24 m.

To display the data on a web browser, you will need to understand some basic HTML code, too. Let's suppose that you have collected data from a table, and you wish to display it in a tabular format on the web browser. For this example, let's assume that you have collected some data from a SQL query and stored it in an object named `results`. You can iterate the results object to get all the records contained within it using a for-each loop. The for-each loop will repeat a given instruction for every item contained within an array. For example, suppose your result has a field called `Continent_Region`. The following code would print each record's `Continent_Region` field to the console:

```
result.forEach(function(Statistics){
    console.log(Statistics.Continent_Region);
}
```

On each iteration, the current record is placed in the `Statistics` object. When you want to refer to the current record, you can use the `Statistics` object. For example, if you want to log the `Continent_Region` field of the current record, you would write `Statistics.Continent_Region`. This gets the current record, which is stored in `Statistics`, and retrieves the `ContinentRegion` field from the record.

Now, suppose you wanted to create a table of all the `Continent_Region` fields. To begin, you would need to create an HTTP server, as discussed in the previous chapter:

```
var http = require('http');
http.createServer(function (req, res) {
res.writeHead(200, {'Content-Type': 'text/html'});
```

From here, you just need to construct your HTML table. An HTML table starts with a `<table>` tag, which indicates where the table starts. Next, there is a `<tbody>` tag, which indicates where the table data is. Following this, you add the rows of the table using the `<tr>` tag, which indicates a row of the table. Finally, you can use either `<td>` or `<th>` to indicate the table data in the row or the table headers in the row, respectively. The following code shows a full table construction:

```
var http = require('http');
http.createServer(function (req, res) {
res.writeHead(200, {'Content-Type': 'text/html'});

string = "<table><tbody>";
result.forEach(function(Statistics){
    string = string + "<tr>";
        string = string + "<td>" + Statistics.Continent_Region
```

```
   + "</td></tr>";
});
    string = string + "</tbody></table>";
    res.end(string);
}).listen(82);
```

In the preceding code, you are constructing an HTML table. The code starts by telling the browser that a table using the `<table><tbody>` tags needs to be built. From here, the `result` object is reiterated to add data to the table. Then, you start by opening a new table row with `<tr>` and adding in some table data using `<td>`. Finally, you close the `</td></tr>` tags to complete the table data. Once the iteration is complete, you add the closing tags for the `<tbody>` tags, along with the `<table>` tags, and send it as a result to the requester. This creates a table that is displayed on the browser through HTML.

One additional note regarding HTML tags that you might want to know is how to adjust the actual content that is displayed in the table. This can be done by adding style code to the tags. For example, if you want your font to be green, you can add `color='green'` to the tag. If you want your font to be size 5, you can add `font size='5'` to the tag. The following example shows how this is formatted:

```
<font size = '5' color='green'>
```

There are a variety of formats that are available for HTML tags, which can help you to customize the style of the output. In this section, you will primarily work with font size and color adjustments. However, you might find it valuable to read further into the options that are available for formatting.

> **Note**
>
> To learn more about the specific HTML syntax for tables, check out the Mozilla HTML documentation at `https://developer.mozilla.org/en-US/docs/Web/HTML/Element/table`.

In the next exercise, you will look at a full example of how to build an HTML table through SQL data.

Exercise 8.05 – formatting data to the web browser

Now your company would like to create a report that shows information about the population of different areas of the world in the browser. Specifically, you have been asked to display the continent, the total population, and the total number of countries in each continent. This data is currently stored in the `world_statistics` database. The population should be formatted in `0.00a` using the `numeral` module. Additionally, the company would like the user to be able to filter the query for a specific year. Currently, they would like to see historical data for the year 2011, so this should be set as the filter for the query. To do this, you decide to create an HTML table to display the results, as this would be easy for users to access and read.

The data for this exercise will come from the `continents` table, which is joined with the `countryalldetails` table. The `continents` table can be loaded from `https://github.com/PacktWorkshops/The-MySQL-Workshop/blob/master/Chapter07/Databases/PracticeDatabaseNoSchema%2020190926a.sql`. Additionally, the `countryalldetails` table was created in *Activity 6.01 – multiple updates*. You will be required to add the population and count the countries in the table.

To complete this exercise, perform the following steps:

1. Create a new Node.js file named `TotalPopulationByContinents.js`.

2. Add the `numeral` module to the current project. You can do this by using the following commands, which are run through the command line:

   ```
   npm install numeral
   ```

3. Connect to the database, import the required `http` and `numeral` modules, and instruct the server to use the `world_statistics` database:

   ```
   var http = require('http');
   var mysqlconnection = require("./mysqlconnection.js");
   var numeral = require('numeral');
   mysqlconnection.query("USE world_statistics");
   ```

4. Prepare the variables that you are required to use in the exercise. Note that `FilterYear` will allow you to change the year of the output, while the remaining variables will be keeping track of the HTML code that is being created and the SQL queries that are being run:

   ```
   var FilterYear = "";    //Filter for year
   var string = "";        //To write the output to
   var banner = "";        //Page banner
   var headings = "";      //Column headings
   ```

```
var temp = "";          //for building output banner
var sql = "";           //For the SQL statement
var tablestyle = "";    //styling for the table
```

5. Build the SQL query to extract the data from the database. To get the population data with continents, add the population from the countrypopulation table, and join it with the continents table:

```
var sql = "SELECT \
continents.Continent AS Continent_Region, \
Sum(countrypopulation.StatisticValue) AS 'Total_
Population', \
countrypopulation.Year, \
Count(country.CountryID) AS 'Total_Countries' \
FROM continents \
INNER JOIN country \
ON country.ContinentID = continents.ContinentID \
INNER JOIN countrypopulation \
ON countrypopulation.'Country Code' = country.'Country
Code' \
WHERE \
countrypopulation.Year = ? \
GROUP BY continents.Continent, countrypopulation.Year \
ORDER BY 'Total_Population' DESC ";
```

This code will find the total population and number of countries in a continent for a given year, sorted by the total population.

6. Create the server to monitor the request. Include the request and response functions:

```
http.createServer(function (req, res) {
    res.writeHead(200, {'Content-Type': 'text/html'});
```

7. From here, the code will be run when the server gets a request from a browser. Set the year for which the report is being generated using the FilterYear variable. To format the data, use the table column headings. Each heading is enclosed in table heading tags, <th> ... </th>:

```
FilterYear = 2011;
banner = "Continent Population " + FilterYear;
```

```
headings = "<th>Continent_Region</th><th>Total
Population</th><th>Total Countries</th>";
```

8. Execute the SQL query with error handling using the following command:

```
    mysqlconnection.query(sql, FilterYear, function (err,
result) {
    if (err) throw err;
```

9. Loop through the records in the `result` variable, and in each loop through, move the record into an object named `statistics`. The output string will be built on each pass:

```
    result.forEach(function(Statistics){
```

10. In each loop through, start by adding a table row tag, `<tr>`:

```
    string = string + "<tr>" //Start table row
```

11. Add the three fields, where each field includes formatting tags and heading tags. Use the `numeral` module to format `Total_Population` for easier reading as the numbers are quite large:

```
    string = string + "<th><font size='3' color='blue'>" +
Statistics.Continent_Region + "</font></th>"
    string = string + "<th><font size='3' color='black'>" +
numeral(Statistics.Total_Population).format('0.00a') +
"</font></th>"
    string = string + "<th><font size='3' color='green'>" +
Statistics.Total_Countries + "</font></th>"
```

12. Close the table row tag and close the loop bracketing:

```
    string = string + "</tr>" //End table row
    });
```

The code will loop to the next record and add it to the string. When all of the records have been read, it will move out of the loop.

13. You are now out of the loop, and all your data is stored in the `string` variable, which also contains the HTML formatting. To improve the output, add some formatting to the `tablestyle` variable:

```
    tablestyle = "<style>table, th, td {border: 1px solid
black;}</style>"
```

14. Add the banner with some formatting tags and a start tag for the table, set the width of the table to 30% of the screen's width, and add the `headings` string, which is now enclosed in table row tags:

```
temp = "<font size='5' color='red'>" + banner + " </
font></br>"
temp = temp + "<table style='width:30%'>"
temp = temp + "<tr>" + headings + "</tr>"
```

15. Now, put them together. First, use `tablestyle`, which starts your table, then the data rows, and, finally, the closing table tag:

```
string = tablestyle + temp  + string + "</
table>";
```

16. Send the entire string to the browser in response to the request:

```
res.end(string);
```

17. Close the switch bracketing. This is the end of the response code:

```
});
```

18. Close the `createserver` bracketing, and tell the server to listen on port `82`:

```
}).listen(82);
```

> **Note**
>
> You can find the entire script at `https://github.com/PacktWorkshops/The-MySQL-Workshop/blob/master/Chapter08/Exercise8.05/TotalPopulationByContinent.js`.

19. Save and run the script. Your console will respond with the following output. However, your prompt won't come back since the code is monitoring port `82` for a request: 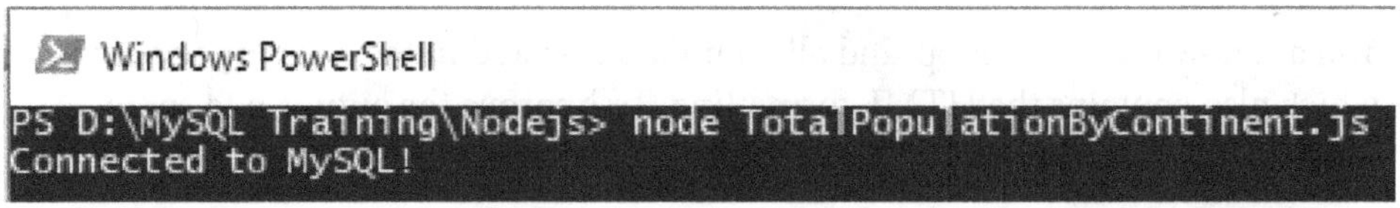

```
Windows PowerShell
PS D:\MySQL Training\Nodejs> node TotalPopulationByContinent.js
Connected to MySQL!
```

Figure 8.14: A connection validation but no cursor since the server is monitoring port 82

20. Enter `localhost:82` in your browser's navigation bar. The browser will send a request to the page, and the code will respond with the following output:

Continent Population 2011

Continent_Region	Total Population	Total Countries
Asia	4.20b	49
Africa	1.06b	53
Europe	706.53m	42
North America	546.51m	31
South America	396.27m	12
Australia/Oceana	37.23m	19

Figure 8.15: The server responds with the formatted browser output

This output shows that your data has been successfully outputted to the browser when you navigated to `localhost:82`.

21. Finally, remember to press *Ctrl + C* in the CLI to stop the script.

You can create some impressive browser output from Node.js and MySQL with little code. Additionally, including HTML tags for formatting improves the appearance. This exercise demonstrated how data can be reported through users and served through an HTML page. This allows any user to type in the address of the web page and instantly receive the data they require. Often, this type of formatting is also used in applications to generate dynamic web pages for a user. For instance, if a user logs into a web application, it might display their name on the page. This name would be added from a database, in the same way that you have done in the exercise.

Aside from showing data through HTML, there are several different options you can take advantage of to display data. Tools such as Microsoft Excel and Access are popular for managing data and databases. To connect to these external programs, you will need to learn more about the connection types they use.

In the next section, we will explore ODBC connections in detail. They are used to provide database connections to applications.

ODBC connections

When you worked with Node.js, you learned that it was possible to connect to a database using a module called `mysql`. Some applications utilize a different method of connection—known as ODBC. An ODBC connection allows a user to connect to a database through a program such as Excel, without needing to create a program to connect to the database. ODBC is the primary method of almost all application/data store connections. ODBC allows many different applications to connect to and use a data store such as MySQL databases. For this reason, you must have a good understanding of ODBC and how to use it. Some applications do not require an ODBC driver to be installed. They either install it themselves as a part of the installation or can communicate with the data store directly. Node.js does not require you to install an ODBC driver. However, Node.js still requires you to provide the connection details at some point before using the data store. With Node.js, you provide these details within the scripts.

ODBC is a method of connecting applications to a database or another data source. Think of them as translators, translating commands from your application language into the database language and back. They are available for all databases and programming languages and are present in all database-orientated applications.

A **Data Source Name** (**DSN**) is a data structure that holds the information required by the ODBC driver to connect to the target database or data source.

An ODBC can connect an application to many types of data sources, including the following:

- A MySQL database
- SQL Server, that is, dBase (`.dbf`), Paradox (`.db`), and FoxPro (`.dbf`)
- MS Access files, with the `.mdb` and `.accdb` extensions
- MS Excel files, that is, `.xls`, `.xlsx`, `.xlsm`, and `.xlsb`
- Text data sources, that is, `.txt` and `.csv`
- Unicode and ANSI

You can store DSNs in several ways, each of which will be addressed in this chapter.

Once the ODBC and the DSN have been set up, you rarely need to do anything more with them unless something changes—for example, if the database moves to another IP address or a user account or password changes. You can think of the DSN as an object that specifies what database it needs to be connected to. The ODBC takes the DSN information and establishes the connection to the database. The ODBC will handle all communications, while the DSN tells it what to communicate with.

Now, in the following section, let's look at the different types of DSN structures.

Types of DSNs

There are several types of DSN structures available to the developer, and which one you use depends on several factors. Ideally, the application developer puts some thought into the use of the application—who would be using it, their location, and the data source's location before development commences. This is so that the appropriate DSN and ODBC can be set up.

There are three types of DSN structures:

- A system DSN
- A user DSN
- A file DSN

A system DSN and a user DSN are both examples of computer DSN types. These DSN types are defined on a computer system, with different levels of accessibility. A system DSN can be accessed by any user on a system, whereas a user DSN can only be accessed by a single user on a system.

A file DSN is defined within a file either on a computer system or on a network drive shared by multiple systems. This allows for an ODBC connection that can be used by multiple users that are distributed.

Now, let's take a more detailed look at DSN types to understand how they can be defined:

- **Computer DSN**: An application will use a DSN to define the connection details of the data store that the application wants to connect to. The DSN is then used to establish an ODBC connection. The ODBC connection handles any data requests between the application and the data store. The following diagram shows the general way in which DSN interfaces are used to connect to a data store:

Figure 8.16: A diagram that shows how a DSN is used

A computer DSN is permanently available on the user's computer. There are different ways of setting it up. A system DSN is available to any user that logs into a computer. A user DSN is available to the user who sets up the DSN on the computer. You cannot set up a user DSN for MySQL in Windows. A computer DSN needs to be set up on each computer running your application and can be used by other applications. It is secure and usually created using an ODBC manager. It can be created with the application's code when it's initially run.

Typically, a computer DSN is defined in the computer's ODBC data sources. The ODBC data source keeps track of all DSN objects that are available to connect to. A connection can be added to the ODBC data source, which involves the user providing the database type and connection information. Once this information has been provided, a user can select any DSN object they wish to connect to. The ODBC will take the stored information and use it to establish a connection for the user.

- **File DSN**: In the following diagram, the typical usage of a file DSN is outlined. The application reads in the file DSN, then establishes an ODBC connection. The ODBC connection helps to facilitate any requests between the data store and the application:

Figure 8.17: A diagram of how a file DSN works

A file-based DSN stores the required connection information for the ODBC. It is saved as a text file and is portable. It can be stored in a shared folder for several users and can be distributed with the application when no shared folder is available. It is not as secure as a DSN.

A file DSN is similar to a computer DSN in terms of functionality; the only difference is how it is stored. Typically, a file DSN is used for companies where a shared drive exists, so a user can read the text file from the shared drive. Once the DSN has been read, it is provided to OBDC, which establishes the connection.

- **DSN-less**: The following diagram demonstrates how DNS-less connections typically work. In this scenario, the application interfaces directly with the ODBC, providing connection details itself without the need for the DSN:

Figure 8.18: A diagram of DSN-less connections

A DSN-less connection is portable in that it does not exist until the application creates it. It is created by the application code on installation or first run and does not require a separate setup.

What do they all have in common?

Regardless of the type of DSN employed, all of them need the relevant ODBC driver to be installed on the client's PC to work. Computer OSs such as Windows come with several of the more common ODBC drivers already installed, but not all. As a developer, you need to determine whether your required driver is installed on the user machine and install it. You might need to do this manually or install it as part of an installation package or MSI.

Determining whether ODBC drivers have been installed

This section demonstrates how to set up ODBC drivers for Windows. For Linux instructions, please refer to `https://docs.microsoft.com/en-us/sql/connect/odbc/linux-mac/installing-the-microsoft-odbc-driver-for-sql-server?view=sql-server-ver15`. For macOS instructions, please refer to `https://docs.microsoft.com/en-us/sql/connect/odbc/linux-mac/install-microsoft-odbc-driver-sql-server-macos?view=sql-server-ver15`.

Often, you will need to install an ODBC driver to access ODBC connections. Many workplaces use ODBC to allow people to establish connections to databases. In these cases, ODBC will already be available on your device. Windows does not provide them as part of the OS, so you need to check whether they have been installed on your system.

To check for ODBC drivers, you need administration rights to use the ODBC connection manager. To accomplish this, perform the following steps:

1. Press the *Windows Start* button on your keyboard and type in ODBC.
2. Select **ODBC Data Sources (32 bit)**. Right-click and select **Run as administrator**.

3. Click on **Yes** when prompted to allow this application to make changes. The **ODBC Data Source Administrator (32-bit)** window will open:

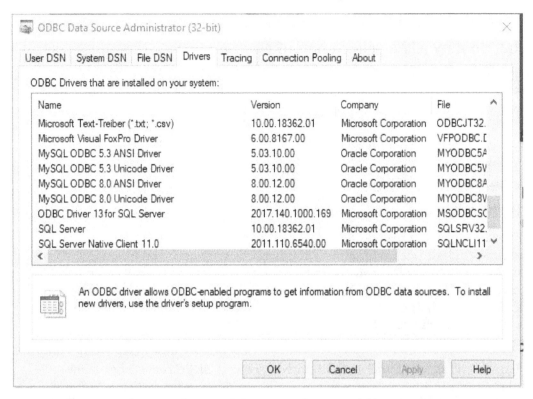

Figure 8.19: The ODBC drivers and the versions that are available on your computer

4. Select the **Drivers** tab, as shown in the preceding screenshot, and scroll through to locate the **MySQL ODBC 5.3** drivers, preferably 5.3 or better.

5. If they are not there, refer to the *Preface* section and follow the instructions to install them. If the drivers are there, you should also check whether **MySQL ODBC Connection Manager** has been installed.

6. Click on the **System DSN** tab, scroll through, and select any **MySQL ODBC driver** instance. Then, click on **Finish**:

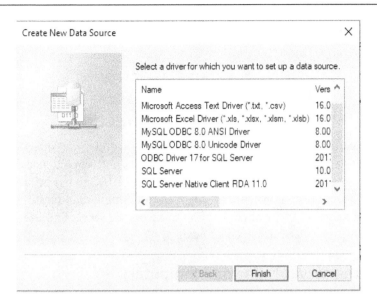

Figure 8.20: The data source selection screen

7.   Check for **MySQL Connector**. If the window that opens has a title reading **MySQL Connector/ODBC** and the dolphin logo, as shown in the following screenshot, then it has been installed and you are good to go. Otherwise, please refer to the *Preface* section of this book to install MySQL Connector:

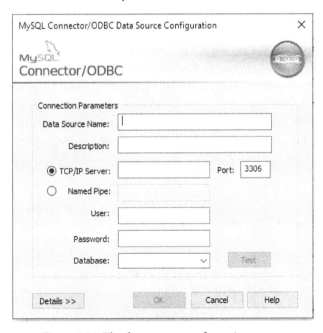

Figure 8.21: The data source configuration screen

8. You can click on **Cancel** to close the window and back out. That's all you need to check.

In the next section, you will explore local, LAN, and remote ODBC connections.

Local, LAN, and remote ODBC connections

You are now ready to create some ODBC connections. However, first, you'll need to consider some of the different types of ODBC connections and when to use them. You have several options and will need to decide what type of ODBC to use. As a developer, you can create connections that are suitable for your development and testing. A user might need to use a different type of connection to access the server and data. Your options, and when to use them, are detailed next.

Local ODBC (the server is on your computer)

Often, these types of ODBCs are used by developers in situations where a server only exists on the computer establishing the connection. In these cases, the address of the server being connected will be `localhost` or `127.0.0.1`. These connections allow for faster connections to be established, as they are not done over the network but are, instead, local to the computer. As such, this connection type should be used in cases where only you need to connect to the database, not anyone else.

LAN ODBC

Often, these types of connections are used in situations where a database is on the same network as your computer. Typically, the IP address will look similar to `192.168.#.#`. These are servers that are used internally by the users, not externally by people outside of the network. Common examples include development servers and internal servers containing customer information. These connections tend to be a bit slower than local servers, but they allow more users to connect since they are available to anyone on the network.

Remote ODBC

A remote ODBC is used for connections that are outside of your current network. Typically, you can access these through the IP address of the server, using either port `3306` or another port specified by the administrator of the server. In this format, any user from any location can connect to the server. This is useful for applications that are distributed over many networks, such as a database of clients used by salespeople in different countries. The speed of this connection is mostly reliant on the internet connection that is being used. This type of connection is mostly used for production databases.

Now that you have a good understanding of the different types of ODBC connections, in the following exercise, you can take a look at how to set up an ODBC connection on your computer.

Exercise 8.06 – creating a LAN or remote DSN/ODBC connection to the world_statistics database

In this exercise, you will create a LAN ODBC connection to the `world_statistics` database that you created in *Exercise 5.07 – creating a new database*. The only difference between a LAN and remote ODBC is the IP address and the port number. The user accounts must be set as either the local network for a LAN connection or everywhere for a remote connection. You need to have administrative rights to use the ODBC connection manager. To accomplish this, perform the following steps:

1. Press the *Windows Start* button on your keyboard and type in ODBC.

2. Select **ODBC Data Sources (32 bit)**, and click on **Yes** when prompted to allow this application to make changes. Now, the **ODBC Data Source Administrator (32-bit)** window will open.

3. Select the **System DSN** tab and click on **Add**:

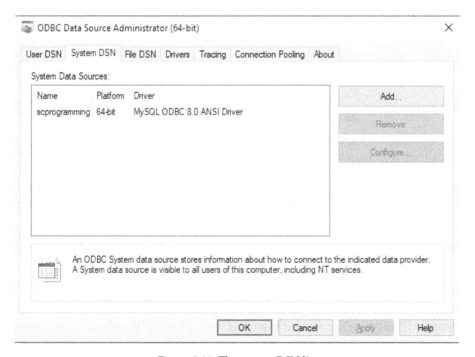

Figure 8.22: The system DSN list

> **Note**
>
> The options you will see in this window will depend on what ODBC connections have already been created on your computer. On a fresh installation, it is likely to be empty.

4. The driver selection window will open. Select the MySQL driver you wish to use and click on **Finish**. For this exercise, use **MySQL ODBC 5.3 ANSI Driver**:

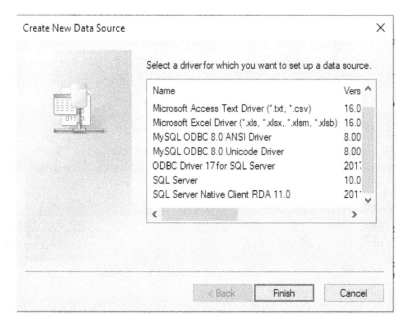

Figure 8.23: The data source selection screen

5. For the configuration windows, enter your connection details. The list of options you can use is as follows:

- `Data Source Name`: Give it a name that is meaningful to your application. Once you start to develop with it, it could be challenging to change it, so give it some thought.

- `Description`: This is an optional field that can be used to provide information about what the connection is used for.

- `TCP/IP Server`: The address of the server. If the server is on your local computer, use `localhost` or `127.0.0.1`. If the server is on your LAN, use its internal IP address. The sample shows an internal LAN IP address. If the server is located somewhere on the internet, use the IP address.

- `Port:` This is already set at 3306. If the server is local to your computer or on the LAN, leave it at 3306. If you changed it during installation, you need to use the port number that you set. If the server is on the internet somewhere, use the port number you have been instructed to use or that you set when you opened the database to the web and mapped the port numbers.

> **Note**
>
> This book does not use named pipes. Oracle states that they can be problematic when shutting down the server with some Windows configurations, and they are slower than TCP/IP. They are not on by default. If you want to read more about named pipes, you can do so at `https://docs.microsoft.com/en-us/windows/win32/ipc/named-pipes`.

- `User:` Enter the username of the account used to connect to the database. By default, the username is `root` and will be the same username you use to connect to your database through Node.js and MySQL Workbench.

- `Password:` This is the password of the account used to connect to the database. This is the same password used in Node.js and MySQL Workbench when connecting to your database.

- `Database:` If the IP and port addresses are valid, a list of databases on the server will be listed. Select the database you want the connection to use, and select the `world_statistics` database, as shown in the following screenshot:

Figure 8.24: Completed details (make sure that you use your own)

6. Test the connection by clicking on **Test**. The manager attempts to connect and, if successful, displays the following result:

Figure 8.25: A successful connection

7. If you get a **Connection Successful** result, click on **OK** to close the test message. Then, click on **OK** again to close the new ODBC window. If the connection has failed, check your values for **TCP/IP Server**, **Port**, **User**, and **Password** and try again.

In this exercise, you created a LAN or remote DSN/ODBC connection to the `world_statistics` database. In addition to a computer DSN, you can also create a file DSN to connect to databases. In the next section, you will learn more about file DSNs, and learn how to create these connections on your systems.

Creating file DSN/ODBC connections

In the previous exercise, you saw how a computer DSN is created. A file DSN will allow you to create a file that contains information about the DSN you are connecting to. This section will discuss how the file DSN is formatted and how one can be written for an ODBC.

When you create a DSN file, it will have the `.dsn` file extension. This file will follow a specific format and contain the following information:

- `DRIVER`: This is the driver that handles the database connection you are attempting to make.

- `UID`: This is the username that you are authenticating with.

- `PORT`: This is the port number that the server is listening on.

- `DATABASE`: This is the name of the database that you are connecting to.

- `SERVER`: This is the IP address of the server you are connecting to.

For example, if you wanted to connect to a `world_statistics` MySQL database that was at IP address `127.0.0.1`, port `3306`, and with the `root` username, you would have the following DSN file:

```
[ODBC]
DRIVER=MySQL ODBC 5.3 ANSI Driver
UID=root
PORT=3306
DATABASE=world_statistics
SERVER=127.0.0.1
```

Figure 8.26: The file ODBC structure for our connection

When you want to connect to this DSN, you will direct your ODBC toward this file. The ODBC will read the file, set each of the properties provided, and establish the connection to the server.

In the next exercise, you will create a file DSN/ODBC connection to the `world_statistics` database.

Exercise 8.07 – creating a file DSN/ODBC connection to the world_statistics database

A file DSN is a simple text file that holds the connection information of the data store. The process is very similar to the standard DSN setup. You need to have administrative rights to use the ODBC connection manager. To accomplish this, perform the following steps:

1. Press the *Windows Start* button on your keyboard and type in `ODBC`.

2. Select **ODBC Data Sources (32 bit)** and click on **Yes** when prompted to allow this application to make changes. The **ODBC Data Source Administrator (32-bit)** window will open.

3. Select the **File DSN** tab. Using **Look in:**, navigate to the folder you want to store the file in. Then, click on **Add**:

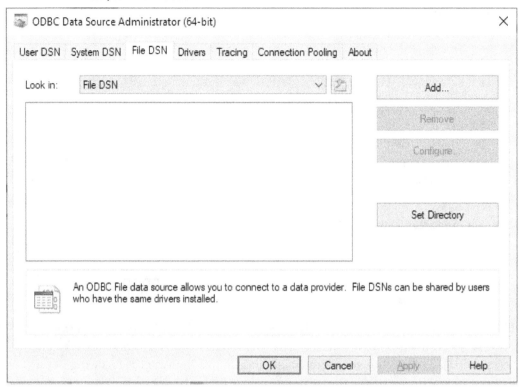

Figure 8.27: The File DSN tab

4. Select the driver you wish to use, just like you did with the LAN and remote connections. When prompted, enter the name you want to give the file DSN—that is, `world_statistics`. The following window will be displayed. Click on **Finish**:

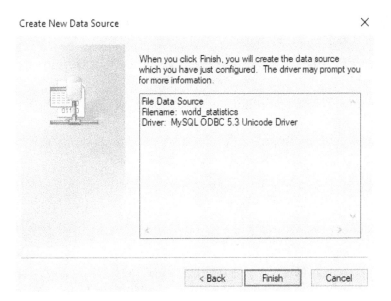

Figure 8.28: Displaying your selections

5. The following window will open so that you can finish entering your options. As you can see, it is very similar to the LAN and remote windows. Enter your details as before, and use the `world_statistics` database:

Figure 8.29: The connection details have been completed

6. Test the connection. If the test is successful, click on **OK**. If it is not successful, check your details and try again.

7. If you cannot find the file for some reason, look inside your **Documents** folder—it is named `world_statistics.dsn`.

8. Open the file with your text editor. The file should look similar to the following:

```
[ODBC]
DRIVER=MySQL ODBC 5.3 ANSI Driver
UID=Tom
PORT=3306
DATABASE=world_statistics
SERVER=192.168.0.3
```

Figure 8.30: The contents of the file ODBC

From opening the DSN file, you can see that it matches the format we discussed. The ODBC Data Source Administrator has automatically created the file, and now you can use it as required.

An ODBC connection is your primary method of accessing the database, and once it has been set up, you won't have to change it unless the server is moved or the account details are changed. Usually, you will have one account set up for all your users so that the connection can be standardized. While there are three distinct types of ODBC connections, you will usually use the fixed system or user types and set them up on each machine.

Now, in the following activity, let's test your knowledge of the skills that you have learned so far in this chapter.

Activity 8.02 – designing a customer database

In *Activity 5.01 – building a database application with Node.js*, you created a database named `MOTDatabase`, which contained a `Customer` table and a `CustomerPurchases` table. The company has since started to acquire customers and sales, and as such, they require data to be added to these existing tables.

Additionally, the company would like an ODBC connection available for the database. You can assume that the database exists on localhost, so the ODBC IP will be `127.0.0.1`.

Currently, the company has the following customers that they would like inserted into the `customers` table:

Customer ID	Customer Name
1	Big Company
2	Little Company
3	Old Company
4	New Company

Figure 8.31: The data to be inserted into the customers table

The company currently has the following purchases that it would like inserted into the `CustomerPurchases` table:

CPID	CustID	SKU	SalesDateTime	Quantity
1	1	SKU001	01-JAN-2020 09:10am	3
2	2	SKU001	01-Jan-2020 9:10am	2
3	3	SKU002	02-Feb-2020 9:15am	5
4	4	SKU005	05-May-2020 12:21pm	10

Figure 8.32: The items to be inserted into the CustomerPurchases table

To do this, you will need to complete the following steps:

1. Create a script named `Activity_6_02_Solution_Populate_Tables.js`.

2. In the `Activity_6_02_Solution_Populate_Tables.js` script, add code that will insert the customer data into the customer table of `MOTDatabase`.

3. In the `Activity_6_02_Solution_Populate_Tables.js` script, add code that will insert the customer purchases data into the customer table of `MOTDatabase`.

4. Open the **ODBC Data Sources** interface on your computer.

5. Using the **ODBC Data Sources** interface, create a system ODBC connection to `MOTDatabase`.

> **Note**
>
> The solution scripts can be found at `https://github.com/PacktWorkshops/The-MySQL-Workshop/tree/master/Chapter08/Activity8.02`. The solution to this activity can be found in the *Appendix*.

Summary

In this chapter, you worked your way through a lot, so take a moment to recap what you have learned so far. Using Node.js, you learned how to insert, read, modify, and delete data from the tables; how to output the data to the console and the browser; and how to format the data to make it easier and more pleasant to read for the user.

With these skills, you should now be able to construct complex applications that work with MySQL databases. Queries to modify and read data are common for applications using MySQL databases, so these skills are essential not only for Node.js but any other programming language that you might use with MySQL databases. Output formatting is an important aspect of working with MySQL databases. When you want to show a user a set of data, it is essential for it to be easily readable. Formats such as HTML tables are a great way in which to display database data to a user, and they are commonly used in the industry.

In the *ODBC connections* section, you learned what an ODBC connection is and what types of connections are available to you, along with how to create each of the connection types. ODBC connections are common in the industry, as they allow for a simple and convenient way to connect many users to a database. These skills are especially useful for programmers and system administrators, as they will allow you to set up clients for database connections.

In the next chapter, you will learn how to use ODBC connections in real-life situations, how to migrate an MS Access database to MySQL, important tips when migrating from MS Access to MySQL, how to convert MS Access SQL into MySQL, how to use pass-through queries to move the processing to the MySQL server and speed up the application, and how to create an unbound data form.

Section 3:
Querying Your
Database

This section covers the various ways you can query data through applications. We will discuss how MS Access and MS Excel can be used to interact with MySQL databases, allowing you to efficiently and effectively work with data.

This section consists of the following chapters:

- *Chapter 9, MS Access Part 1*
- *Chapter 10, MS Access Part 2*
- *Chapter 11, MS Excel VBA and MySQL*
- *Chapter 12, MS Excel VBA and MySQL Part 2*

9
Microsoft Access – Part 1

Microsoft (**MS**) Access is still a very popular database application that has a lot of components. Due to this, we will cover it over two chapters. In this chapter, you will learn about the MS Access application and its database architecture, the problems associated with the architecture, and how and why to improve on the architecture by migrating to a MySQL backend. You'll also learn how to provide more stability and longer life to the MS Access database application. You will start by upsizing an MS Access database to MySQL and setting up the ODBC connections to the database. After that, you will learn about some of the issues you may face when migrating databases to MySQL and how to fix or avoid them.

Finally, you will convert a sample application to use MySQL data using passthrough queries before learning how to convert an MS Access table-reliant form into an unbound form that doesn't rely on local or linked tables. By the end of this chapter, you will be able to remove all linked tables and check that the application still works.

In this chapter, we will cover the following topics:

- Introduction to MS Access
- MS Access database application configurations
- Upsizing an MS Access database to MySQL

- Manually exporting MS Access tables
- Adjusting field properties
- Migrating with wizards
- Linking to your tables and views
- Refreshing linked MySQL tables

Introduction to MS Access

In the previous chapter, we learned how to use MySQL with Node.js to manipulate a database and read and output data to several common data destinations. Now, we will learn about MS Access.

Before we begin, let's discuss what MS Access is all about. It is a **Relational Database Management System** (**RDBMS**) that was released by Microsoft in late 1992. It provides a **Graphical User Interface** (**GUI**) so that you can easily and interactively develop queries, forms, and reports. It provides the **Visual Basic for Applications** (**VBA**) programming language, which was specifically designed for database development, as well as a host of libraries that add programming features that aren't included in the basic installation. These libraries are provided by Microsoft as well as third-party applications, and they can be integrated into MS Access applications. MS Access has had 11 version releases since 1992, with the current version being MS Access 2019. Unlike most RDBMSs, MS Access uses a single file to hold the entire system, including its tables, data, forms, reports, queries, and VBA code. MS Access also stores the temporary data it generates during its data query operations in the same file. Since version 2000, the maximum file size has been increased from 1 GB to 2 GB. The tables and data can be separated into a separate MS Access file known as a backend, and the tables are linked back to the application file to share the database with other users, as well as increasing the file size limitation as both files will have the 2 GB size limit. The backend data file has no processing power and is only a container for the data. The single-file architecture of MS Access permits entire database applications to be easily transferred electronically to other users.

Due to the speed, ease, and cost-effectiveness of creating database applications in MS Access, it has become widely popular for both personal and business use, with many businesses relying heavily on these systems to function.

Most MS Access database applications start out working well; they are fast and responsive and do their job well. The longer they are used, the more that businesses rely on them. After a while, as they fill up with data and more users need access to them or need remote access, they can become slow and unreliable. This often leads to crashing and can be very frustrating for the users, the business, and those who maintain them.

In this chapter, we will be using the objects we learned about in *Chapter 4*, *Database Objects*, to improve the MS Access experience for the users, businesses, and developers maintaining the database. In the next section, we will explore several configurations of the MS Access database application.

MS Access database application configurations

There are several configurations an MS Access database application can have, such as the following:

- **Single-file application/data**: The application logic, forms, queries, reports, and data are all contained in a single MS Access file. This works fine if there's only one user and not a lot of data. Usually, we can access data quickly. If the file gets corrupted, then you may lose both the application and your data permanently, especially if a regular backup regime is not adhered to.

  A small home-based company may use this configuration for the inventory and sales data when they are a new start-up, often using ready-made database templates that are available for free as part of the MS Access installation.

- **Multi-user/single-file application and data**: A single MS Access file contains both the application and data, as described earlier; however, there are two or more concurrent users using it at the same time. It is never good to share a common single-file database between users simultaneously because there is a high chance of both the application and data becoming corrupted, especially if users do not exit the application correctly.

  The home-based start-up mentioned previously has now employed one or more people to assist with sales and inventory updates and is still using the single-file MS Access configuration.

- **Split access frontend/access backend**: The data has been separated into a separate MS Access file. This is a better option since the data is protected from corruption if the application file gets corrupted. Each user should have a version of the frontend on their computer. The backend needs to be in a LAN location that all users have access to. As MS Access still does all the data processing work, the network resources could get overloaded.

  The home-based startup now has up to eight users on the system concurrently and has split the data into a separate MS Access file due to corruption issues.

Out of these three configurations, the split frontend/backend configuration is the best option. Even for a single user, it offers more protection for the data; if the frontend application gets corrupted, usually due to incorrect shutdown procedures by the users, the backend data file will rarely get corrupted. Depending on several factors, however, even the split configuration may struggle after some time, and moving the data to a more robust backend needs to be considered. There are several good reasons you should move the backend to a MySQL database. A few of them are as follows:

- The system now has multiple users, possibly at multiple sites.

- The business has grown and the system is not coping with the data growth.

- Inefficient data handling over networks often arises as the amount of data increases.

- Once the backend data file starts to approach the MS Access file size limit, corruption could occur and the application will stop working. MySQL does not have a limitation on file size.

- The 2 GB size limit also includes all of the temporary data that Access generates internally while processing queries. It is not cleaned up until a compact and repair procedure is done. Even with the database split, the frontend can be affected by this limit.

- Access backends are just containers and have no processing power, which means all the data must be transferred over the network to the frontend for processing, which makes it slower, especially when there is a lot of data. With MySQL, you can transfer much of the processing requirements to the MySQL server and only transfer the results back to Access.

- All that data transfer can slow the LAN down, especially for long-running and complicated queries that are using several tables, whereas MySQL can only pass back the results.

- Remote access via the internet is slow to unusable. MySQL will work well for local or remote access when the MS Access application's data handling is designed to use MySQL.

Let's look at what MySQL will do for an MS Access application:

- MySQL has the necessary processing power. So, moving to MySQL and modifying the application's data handling means we can simply request some data; here, MySQL will process the request and return only the results.

- MySQL will protect the data from unauthorized access better than MS Access can, especially over remote connections.

- MySQL will reduce network traffic when the application is tuned to use it properly.

Taking these points into consideration, we can put the data into MySQL and link the tables. Unless you optimize your data handling to leverage MySQL's processing power, you won't see any real difference; it may even be slower. This chapter will show you how to leverage the power of MySQL. In the next section, we will explore migrating, also known as upsizing, an MS Access database to MySQL.

> **Note**
>
> In this chapter, we assume that you are familiar with MySQL and Workbench, so references to them will be high-level only.

Upsizing an MS Access database to MySQL

In this section, we are going to set up a database to work with, export its tables to MySQL, and relink the tables back to MS Access. The training database for this chapter is a single-file application/database. We will be migrating the data tables to MySQL and linking them back, ensuring that the application will work with linked tables. By the end of this chapter, all data access to MySQL will be in VBA code with no linked tables. In the next section, we will complete an exercise where we will set up our training database.

First, we need an MS Access application and database to work with. One has been provided with this book's resources. This is a database template that can be reused. First, let's learn how to set up our Access database:

1. Create a work folder.

2. Double-click on the `MySQL Training Database.accdt` file in this book's resources. The template will open.

> **Note**
>
> The `MySQL Training Database.accdt` file can be found here:
> `https://github.com/PacktWorkshops/The-MySQL-Workshop/tree/master/Chapter09/Exercise9.01`.

3. When prompted, enter `MySQL Training DB.accdb` as the name for the database and select your work folder to save it in. Then, click **OK**. The training application will be copied into your work folder and will open:

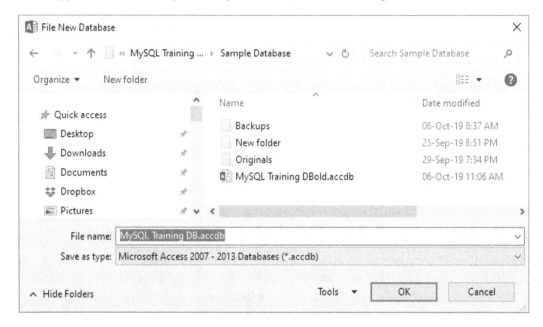

Figure 9.1 – The template database will prompt you for a name and a location to save the database to

4. When the training application starts, you will be presented with a dialog for locating the backend data file – that is, `MS Access Training Data.mdb`.

> **Note**
>
> The `MS Access Training Data.mdb` file can be found at
> `https://github.com/PacktWorkshops/The-MySQL-Workshop/tree/master/Chapter09/Exercise9.01`.

5. Locate the file and select it. The file will then be copied into your work folder and the frontend application will link to its tables. If you are asked whether to trust the database, answer **Yes**.

Your training database is now ready. Take a moment to look through it and try out the main form. However, don't try **Open Users** yet as it has been set up for a later exercise and you will get a prompt to select an ODBC connection. The training database has the following features:

- Forms containing drop-down combo lists, text boxes, drop-down lists, and graphs.

- All the text box controls are populated with SQL statements by VBA code to access the data tables.

The following screenshot shows these features:

Figure 9.2 – The opening screen after the training database has been linked

The forms and VBA have features that you can find in a typical application database. In the upcoming sections, we will focus on migrating our training database. But before that, we have to prepare a MySQL database along with the ODBC, which we will look at in the next exercise.

The data we are using has been sourced from World Bank Open Data. It is a small subset of statistical information from between 2004 and 2018 that represents three groups of data – jobs, gender, and capacity indicators – for all countries worldwide. To reduce the size of the database, we will cover three to five specific series of data for the three groups. The database consists of one country table listing the world's countries, three group tables holding the statistical data and the series and country links, and one series table listing the series and the groups they belong to. There are also three other tables to use for various exercises that are not related to the statistical data.

> **Note**
>
> World Bank Open Data can be found here: `https://data.worldbank.org/`.

To connect Access to MySQL, there are a few important steps we must take. First, we must set the collation of the database to one that Access can parse. For this book, we will use `utf8 - utf8_unicode_ci`. This collation can be set in the MySQL database schema when the database is created.

To be able to reach the database from Access, we will also need to create an ODBC for our database. This process was discussed in *Chapter 8*, so we will follow the same process for our Access ODBC.

Exercise 9.01 – preparing your MySQL database and ODBC

In this exercise, we will create a MySQL database that can be accessed from Access. This will allow external users to easily access the database. We will start by creating a database with the appropriate collation, then create an ODBC for the database. Follow these steps:

1. Create a new MySQL database schema.

2. Name it `ms_access_migration`.

3. Ensure that you set **Collation** to **utf8 - utf8_unicode_ci**, as shown in the following screenshot:

Figure 9.3 – Entering a name for the MySQL database and the Collation type

> **Note**
>
> Collation is very important. If it is not set as described here, then you will not be able to read data from the database with MS Access.

4. Once the database has been created, you can find it in the **SCHEMAS** section:

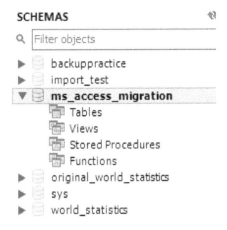

Figure 9.4 – The training database in the SCHEMAS section

5. Now, let's create an ODBC for the database.

> **Note**
>
> All MySQL ODBC drivers from version 5.3.11 and later have a well-known bug that can cause the table fields to incorrectly display **#DELETED** under certain circumstances when linked to MS Access. This is a well-documented bug and there is no workaround at the time of writing. The **MySQL ODBC 5.3.10 driver** does not have this issue. When working with MS Access, you should use the MySQL ODBC 5.3.10 driver. If required, download and install the MySQL ODBC 5.3.10 driver and recreate the ODBC connection. Both the 32-bit and 64-bit versions can be found in this book's resources so that you can download and install them.

6. Name it `ms_access_migration`:

Figure 9.5 – ODBC connection screen – use your details

Now, we are ready to begin the migration process. In this exercise, we created the database and ODBC that are required for migration. In the next section, we will learn how to manually export MS Access tables.

Manually exporting MS Access tables

Before we start, let's look at some information that will help us decide what tables to move into MySQL and what tables to keep in MS Access. If you are thinking of migrating tables to a MySQL database, then it depends on the application. You also need to consider where your users are accessing the database, as well as the purpose of the tables. In the case of remote users, we only need to list those tables that feed drop-down lists. It may be better to keep these in the application as local tables. On the other hand, tables with MS Access-specific field types such as multivalued fields and attachment fields cannot be migrated. MySQL does not have a comparable field type, so it cannot use them.

Please note that it is never a good idea to store files and images in any database as you can with attachment fields or MySQL BLOB fields. They will make the database grow very large very quickly. It is better to store the path names in the files and store the files separately on a server. Other than these exceptions, all the tables should be migrated to a MySQL database. In the next exercise, we will manually migrate (or upsize) a single table.

Exercise 9.02 – manually upsizing a table

Suppose that you have found that your **Users** table has now grown too large for MS Access. To remedy this, you will need to upsize the table. In this exercise, we will manually upsize a single table from an MS Access database into a MySQL database. We will start with the **Users** table. Follow these steps to complete this exercise:

1. In the MS Access navigation bar, right-click on the **Users** table.

2. Select **Export** and then **ODBC Database**:

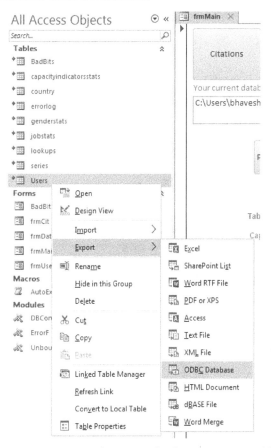

Figure 9.6 – Locating the ODBC database to export a table

3. An input box will appear that will allow you to set a name for the exported table; it will display the table's name. Keep the original name.

> **Note**
> If you change the name of the table, any code, queries, or objects that are using the table when it is linked back to Access later using the new name will not work.

4. You can give a linked table an aliased name when linking it. Click **OK** to accept the default name provided:

Figure 9.7 – Changing the name of the table

5. After a while, the DSN data source window will open. Select the `ms_access_migration` ODBC you created in the preceding exercise and click **OK**:

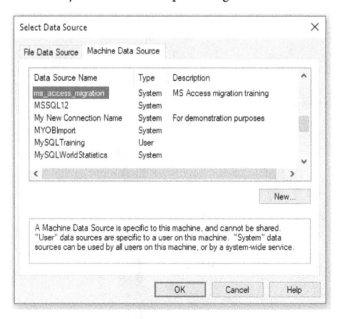

Figure 9.8 – Selecting the named DSN for the destination database

6. The table will now be exported to MySQL and the following window will open to confirm the export. Click **Close** to close the window:

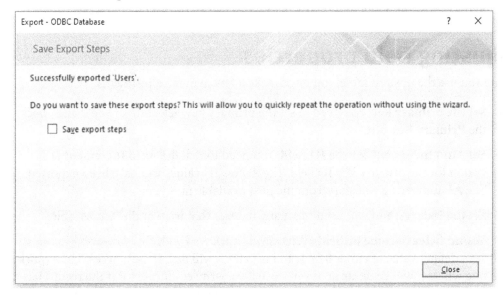

Figure 9.9 – Screen indicating the success or failure of the table being exported

> **Note**
> The time it takes to export any given table will depend on how much data is present in the table.

7. Open **Workbench**. Refresh the schema; you should see the **users** table present inside the **ms_access_migration** database:

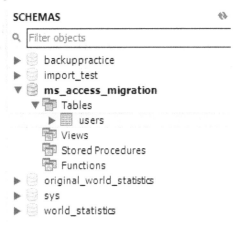

Figure 9.10 – The exported table in the MySQL database

With that, your table has been migrated to MySQL. That was easy! Exporting manually like this will copy the fields and data, but there are several things it does not do or change that you need to know about and rectify before you can start using the table.

Adjusting field properties

Before you use the upsized table, you must make a few manual adjustments:

- **Set the primary key**: You need to set the primary key. Select the ID field and check the **Primary Key** box.

- **Set Auto Increment for the ID field**: You need to tick the **Auto Increment** (**AI**) option for the primary key ID field. The Access ID values will have been exported. The AI numbering will start from the next available number.

- **Set the indexes**: You will need to set any indexes that were in the Access table.

- **Yes/no fields become bitfields**: Access will work well with this; however, you *must* set a default value of either 0 or 1, and usually, it will be 0 (false). Access has a quirky bug where it will generate an error – usually a write conflict error if the record has a bitfield with a NULL value. You will also need to ensure there are no NULL values in any bitfields in MySQL, which is what causes the error. Access does not like NULL values in bitfields. You can check whether there are any NULL values in a large number of records by checking **Not Null** for the bitfield and clicking **Apply** if there are any; then, it will not permit the change. Setting a default value for bitfields will ensure there will never be a NULL value.

- **Set default values in MS Access**: You need to reset any default values Access had. Generally, it will not cause any issues unless the field is a bitfield, as mentioned previously.

- **Hyperlink fields**: Hyperlink fields are changed to **MEDIUMTEXT** if a URL is going to be longer than 255 characters. It will not behave as hyperlinks do in MS Access; however, this can be fixed in the controls for MS Access by setting the **IsHyperlink** property to **True**. The link is encased in **#**.

- **Field description/comments**: They are not migrated in Access. You will need to reset these if required.

Once you've made these field adjustments, the tables will be migrated. These adjustments will look as follows:

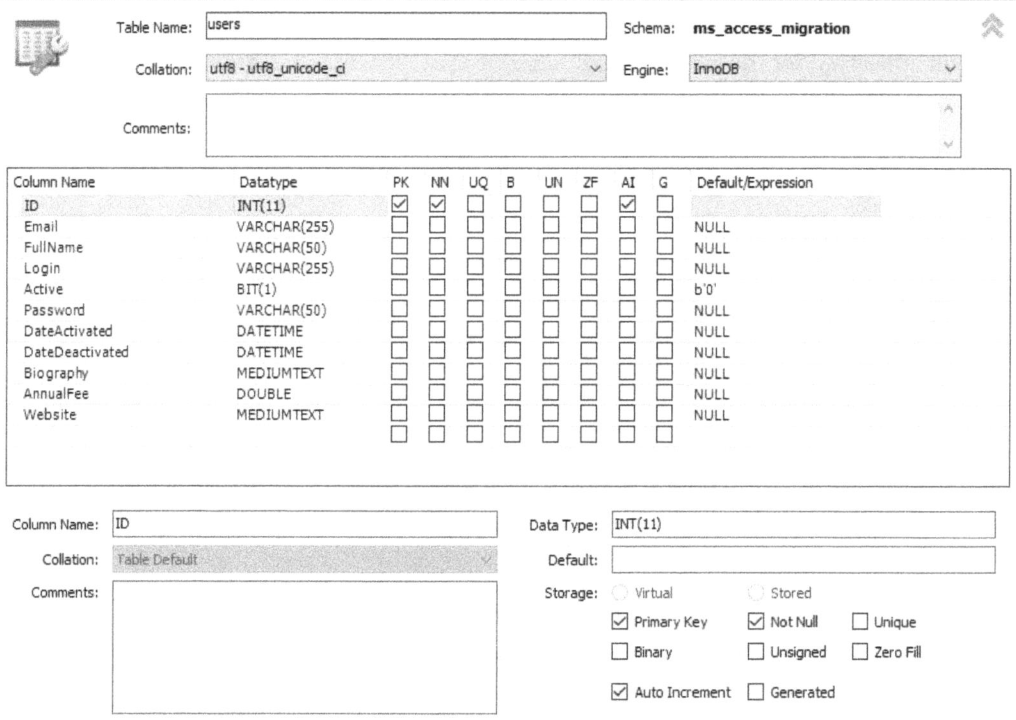

Figure 9.11 – The Alter Table view of the new table. The highlighted areas will need reviewing as described

Now that you have clarity on how to make adjustments to the field properties, in the next exercise, we will migrate more tables and adjust their field properties.

Exercise 9.03 – manually migrating tables and adjusting their field properties

There are several additional Access tables that we would like to upsize. While doing this, we have found that certain field properties haven't been exported properly. In this exercise, we will be migrating series, errorlog, and badbits tables and adjusting their field properties.

Follow these steps to complete this exercise:

1. To migrate the **series, errorlog**, and **badbits** tables, you must complete *Exercise 9.02*, where we migrated the **users** table. Once you've done this, the three migrated tables will appear under the `ms_access_migration` schema in Workbench:

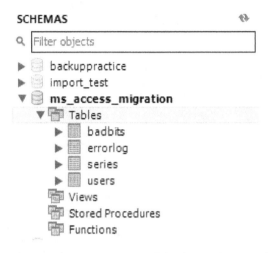

Figure 9.12 – The SCHEMAS panel displaying the new tables

2. Using Workbench, alter the series and errorlog tables and fix the primary key, AI, and any bitfields.

    The changes in the series table must look as follows:

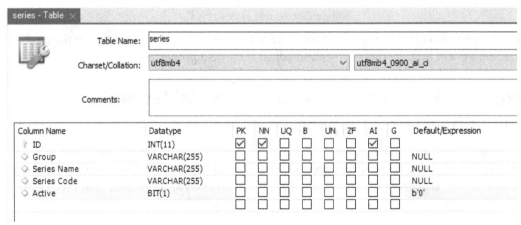

Figure 9.13 – Setting adjustments for the series table

The changes in the errorlog table must look as follows:

Column Name	Datatype	PK	NN	UQ	B	UN	ZF	AI	G	Default/Expression
SQLID	INT(11)	☑	☑	☐	☐	☐	☐	☑	☐	
ErrNumber	VARCHAR(25)	☐	☐	☐	☐	☐	☐	☐	☐	NULL
ErrDescription	MEDIUMTEXT	☐	☐	☐	☐	☐	☐	☐	☐	NULL
CallingProc	VARCHAR(100)	☐	☐	☐	☐	☐	☐	☐	☐	NULL
ErrDate	VARCHAR(40)	☐	☐	☐	☐	☐	☐	☐	☐	NULL
UserName	VARCHAR(25)	☐	☐	☐	☐	☐	☐	☐	☐	NULL
ShowUser	INT(11)	☐	☐	☐	☐	☐	☐	☐	☐	NULL
Parameters	MEDIUMTEXT	☐	☐	☐	☐	☐	☐	☐	☐	NULL

Figure 9.14 – Setting adjustments for the errorlog table

3. The field properties of the badbits table must remain unchanged because we need this table to remain migrated for a later exercise. It looks like this:

Column Name	Datatype	PK	NN	UQ	B	UN	ZF	AI	G	Default/Expression
ID	INT(11)	☐	☐	☐	☐	☐	☐	☐	☐	NULL
TextData	VARCHAR(255)	☐	☐	☐	☐	☐	☐	☐	☐	NULL
BitField1	BIT(1)	☐	☐	☐	☐	☐	☐	☐	☐	NULL
BitField2	BIT(1)	☐	☐	☐	☐	☐	☐	☐	☐	NULL

Figure 9.15 – Don't make any changes to the badbits table

Manually migrating tables using the single-table ODBC method is fine for a few tables. It is fast and easy, although having to reset the table properties can be a little frustrating. If you need to migrate many tables, you will need a more automated approach that will also migrate as many of the properties as possible, if not all of them. In the next section, we will learn how to use wizards that are designed to migrate tables.

Migrating with wizards

Using wizards to migrate your tables has the following advantages:

- They let you select some or all tables to migrate to MySQL.

- They will set most properties on the tables, as described in the *Adjusting field properties* section, while migrating so that you don't have to adjust them.

- There are dedicated applications available to migrate from MS Access to MySQL. However, all the ones I have tried are slower than the Workbench wizard.

However, this is where things may get a little tricky. MS Access versions 95 to 2010 included the Upsizing Wizard, which always handled upsizing well, but Microsoft removed it in version 2013.

How you approach the data migration process will depend on your specific MS Access version and MySQL setup. Let's see what we can do with a variety of setups:

- **MS Access 2010 or earlier**: You can use the Upsizing Wizard on the ribbon to access **Database Tools**, **Move Data**, and **SQL**.

- **MS Access 2013 or later**: Other than the single-table export ODBC we just covered, there is no longer any mechanism in MS Access to migrate more than one table at a time.

- **MS Access 32-bit and Workbench 32-bit**: You can use the Workbench Migration Wizard.

- **MS Access 64-bit and Workbench 64-bit**: You can use the Workbench Migration Wizard.

- **MS Access 32-bit and Workbench 64-bit (or vice versa)**: You cannot use the Workbench Migration Wizard. They both must have the same bit architecture. If this is your setup, your options are as follows:

- **Access 32-bit**:

  I.   Download `mysql-workbench-community-6.3.8-win32.msi`.

  II.   Uninstall Workbench 8.0.xx from your computer.

  III. Run and install Workbench from `mysql-workbench-community-6.3.8-win32.msi`.

- **Access 64-bit**:

  I.   Download `mysql-installer-web-community-8.0.17.0.msi`.

  II.   Uninstall Workbench from your computer.

  III. Run and install Workbench from `mysql-installer-web-community-8.0.17.0.msi`.

Obtain a third-party application such as Bullzip MS Access to MySQL and use it to migrate. Now that we have sorted that out, let's continue. Regardless of which wizard or application you use, the process will be similar, and by this stage, you have attained all the knowledge you need to work your wizard out. The basic steps for all wizards are as follows:

1. Select the source database (`MS Access Training Data.mdb`).

2. Select the target ODBC or enter the necessary connection details.

3. Select the tables to migrate.

4. (Optionally) Select some options related to the tables.

5. Start the migration and have a break while it runs – you deserve it.

In the next exercise, we will use the Workbench Migration Wizard to upsize the table.

Exercise 9.04 – using the Workbench Migration Wizard to upsize the table

We will be using Workbench 6.3.8 (32-bit) for this exercise since Workbench 6.3.8 will only upgrade from an MS Access .mdb file. Follow these steps to complete this exercise:

> **Note**
>
> In this exercise, we will be using the `MS Access Training Data.mdb` database that's linked to the frontend sample application.

1. Start Workbench and connect to your MySQL server.

2.   From the **Database** tab, select **Migration Wizard**. The wizard will open:

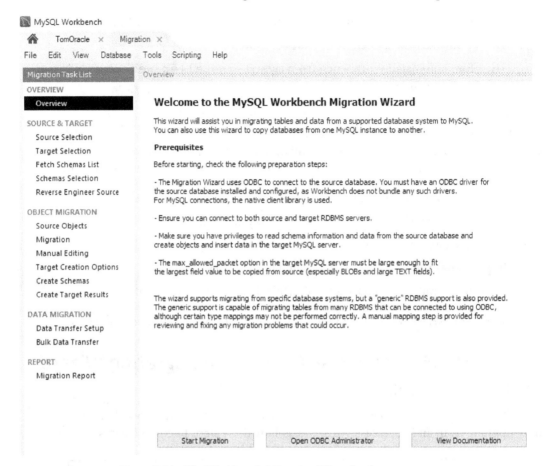

Figure 9.16 – The Workbench Migration Wizard welcome screen

3.   Now, we need to check whether we can import the database by checking whether the relevant drivers are available. Click the **Open ODBC Administrator** button – that is, the center button at the bottom of the screen. You will be prompted to allow the program to make changes. Click **Yes**.

4.   Click on the **Drivers** tab and locate **Microsoft Access Driver (*.mdb, *.accdb)** or **Microsoft Access Driver (*.mdb)**, as shown in the following screenshot. If you only see three SQL drivers, then you are using the wrong version of Workbench and cannot continue. If you found the driver that's shown in the following screenshot, you can continue. Stay in ODBC Administrator:

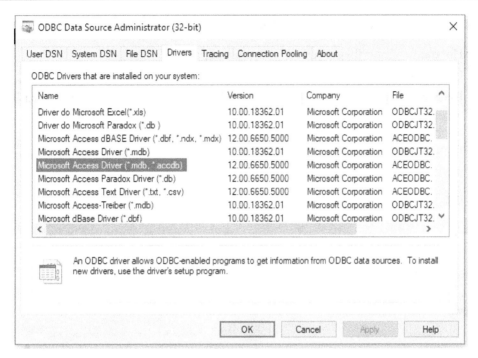

Figure 9.17 – Check that the .mdb drivers are installed

5. Create a system DSN ODBC connection to the `MS Access Training Data.mdb` source in your work folder using the MS Access driver. Name it `MSAccessForUpsize` and click **Test Connection** to make sure it connects successfully. Close the ODBC connection window if successful.

6. Click **Start Migration**; the **Source Selection** screen will appear. Select **Microsoft Access** for **Database System**. If Microsoft Access is not available in the list, close and restart Workbench. Select **ODBC Data Source** for **Connection Method** and **MSAccessForUpsize** for **DSN**:

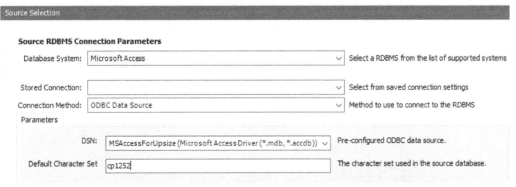

Figure 9.18 – Parameters for the source database

7.   Click **Next**. The **Target Selection** screen will show the following:

- **Hostname**: Enter the server's IP address and port.
- **Username**: Enter the MySQL user's account name.
- **Password**: Click **Store in Vault...**. You will be prompted to enter the password. The wizard will use this later when migrating the data. If you do not enter it now, it will not be able to connect.
- **Default Schema**: Type `ms_access_migration`, although the wizard still won't use it, as we will see later.

8.   Click **Test Connection** to test it:

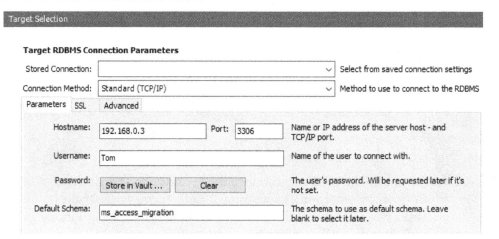

Figure 9.19 – Parameters for the target database. Use your host and login details

9.   Click **Next**. The **Fetch Schema List** window will open with the following results:

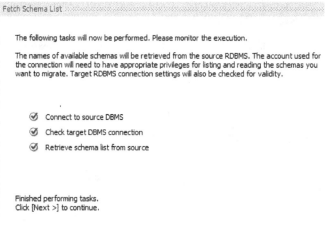

Figure 9.20 – Ensure all three options are checked

10. Before we continue, in the next step, the wizard will attempt to read from the Access **MSysRelationships** table to retrieve the foreign keys. This is a system table that is usually hidden, and Access will not permit the wizard to read it. At this point, you will need to grant permission.

11. If you have the frontend MySQL Training DB.accdb file open, close it.

12. Open the MS Access database we are importing – that is, MS Access Training Data.mdb.

13. Open a module so that you can access the VBA development screens. If there is no module, then create one.

14. Open the **Immediate** pane and select **View | Immediate Window** from the ribbon.

15. Test that you have admin rights by typing ? CurrentUser and pressing *Enter*.

    If the response is *Admin*, then type the following:

    ```
    CurrentProject.Connection.Execute "GRANT SELECT ON
    MSysRelationships TO Admin"
    ```

16. Press *Enter*; there will be no response, as shown in the following screenshot:

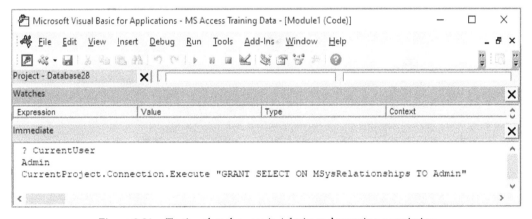

Figure 9.21 – Testing that the user is Admin and granting permission
to read the system table for the wizard

17. Close the Access database so that you can continue in the wizard. Click **Next**. The **Reverse Engineer Source** screen will be displayed. It will look as follows:

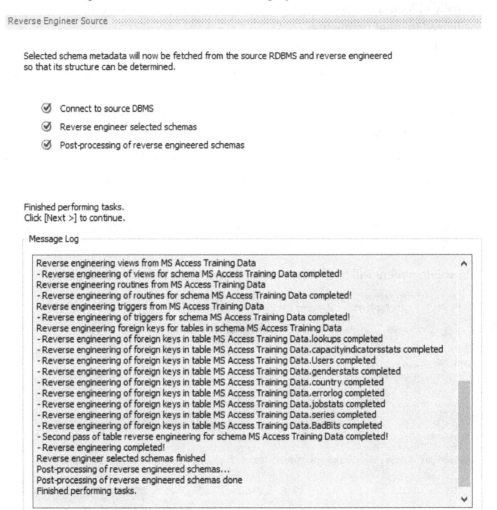

Figure 9.22 – Output after connecting and performing reverse engineering

18. Click **Next**; the **Source Objects** screen will appear. Click **Show Selection**. All the tables will be displayed on the right pane. Using the arrowheads, select and move the objects we have already migrated back to the left pane. Your screen should look as follows:

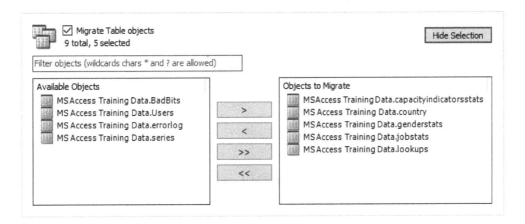

Figure 9.23 – Selecting the tables to migrate; only the tables in the right pane will be migrated

19. Click **Next**; the **Migration** screen will appear. The wizard will have generated scripts to create the objects:

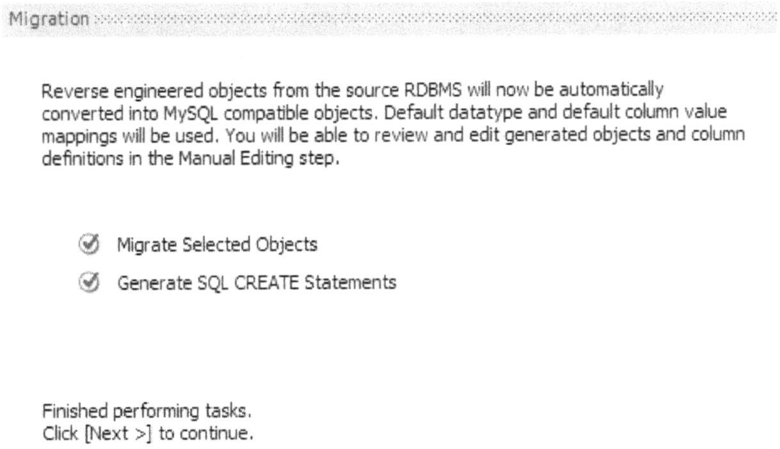

Migration

Reverse engineered objects from the source RDBMS will now be automatically converted into MySQL compatible objects. Default datatype and default column value mappings will be used. You will be able to review and edit generated objects and column definitions in the Manual Editing step.

✓ Migrate Selected Objects

✓ Generate SQL CREATE Statements

Finished performing tasks.
Click [Next >] to continue.

Figure 9.24 – The Migration screen indicating that the selected tables have been generated successfully

20. Click **Next**; the **Manual Editing** screen will appear. It will report any issues and will also allow you to manually edit the proposed settings using the **View** dropdown.

21. In the **View** dropdown, select **All Objects**; you will see the following output. You may notice that the **Target Object** column contains **MS Access Data Training**. If it does, change it to ms_access_migration, as follows:

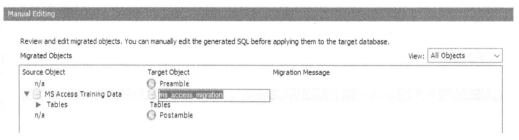

Figure 9.25 – The initial Manual Editing screen

22. Select **Column Mappings** from the **View** dropdown. All the tables and columns to be migrated will be listed. The **Target Schema** column still says **MS Access Training Data**. We need to check whether the table is going to go to the database we changed it to in *Step 15*:

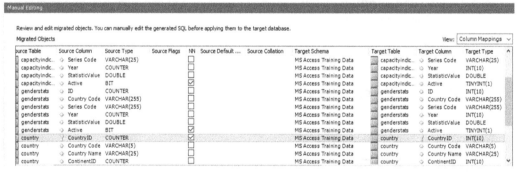

Figure 9.26 – Making adjustments to the columns and their mappings. The target schema may not match the new name that we changed it to in Step 15

23. Select the CountryID row, as shown in the preceding screenshot, and click **Show Code and Messages** (at the bottom left). The following script will be displayed:

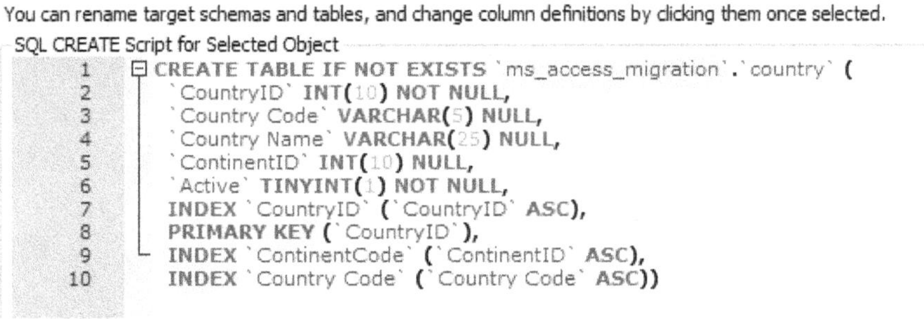
You can rename target schemas and tables, and change column definitions by clicking them once selected.
SQL CREATE Script for Selected Object
1 CREATE TABLE IF NOT EXISTS `ms_access_migration`.`country` (
2 `CountryID` INT(10) NOT NULL,
3 `Country Code` VARCHAR(5) NULL,
4 `Country Name` VARCHAR(25) NULL,
5 `ContinentID` INT(10) NULL,
6 `Active` TINYINT(1) NOT NULL,
7 INDEX `CountryID` (`CountryID` ASC),
8 PRIMARY KEY (`CountryID`),
9 INDEX `ContinentCode` (`ContinentID` ASC),
10 INDEX `Country Code` (`Country Code` ASC))

Figure 9.27 – The SQL that will run to create the selected table. Check that the database name (highlighted) in line 1 matches the new name we entered in Step 15

24. Take note of what database the code will create the table in. Also, note that the primary key and indexes will be set. Check the other tables to make sure. At this point, you can make changes to the structure of the table that will be created if you feel it is necessary. Let's move on.

25. Click **Next**; the **Target Creation Options** screen will appear. There are three checkboxes. **Create schema in target RDBMS** is already checked; we want this, so leave it as is. **Create a SQL script file** also needs to be selected so that a script file is created. You can select a folder and name for the file.

    Keep schemas if they exist should be checked; otherwise, the existing schema will be dropped. You will get a warning before this happens:

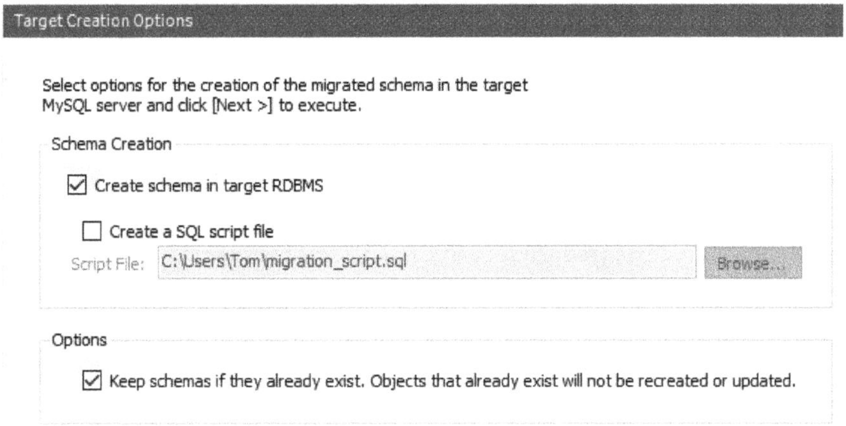
Target Creation Options
Select options for the creation of the migrated schema in the target MySQL server and click [Next >] to execute.
Schema Creation
Create schema in target RDBMS
Create a SQL script file
Script File: C:\Users\Tom\migration_script.sql
Browse...
Options
Keep schemas if they already exist. Objects that already exist will not be recreated or updated.

Figure 9.28 – Schema and SQL creation options

26. Click **Next**. The scripts will be run to create the schema tables. The result will be as follows:

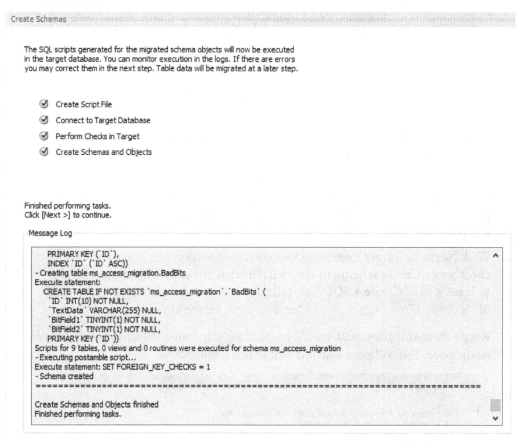

Figure 9.29 – Output after the database schema is created in the target

27. Click **Next**; the **Create Targets Results** screen will appear. If there were any errors, they will be displayed. Here, you can click on the object, view the script, and correct it:

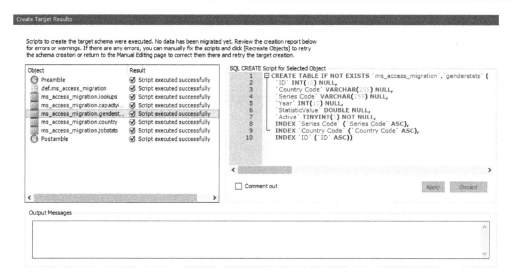

Figure 9.30 – The results of creating the target schema

28. Click **Next**. We are now ready to set up the data transfer. You can choose to do an online transfer (transfer now), which is the default, or create a script to transfer the data later. We want to copy the data now:

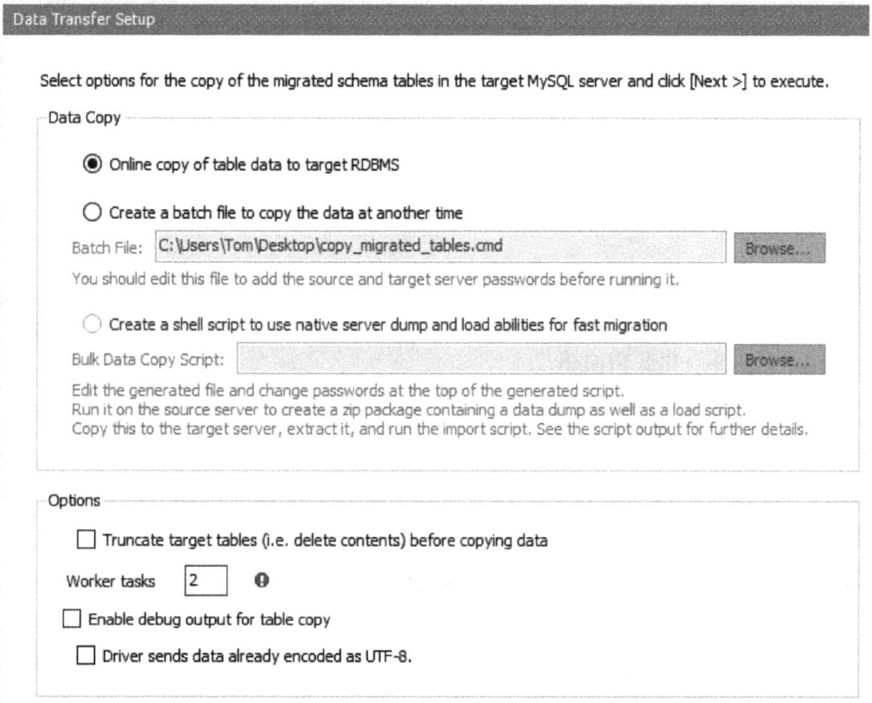

Figure 9.31 – Options for the data transfer

29. Click **Next** to start the transfer. You will see a progress bar. At the end, you will see the following screen:

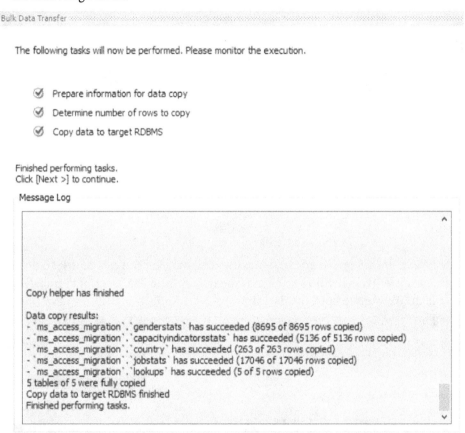

Figure 9.32 – Results displayed after the data transfer is completed

30. After clicking the **Next** button, you will be presented with a report of the whole operation. Finally, click **Finish**.

31. Remember when we granted **SELECT** access to **Admin** on one of the system tables back in *Step 10*? We need to revoke this to tidy things up.

32. Workbench will retain a connection to the source database, so close Workbench.

33. Open the source Access database.

34. Create a module if required. In the **Immediate** panel in the source database, type in the following; again, there will be no response:

```
CurrentProject.Connection.Execute "REVOKE SELECT ON
MSysRelationships FROM Admin"
```

35. Close the Access database.

36. Now, we are done with the migration. With that, the table has been migrated to MySQL. Examine the tables and data in Workbench.

> **Note**
>
> The `badbits`, `jobstats`, `genderstats`, and `capacityindicatorsstats` tables do not have primary keys; this is by design for the upcoming exercises. Please don't add them or modify the tables at this stage.

There are a lot of steps involved, and when you are more comfortable with the process, it doesn't take that long. Using the wizard will save you a lot of time and ensure the data is migrated correctly. The Workbench Migration Wizard is also the fastest in terms of performing the data transfer compared to the third-party applications we have tested to date. Here are a few things you should keep in mind for smoother migration:

- Try to ensure that your data is in good order before you attempt to migrate it. Embedded characters such as backticks can cause problems because they are used in MySQL as delimiters for tables and fields.

- MySQL has difficulties importing from MS Access `accdb` files. If your source data is in an `accdb` file, create an Access `mdb` file and export your data from `accdb` to `mdb` first (you cannot import from `accdb` to `mdb`) and then migrate the `mdb` file into MySQL.

- Commercial or other third-party migration tools may deal with these issues better, but they may be slower in the actual data transfer, which isn't a problem for smaller databases. Larger databases may take a long time.

There is still one more step. Before we can use our shiny new MySQL database, we need to link the tables to the Access frontend, which we will explore in the next section.

Linking to your tables and views

Some things can cause issues when you're linking tables to MS Access. Some of the tables you just migrated have been set up to highlight these problematic situations, and we will show you how to get around them. It is not difficult, but first, we will start with a table with no problems so that you can see how it should happen.

Exercise 9.05 – linking a good MySQL table to Access

Previously, you exported the User table from Access into MySQL. Although it is no longer in Access, your company's business analysts would still like to be able to view it in Access. To achieve this, you will need to link the table in Access. Follow these steps to complete this exercise:

1. Open the frontend **MySQL Training DB.accdb** application; we no longer need the old backend database.

2. If any forms are open, close them.

3. Rename the **Users** table in MS Access to **zUsers**. This way, we can have both tables available for comparison before we remove the old table.

4. In Access, click **External Data | ODBC Database**. The following window will open:

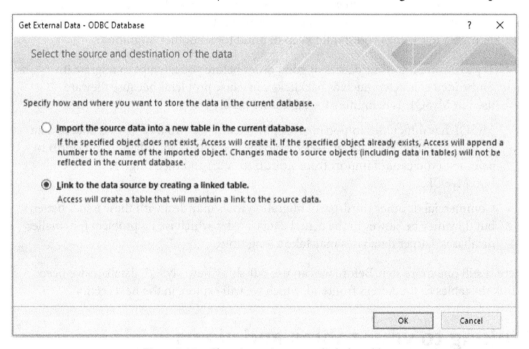

Figure 9.33 – Choosing to import or link the table

5. Select the **Link to the data source by creating a linked table** option.

6. Click **OK**; the **Select Data Source** window will open. Select the **ms_access_ migration** data source you created earlier:

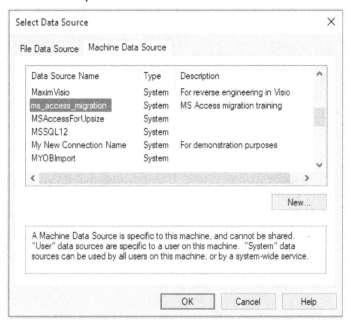

Figure 9.34 – Selecting the named DSNs for the data source

7. Click **OK**; the list of tables in MySQL will be displayed:

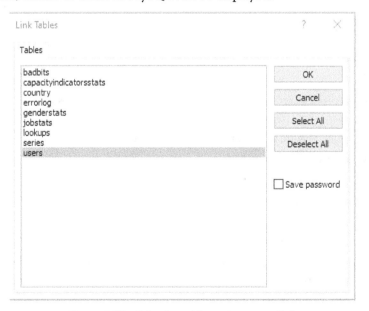

Figure 9.35 – Selecting tables to import or link

8. For now, just select **users** and click **OK**. The table will be linked and displayed in the table list in Access:

Figure 9.36 – Users table linked to Access

9. Notice that it has a globe icon next to it. This indicates that it's an ODBC linked table. If you hover your cursor over the table, you will see its source:

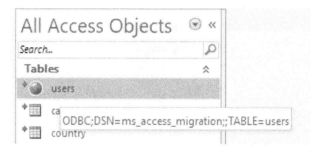

Figure 9.37 – The table's source is displayed when you hover your cursor over it

10. *Double-click* on the new **users** table. It should open and display its data.

    This is good:

Figure 9.38 – A perfect link; the data is displayed

    This is not so good:

Figure 9.39 – The data not appearing indicates a problem

We know that the ODBC works because the table is linked. The issue here is if your data looks like it does in *Figure 9.39* and the collation is incorrect. The following steps will show you how to correct this.

11. Examine the table in Workbench and check the collation that the tables and text fields have been set at. They should be `utf8-utf8_unicode_ci`. If any other collation is set, correct it now. Don't worry – you won't have to import the database again. By following these steps, we can ensure the schema, tables, and fields are all set correctly.

> **Note**
>
> Collation is only used for text field types such as **VarChar** and **MediumText**. Date and numeric fields do not use collation, but all fields will show **$Name?** in MS Access if the incorrect collation is set.

12. Open the `Convert ms_access_migration to UTF8.sql` file in a new **Query** tab in Workbench:

> **Note**
>
> The `Convert ms_access_migration to UTF8.sql` file can be found here: `https://github.com/PacktWorkshops/The-MySQL-Workshop/tree/master/Chapter09/Exercise9.06`.

```
SQL File 2*  x
       /*Set the default collation on the database schema*/
  1 ☐
  2 ●   ALTER DATABASE `ms_access_migration` CHARACTER SET utf8 COLLATE = utf8_unicode_ci ;
  3
  4 ☐   /*Set the default collation on the tables, using convert will also set it for the fields in each table*/
  5 ●   ALTER TABLE ms_access_migration.badbits CONVERT TO CHARACTER SET utf8 COLLATE utf8_unicode_ci ;
  6 ●   ALTER TABLE ms_access_migration.capacityindicatorsstats CONVERT TO CHARACTER SET utf8 COLLATE utf8_unicode_ci ;
  7 ●   ALTER TABLE ms_access_migration.country CONVERT TO CHARACTER SET utf8 COLLATE utf8_unicode_ci ;
  8 ●   ALTER TABLE ms_access_migration.errorlog CONVERT TO CHARACTER SET utf8 COLLATE utf8_unicode_ci ;
  9 ●   ALTER TABLE ms_access_migration.genderstats CONVERT TO CHARACTER SET utf8 COLLATE utf8_unicode_ci ;
 10 ●   ALTER TABLE ms_access_migration.jobstats CONVERT TO CHARACTER SET utf8 COLLATE utf8_unicode_ci ;
 11 ●   ALTER TABLE ms_access_migration.lookups CONVERT TO CHARACTER SET utf8 COLLATE utf8_unicode_ci ;
 12 ●   ALTER TABLE ms_access_migration.series CONVERT TO CHARACTER SET utf8 COLLATE utf8_unicode_ci ;
 13 ●   ALTER TABLE ms_access_migration.users CONVERT TO CHARACTER SET utf8 COLLATE utf8_unicode_ci ;
 14
```

Figure 9.40 – The script will set all the tables to the correct collation for use with MS Access

13. Execute the query.

14. Delete the **users** table you just linked so that we can relink it.

15. Go back to *Step 1* and try again. This time, the data should appear, all fixed. That little trick will save you a lot of time and frustration trying to figure out what went wrong. There is another script named `Create Collation Conversion commands.sql` that will create the commands to fix each table in the schema, and you can run this against any future database you create or migrate. Should you forget about the collation, just change the schema name as appropriate.

16. If you are happy that the data has been migrated correctly and matches the original **users** table, you can delete the **zUsers** table from the frontend.

Linking the tables back to MS Access is not difficult, and if their properties have been set correctly, it will not cause many issues. Even a successful migration may cause issues, as you will have found out if you got **#Name?** in *Step 10*, but most issues can be fixed easily.

Exercise 9.07 – linking a problematic MySQL table to Access

The only issue you may find when linking a table from a MySQL database to MS Access is usually because the primary key has not been set. We have some of them in our database, so let's try and link one. Follow these steps to complete this exercise:

1. Rename the `capacityindicatorsstats` table in MS Access to `zcapacityindicatorsstats`.

2. Click **External Data | ODBC Database** and select the link to the data source by creating a linked table option.

3. When you're presented with the table list, select `capacityindicatorsstats`. This time, you will get the following window:

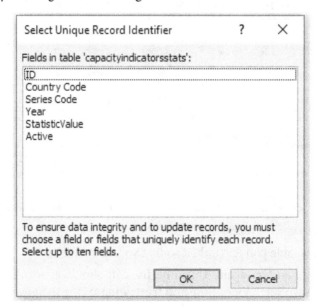

Figure 9.41 – This window only appears if no primary key is set in the table

4. When this happens, this means that the primary key hasn't been set for the table. You have three options to fix this.

   If you do not make a selection and click **Cancel**, the table will be linked but will be read-only. This may or may not be a problem, depending on the table and how it is being used – that is, if it is just populating drop-down lists and is not expected to be updated, then there will be no problems.

5. Select up to 10 fields to ensure the uniqueness of the record. You will usually pick one or two. Select the fields and click **OK**; the table will be linked and can be updated. It is better to fix it in the backend immediately so that you don't get the message because it will pop up every time you refresh the links and can get tedious. If you are linking tables by VBA code, this message will not appear, and the table will be linked as read-only.

6. Click **Cancel** and delete the linked table. Go to Workbench and fix the issue by setting the primary key and auto-incrementing. Then, retry linking. This is the best option.

7. To check whether the table has been linked correctly and is writable, open it to view the records and move to the last record. If the bottom line is blank with an * at the start, then it is writable, and all is well:

57689	ZWE	5.51.01.09.wat	2017	1	0
57690	ZWE	5.51.01.09.wat	2018	1	0
*					

Figure 9.42 – A linked table that can be edited

Tables that have no primary keys are the only possible issue you will have when linking tables from MySQL, and that is more of a nuisance than anything else. In the next exercise, we will refresh linked MySQL tables.

Refreshing linked MySQL tables

Often, during migration and even into further development, you must make adjustments to the tables and fields in the database. MS Access will not pick up these changes until you refresh the links. Access provides you with a tool to do this easily, without having to remove the table and relink it. To be able to refresh tables, follow these steps:

1. In the MS Access frontend application, select **External Data | Linked Table Manager** from the ribbon. You will be presented with the following screen, which displays all the linked tables and their data sources:

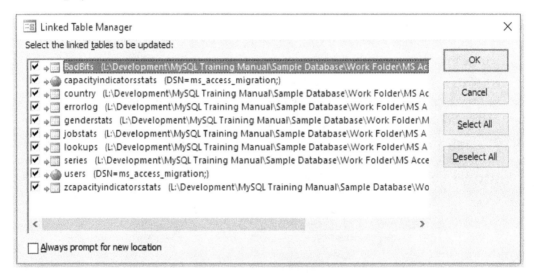

Figure 9.43 – Selecting tables and options to refresh the links

2. Select the tables you want to refresh the links of. You can select any or all of them as required.

3. **Always prompt for new location**, if checked, will ask for the table's source. If it's not checked, the existing source will be used. Use this if the backend database has been moved or changes IP.

4. Click **OK**; the tables will be refreshed and any field changes will now be available.

> **Note**
>
> You can have tables linked from multiple sources. If you're refreshing from only one of the data sources, only select the source tables you wish to refresh.

Refreshing table links is a very fast process and will update the internal MS Access data that's related to the tables. This should be done whenever you're making changes to the data source's structure or moving the backend data location or IP address. All data sources, including MySQL, Access, and Excel, need to be refreshed if changes are made.

We're almost there! Let's complete another exercise and wind this data migration process down.

Activity 9.01 – linking the remaining MySQL tables to your MS Access database

We need to link the remaining tables from MySQL to the MS Access frontend so that we can continue with the conversion process. In this activity, we will complete the linking process for the remaining MySQL tables. The steps to complete the table links should be followed in order. Please refer to *Exercise 9.04* and *Exercise 9.05* if required. Follow these steps to complete this activity:

1. Rename all of the remaining original local tables in MS Access.

2. Check and set the primary keys of all the remaining MySQL tables in Workbench if necessary.

3. Set the AI property of all the primary key fields in the remaining tables in MySQL.

4. Link all the remaining tables to MS Access.

5. Validate that the data is correct in all the MySQL linked tables compared to the MS Access tables.

6. Finally, remove all of the old MS Access linked tables.

After performing these steps, you should see the following output:

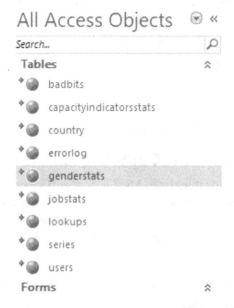

Figure 9.44 – The linked tables once all the tables have been linked

Now, we get to start with the fun stuff. In the next chapter, we are going to transform the MS Access application to leverage the power of MySQL. Come with us and we shall work some magic that will mystify you!

> **Note**
> The solution to this activity can be found in the Appendix.

Summary

In this chapter, we migrated a backend database to MySQL using manual techniques. We learned when we can and cannot use automated methods directly from MS Access, depending on the versions and bit versions of MS Access and MySQL. Then, we learned how to migrate an MS Access database to MySQL using MySQL Workbench. Finally, we linked the MS Access application to the new MySQL server tables and proved that the application still works.

Now that we have migrated our test database to MySQL, in the next chapter, we will migrate the MS Access application to MySQL to leverage the power of MySQL.

10
Microsoft Access – Part 2

In this chapter, you will convert a sample application to use MySQL data using passthrough queries, and then you will learn how to convert a **Microsoft Access** (**MS Access**) table-reliant form to be an unbound form that doesn't rely on local or linked tables. By the end of this chapter, you will be able to remove all linked tables and check that the application still works.

This chapter covers the following topics:

- Introduction to MS Access
- Migrating an MS Access application to MySQL
- Activity 10.01—Converting gender and job statistics
- Calling MySQL functions
- Activity 10.02—Creating a function and calling it
- Calling MySQL stored procedures
- Activity 10.03—Creating MySQL stored procedures and using them in **Visual Basic for Applications** (**VBA**)
- Using parameters
- Activity 10.04—Parameterized stored procedure (series list)
- Activity 10.05—Multiple parameters stored procedure (date lists)
- The **Bad Bits** form

Introduction to MS Access

In *Chapter 9*, you learned how to convert VBA **Structured Query Language** (**SQL**) statements designed to work with linked tables. Using MySQL functions will simplify retrieving results from the data while keeping the processing on the MySQL server and therefore speeding up the MS Access application. In this chapter, you will migrate the MS Access data processing over to the MySQL server to speed up the application. You will do this using passthrough queries. You will learn how to integrate MySQL-based functions, procedures, and views with MS Access VBA code and passthrough queries. You will learn how to create parameterized procedures, how to pass parameters into them for filtering, and how to use the returned data.

Migrating an MS Access application to MySQL

Migrating an MS Access database to MySQL is only half the job. A lot of people think that if you put the data into MySQL, everything is going to be super-fast. But no; you will usually see some improvement in data access speed in some areas, but in others, it may even be slower than before.

Unless you modify the application to properly leverage the data processing power of the MySQL server, you still only have a container for the data, and MS Access is still processing the data. In this section, we are going to move the processing of data to the server by sending requests for data and getting the results only, which we will then use in the application.

You do not have to completely migrate an application before it can be used. You can do parts of it as required, so you will usually concentrate on specific areas that are slow; maybe a report is taking too long to run, a screen is slow, or updating records on a specific form is frustrating the users. Let the application's users direct you to areas that they would like to see improved immediately and concentrate your efforts there for a quick win, and work on other aspects as required or as time permits.

The assumption in this section is that you are comfortable with using the MS Access VBA **integrated development environment** (**IDE**), you can create queries and SQL statements for MS Access, and you have some understanding of the VBA programming language for Access. Let's get started.

Passthrough queries

What is a passthrough query? A passthrough query will pass SQL statements directly to the server for execution, totally bypassing the MS Access data processing engine. They can be used for the following actions:

- Running MySQL console commands
- Running stored procedures and functions

- Running SQL statements
- Retrieving and modifying data

How do we use a passthrough query? Depending on its specific function, it can be used like any other query in Access to populate lists, provide recordsets, update data, and so on. In several of the upcoming exercises, we want to populate several drop-down lists with data based on the options selected in other drop-down lists. We will be designing a SQL statement in VBA (using the users' selections) to be run on the server and generating a passthrough query to pass the statement to the server, which will then run it and pass back the filtered list via the passthrough query so that we can use the data to populate the drop-down list. Passthrough queries can be designed in the Query Designer (text only— not with the **graphical user interface** (**GUI**)) and saved; however, they are not dynamic as the connection details are fixed, and if the server changes name or **Internet Protocol** (**IP**) address, then they all need to be updated. Instead, we will be generating queries dynamically using VBA code provided as a callable function. This will allow us to pass parameters and will also deal with any server IP changes. You are welcome to use this code in any of your future development.

Exercise 10.01 – Passthrough (simple SQL conversion)

In this exercise, we are going to start with a simple query to count records. The code is behind the **Populate Lists** button, and the result will go into the **Capacity Indicators** textbox, as you can see in the following screenshot. This is on the main form, and there are seven database calls in total that we will be converting on this form. At the moment, they are all processed by Access. Each piece of SQL code is numbered. The following screenshot shows number 1, where we will start. The highlighted code is what we are going to replace, and the new code will go into the blank space. Explanations will come after this exercise:

Figure 10.1 – Location of code and structure of code blocks

Follow these steps to complete this exercise:

1. If `frmMain` is open, *right-click* on its tab and select **Design View**. If not, *right-click* on it in the **Navigation Pane** and select **Design View**. The form should open in the design view, as illustrated in the following screenshot:

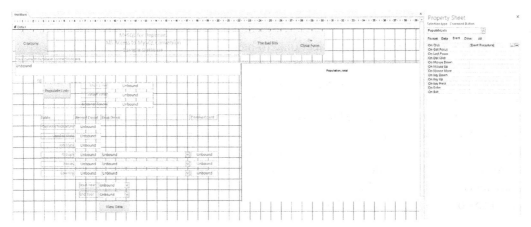

Figure 10.2 – Main form in design view showing the property sheet and activated events for the Populate Lists button

2. If the properties panel is not visible, *right-click* on the **Populate Lists** button and select **Properties**.

3. Where it reads `[Event Procedure]`, click the button with the *three dots*. The code window will open at the code shown in the preceding screenshot.

> **Note**
>
> Unless instructed otherwise, this is how you get to the code we will be working with. The SQL code is numbered and will be referred to by the number.

4. Comment out the two lines of code indicated in *Figure 10.1* by placing an apostrophe at the start of each line.

5. Enter the following code between the two lines:

```
SQL = "SELECT Count(capacityindicatorsstats.ID) AS
RecCount FROM capacityindicatorsstats;"
```

```
Call CreatePassThrough(SQL, "CISCount", True, False)
```

```
Set RS = CurrentDb.OpenRecordset("CISCount",
dbOpenDynaset)
```

Your code for SQL 1 should now look like this. If not, correct it:

```
'SQL 1
'----------------------------------------------------
'When converting, comment out the code lines below (between the lines)
'----------------------------------------------------
    'SQL = "SELECT Count(capacityindicatorsstats.ID) AS RecCount FROM capacityindicatorsstats;"
    'Set RS = CurrentDb.OpenRecordset(SQL, dbOpenDynaset)
'----------------------------------------------------
'Enter your new code between the lines below
'----------------------------------------------------
    SQL = "SELECT Count(capacityindicatorsstats.ID) AS RecCount FROM capacityindicatorsstats;"
    Call CreatePassThrough(SQL, "CISCount", True, False)
    Set RS = CurrentDb.OpenRecordset("CISCount", dbOpenDynaset)
'----------------------------------------------------
    RS.MoveFirst
    Me.cntCIS = RS.Fields("RecCount")
    RS.Close
```

Figure 10.3 – SQL 1 code block with old code commented out and new code added

6. Click **Save**.

7. Return to the form. *Right-click* on its tab and select **Form View**. The form will open in the form view.

8. Click **Populate Lists**, and the data will be populated. **Capacity Indicators** should hold a value of 5136, as illustrated in the following screenshot:

Figure 10.4 – Onscreen results after the SQL 1 code is converted

Let's do a quick analysis of what we just did in the preceding exercise by stepping through the code, as follows:

- The SQL statement is identical to the original and will run in MySQL without modification.

- SQL is passed into a function that creates a passthrough query. The following parameters are passed into the function:

  - SQL statement

  - Name of the passthrough query to create or change

  - `True` or `False` to indicate if the passthrough query returns values or not

  - `True` or `False` to indicate if the passthrough query is to be deleted before being recreated

Because we are returning values, the query is assigned to a recordset. If this were an action query, we would have executed it.

The `CreatePassThrough` function is well documented in code comments, and the code is compact. You can view it in detail at your leisure, but to be brief, it will create a passthrough query with a connection to the MySQL server to execute the SQL statement on it.

As this is a small database and a small query, the improvement is not as immediately noticeable here as it would be in a larger database and a more complex query. The key thing here is that the MySQL server received a command in the form of a single passthrough query, executed it, and returned the value only. We did not pass thousands of records across the network and Access did not process it. Let's look at the passthrough query that was created.

9. Go back to the main Access window.

10. In the **Navigation Pane**, select **Queries** from the drop-down list. You will see one query named `CISCount`. *Double-click* on it to run it (this one is safe, but always check what a passthrough query is doing before you run it). You will get the following result:

Figure 10.5 – Passthrough query (the globe icon indicates a passthrough query) and its result

11. *Right-click* on the CISCount query and select **Design View**. The design view will open. It is not the graphical view that you may be used to as it will be opened in a SQL view, as illustrated in the following screenshot:

Figure 10.6 – The SELECT statement to be passed to the server, the connection details, and the indicator that it will return records

The main panel has the SQL statement we passed in. **ODBC Connect Str** has a **Data Source Name (DSN)** reference to the **Open Database Connectivity (ODBC)** connection that we use to connect to the server. This was assigned in the CreatePassthrough function and has been preset for you in the **lookup table (LUT)**. **Returns Records** is set to Yes, indicating that the query will return results in the form of one or more records.

The rest of the code for the code tagged as SQL 1 ensures that we are positioned to the first record. It assigns the value to the cntCIS textbox for display and closes the recordset.

> **Note**
> Before modifying any SQL code you are about to convert, run it in a Workbench query tab first. If it works, great—one less thing to do. Otherwise, you will need to convert it to the MySQL syntax.

Activity 10.01 – Converting gender and job statistics

Your manager would like to convert the remaining GenderStats and JobStats queries to passthrough queries to allow them to be processed more efficiently. In summary, the following tasks will need to be completed:

1. Convert the SQL for GenderStats (SQL 2) and name the passthrough query GENCount.

2. Convert the SQL for JobStats (SQL 3) and name the passthrough query JOBCount.

3. Convert the SQL for Country (SQL 4) and name the passthrough query CTRYCount.

    After implementing these steps, the expected output should look like this:

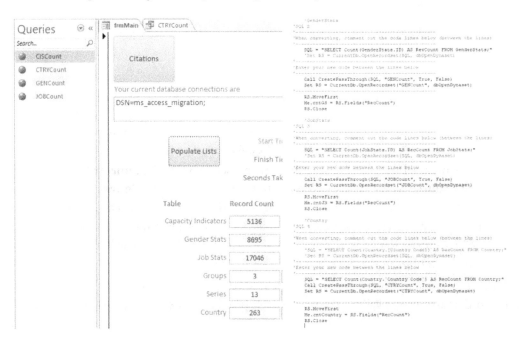

Figure 10.7 – Changes to the code and the affected onscreen controls

> **Hint**
>
> If a field name has spaces, Access encloses the field name in square brackets, whereas MySQL encloses them in backticks. Backticks are located in the top-left corner of your main keyboard, next to the *1* key.

We did not modify or move two of the original SQL statements. We tested them in Workbench, and they worked, so there was no need to modify them. Country, however, had a space in the field name, and the brackets Access uses had to be changed to backticks—our first SQL modification. Always try to make as few changes as possible to achieve the conversion.

So far, we have reduced the dependencies on the linked tables by four queries. We have reduced MS Access from counting 31,140 records to reading only 4 with minimal changes—a good start.

> **Note**
>
> The solution to this activity can be found in the Appendix.

Calling MySQL functions

It is possible to call MySQL functions using passthrough queries. This can help to generate results without having to write additional code. To do this, you simply need to create a passthrough query and use it to call functions as you would in MySQL.

Exercise 10.02 – Passthrough (calling MySQL functions)

You would like to be able to count the values in the series table in order to use the values in analytics for reporting purposes. You currently have a function to do this, called fnCountSeries. To be able to count the values, you can call this function from Access. The following steps will demonstrate how this is done:

1. We are working on SQL 5, the Series count. Locate the code in MS Access.

2. Load the Create Function fnCountSeries.sql file into a query tab in Workbench and run it. This will create a function to count and return the records in the series table, as illustrated in the following screenshot. Verify the function that was created:

Figure 10.8 – New function in the schema panel

> **Note**
>
> The Create Function fnCountSeries.sql file can be found here:
> https://github.com/PacktWorkshops/The-MySQL-Workshop/tree/master/Chapter10/Exercise10.02

3. Calling a function simplifies our SQL statement. We no longer need the original SQL statement in the code, so comment it out.

4. Our new SQL statement is shown in the following code snippet; the value the function returns is stored in a derived field named `SeriesCount`:

```
SQL = "SELECT fnCountSeries()as SeriesCount"
```

5. Add the call to the `CreatePassThrough` function, as follows:

```
Call CreatePassThrough(SQL, "CntSeries", True, False)
```

6. Open the query in a recordset, like this:

```
Set RS = CurrentDb.OpenRecordset("CntSeries",
dbOpenDynaset)
```

7. Position the first record. Assign the value to the textbox for display. Note in the following code snippet that the field name has changed:

```
RS.MoveFirst
Me.cntSeries = RS.Fields("SeriesCount")
RS.Close
```

8. Save and view the results on the form.

Functions are handy when you need to return a single value such as a record count or a calculation result. Calling them from VBA is not difficult. All the previous exercises could have used functions instead of SQL statements to achieve the same results; however, we are demonstrating various ways to achieve results and get the bulk of the processing away from MS Access and onto the MySQL server. In the next section, we will do an activity wherein we will create a function and then call it.

Activity 10.02 – Creating a function and calling it

As part of your project to convert the MS Access application to MySQL, you have reviewed the `SQL 6` SQL statement and have determined that the statement should be converted to a MySQL function to force the processing to the MySQL server, simplify the VBA code, and ensure there is only a single value returned.

You will be working with the code tagged as SQL 6. In this activity, you will create a function to count and assign the total groups to the cntGroups textbox. Follow these steps to complete this activity:

1. Copy the Create Function fnCountSeries.sql file used in the previous exercise and name the new file Create Function fnCountGroups.sql.

> **Note**
>
> The Create Function fnCountSeries.sql file can be found here:
>
> https://github.com/PacktWorkshops/The-MySQL-Workshop/tree/master/Chapter10/Exercise10.02

2. Modify the new file to create a function named fnCountGroups.
3. Use the original SQL statement from VBA. Make a slight adjustment to the SQL for it to work as a MySQL SQL statement.
4. Run it in a Workbench query tab to create this new function.
5. Call the function from VBA.

    After implementing the preceding steps, the expected output should look like this:

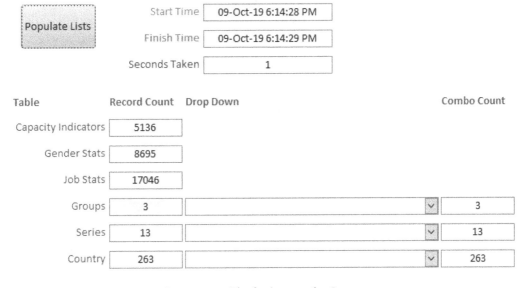

Figure 10.9 – The final output for Groups

Converting the application's SQL code to a function will remove the processing from Access to the MySQL server and reduce the VBA code to a minimum. It will also speed up execution, and if the function's processing needs to be changed in the future, no VBA using the function will need to be changed. In this activity, we created a function using a SQL script file, but we could easily have created it using Workbench. Another advantage of functions and the upcoming stored procedures is that they can be used by any application capable of using them, including MS Excel.

> **Note**
> The solution to this activity can be found in the Appendix.

Calling MySQL stored procedures

Stored procedures are similar to functions, except they can return a recordset. You cannot modify the returned records, but they are ideal for populating ListBoxes, ComboBoxes, and VBA read-only recordsets. Let's populate some dropdowns using stored procedures in the next exercise.

Exercise 10.03 – Calling a MySQL stored procedure

We are working on SQL 7 for the next exercise and activity. SQL 7 comprises three separate queries populating the three dropdowns on the main form. This exercise will work through one of them, the **Series** dropdown. Follow these steps to complete this exercise:

1. Locate the VBA code for SQL 7.
2. Create a SQL file and name it Create Procedure spSeriesList.sql.
3. Type the following code to use a target database:

```
USE ms_access_migration;
```

4. Delete the stored procedure if it exists, as follows:

```
DROP PROCEDURE IF EXISTS  spSeriesList;
```

5. Set up a custom delimiter. This tells MySQL that everything between the custom delimiter is to be treated as one procedure. The code is illustrated in the following snippet:

```
DELIMITER //
```

6. Create and name the stored procedure, like so:

```
CREATE PROCEDURE spSeriesList()
```

7. Add BEGIN to indicate where our procedure code starts, like this:

```
BEGIN
```

8. Add procedure statements. These are the same SQL statements from VBA, so you might just want to copy that. The bracketing on the field names is changed to suit MySQL's requirements. The code is illustrated in the following snippet:

```
SELECT DISTINCT ms_access_migration.series.'Series Code',
ms_access_migration.series.'series Name'
FROM ms_access_migration.series ORDER BY ms_access_
migration.series.'series Name';
```

9. Indicate the end of the code and also the delimiter, as follows:

```
END//
```

10. Reset the delimiter back to its default, like so:

```
DELIMITER ;
```

11. Save the script.

12. Load the script into a Workbench query tab and run it. You should now have a new stored procedure in the schema list, as illustrated in the following screenshot. Don't forget to refresh the list:

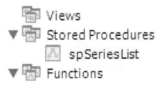

Figure 10.10 – New stored procedure in the schema panel

13. To test the stored procedure, type the following code into a new query tab:

```
call spSeriesList
```

You should get the following result:

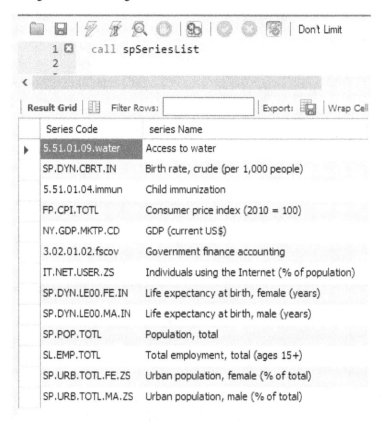

Figure 10.11 – Calling the stored procedure and the output

14. Now, for VBA, go to the VBA window. Insert some blank lines to separate `Me.cmbSeries.RowSource` from the other two lines to give yourself some room to work.

15. Working above the original line of code, add the following line to prepare the SQL statement for the passthrough query:

```
SQL = "call spSeriesList;"
```

16. Add the following line to call the function to create a passthrough query:

```
Call CreatePassThrough(SQL, "spSeriesList", True, False)
```

17. And finally, modify the assignment to `cmbSeries.RowSource`, as follows:

```
Me.cmbSeries.RowSource = "spSeriesList"
```

18. Click **Save**, and we are done. Click the **Populate Lists** button on the main form in Access, and the list should be populated, as shown here:

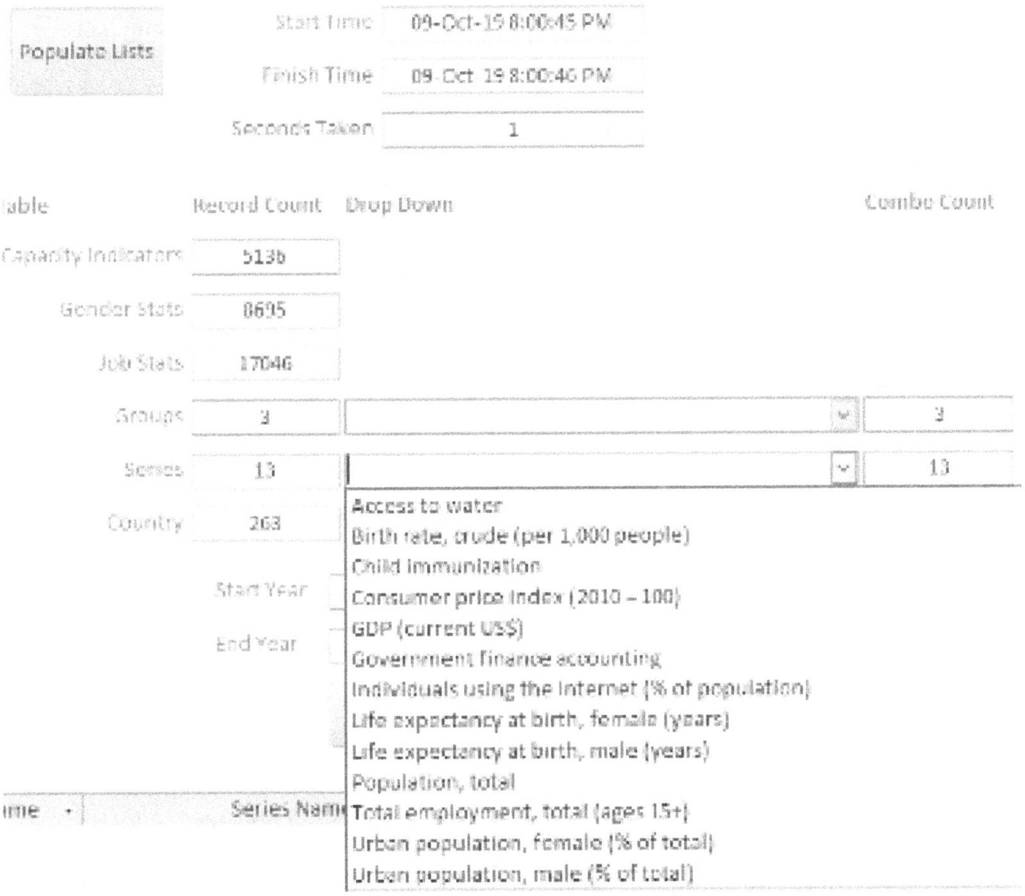

Figure 10.12 – Drop-down list after the passthrough query using the stored procedure is assigned

Simple stored procedures are no more difficult to create than functions but return a lot more. Having a stored procedure return a list makes it very easy to populate a listbox control, and we can assign the passthrough query directly to the list as the row source. Stored procedures or functions can, of course, be much more complex than we have shown here. They can run multiple SQL statements and can have their own built-in logic flow. So, learning more about them will greatly enhance your employability as they are a much sought-after skill in the job market.

Activity 10.03 – Creating MySQL stored procedures and using them in VBA

Continuing with your conversion project, you have noticed two dropdowns using lists provided by the VBA code. You also noticed the lists are not filtered, so you have decided the best way to handle these two lists is to convert them to stored procedures because they can return a recordset. In this activity, we will be creating MySQL stored procedures and using them in VBA. Follow these steps to complete this activity:

1. Create two new stored procedures named `cmbGroups` and `cmbCountry`.

2. Refer to *Exercise 8.11* for the specific steps, if required. Be sure to change the names and SQL as required for each list.

3. Take note of field names with a space. Remember to change the square brackets [ ] to backticks ' '.

4. Modify the VBA code to use the new stored procedures with a passthrough query.

After performing the steps, the expected output should look like this:

Table	Record Count	Drop Down	Combo Count
Capacity Indicators	5136		
Gender Stats	8695		
Job Stats	17046		
Groups	3	Gender Statistics	3
Series	13	Capacity Indicators / Gender Statistics	13
Country	263	Job Statistics	263

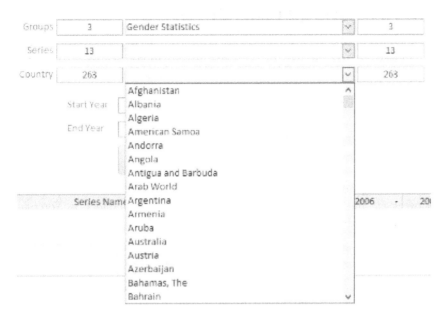

Figure 10.13 – Dropdowns displaying the lists

Stored procedures can return a recordset of data. They can be a single column of data or multiple columns. They can also be assigned directly to data-consuming controls such as drop-down boxes—as we have done here using passthrough queries, lists, or even a single field. As with functions, they can be used by applications such as Excel, so they are ideal when you wish to use the same data across multiple applications.

> **Note**
> The solution to this activity can be found in the Appendix.

Using parameters

Up to now, we have only been dealing with extracting results from the database as either single values or as complete, unfiltered lists as defined by the SQL statements, functions, and stored procedures. However, we often need to filter the data to get the results required for processing in VBA or to populate controls such as drop-down lists. We filter the data by passing in parameters to the SQL statements, stored procedures, or functions. The following exercise will step through creating a stored procedure to accept a single parameter—the group that the user has selected. The stored procedure will then query the database using the filter to return a list of series relating to the group and pass back the results to VBA to display the list in the **Series** dropdown.

Parameterized stored procedures

Most SQL statements in MS Access that you are converting to MySQL have parameters that make them flexible. You can use the same parameters in your stored procedures when you convert them, and most will be of the IN type. The following exercise concentrates on the IN type.

Exercise 10.04 – Parameterized stored procedure (series list)

You have found code generating a list for another dropdown that uses the selections from other dropdowns as a filter to make the data relevant to the user's selection. You have decided to use a parameterized stored procedure to get the relevant list.

We will be working with the code tagged as SQL 8 for this exercise. You can find it in the After_Update event of cmbGroups. The purpose of this SQL is to provide a Series list for the cmbSeries dropdown filtered to the selected group that is used as a parameter. Follow these steps to complete this exercise:

1. Create a new file named Create Procedure spSeriesList_par.sql.

2. Enter the following code. This is similar to what we used earlier to create a stored procedure, except for the name:

```
USE ms_access_migration;
DROP PROCEDURE IF EXISTS  spSeriesList_par;
DELIMITER //
```

3. Enter the following code to create a procedure named sp_SeriesList_par:

```
CREATE PROCEDURE spSeriesList_par(  IN GroupName
VARCHAR(25)    )
BEGIN
```

The IN parameter declaration is within brackets. Here, we are declaring an IN parameter with a GroupName variable name of type VARCHAR(25). With this declaration, the calling program will be required to pass in a parameter. BEGIN indicates the stored procedure code is to follow.

4. Our SQL statement is separated into four lines for readability. Type the following code:

```
SELECT DISTINCT ms_access_migration.series.'Series Code',
ms_access_migration.series.'series Name'
FROM ms_access_migration.series
```

This is the first part of the SQL statement used in VBA, with [] replaced by ''.

5. In VBA, the filter was inserted by referencing the value in cmbGroups; here, it is passed in. Enter the following line of code. Notice we use GroupName for the filter, and we do not need to wrap it in backticks or quotes:

```
WHERE series.'Group' = GroupName
```

6. Finish off the code with ORDER BY, as it was defined in VBA, the END parameter (of our BEGIN parameter). // is the end of the modified delimiter, and DELIMITER; is to set MySQL back to the default delimiter character. The code is illustrated in the following snippet:

```
ORDER BY ms_access_migration.series.'series Name';
END//
DELIMITER ;
```

7. Save and run the SQL in a query tab in Workbench.

8. Test it by typing the following command into a query tab to get the results, as shown here:

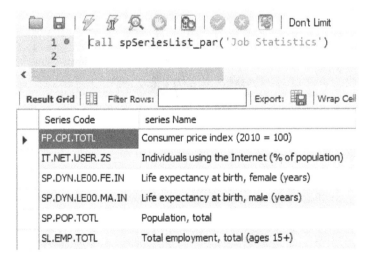

Figure 10.14 – Testing the parameterized stored procedure and its results

Now you have created a stored procedure, we will use it in the next activity to move processing over to the MySQL server.

Activity 10.04 – Parameterized stored procedure (series list)

You have been asked to modify the code tagged as `SQL 8` to call `spSeriesList_par()` from a passthrough query and assign it to the `cmbSeries` row source. Perform the following steps to implement this activity:

1. Locate the SQL code in VBA marked `SQL 8`.

2. Comment out the existing SQL statement. *Hint*: When you are modifying code, always comment out the original lines of code before creating a new line. This gives you *a)* a reference to the original code and *b)* an easy way to reinstate the original code if required. You can remove the line after you have tested and confirmed the new code is working.

3. Create a SQL statement for the passthrough query. This time, pass the filter value to the stored procedure in the brackets and, as we are passing in a string, ensure it is enclosed in single quotes.

4. Assign the resulting passthrough to the dropdown.

After implementing these steps, the expected output should look like this:

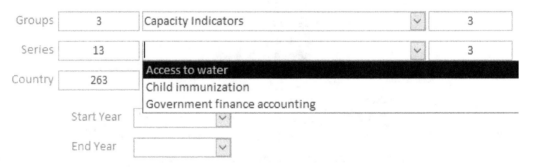

Figure 10.15 – Changing the group will change the series list values

The key to this exercise and activity is the ability to pass parameters into stored procedures. Building the SQL requires a call to the procedure, and passing in the parameters is no different from VBA's original SQL version, so in a lot of cases, you may be able to simply copy parameters from VBA's existing statement. The inclusion of parameters makes the code much more flexible.

The original SQL statement was executed by Access. This means Access pulled all the series data from the server and then applied a filter to get the rows we wanted, whereas the new SQL statement encoded the parameter values into a MySQL-styled SQL statement to call the stored procedure and, by sending it to the server as a passthrough query, MySQL executed the stored procedure, filtered the list based on the parameter value, and returned the results only.

> **Note**
>
> The solution to this activity can be found in the Appendix.

Exercise 10.05 – Multiple parameters stored procedure (country list)

In this exercise, we will be working with multiple input parameters to get a country list for the cmbCountry dropdown. We will be working with the code tagged as SQL 9, which is located in the cmbSeries_AfterUpdate() event.

The purpose of the code is to provide a list of valid countries that have a statistic for the chosen group/series combination. The group data comes from different tables, so we need to pass in two parameters: the table we are looking at and the series. We have some VBA to determine which table we need to include in the SQL based on the selection in the group combo.

Here is the code at the start of the cmbSeries_AfterUpdate() event:

```
Select Case Me.cmbGroups
    Case "Capacity Indicators"
        TableName = "capacityindicatorsstats"
    Case "Gender Statistics"
        TableName = "genderstats"
    Case "Job Statistics"
        TableName = "jobstats"
End Select
```

This will store the table name in the TableName string variable. The series value can be read directly from the cmbSeries dropdown.

We cannot use the TableName value as we did for the normal filtering values in the last exercise. We have to prepare the SQL for the stored procedure differently. We will be dynamically building the SQL based on the TableName value passed in.

The CONCAT() command will help us piece together the SQL statement. Type or cut and paste the following line of code into a query tab in Workbench and run it to see what CONCAT() does:

```
SELECT CONCAT('This ','is ','an ', 'example ','of ','string
','CONCATenation')
```

When you run it, the result will be all the text joined into one string, as shown here:

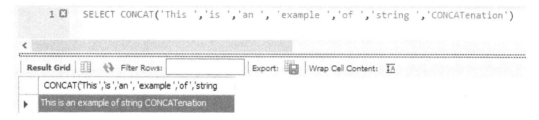

Figure 10.16 – Sample of a concatenated string in SQL and its output

Let's get started with building the stored procedure, as follows:

1. Create a new text file named Create Procedure spCountryList_par.sql.

2. Enter the following code, which is the same as in the last exercise, except for the stored procedure name:

```
USE ms_access_migration;
DROP PROCEDURE IF EXISTS  spCountryList_par;
DELIMITER //
```

3. Add a CREATE PROCEDURE command. This time, we have two IN parameters separated by a comma, as we can see in the following code snippet:

```
CREATE PROCEDURE spCountryList_par(IN TableName
VARCHAR(25), IN TheSeries VARCHAR(25) )
BEGIN
```

4. Start the concatenation, and the resulting string will be stored in the @t1 variable, which we will use later. The code is illustrated in the following snippet:

```
SET @t1 = CONCAT(
```

5. Start building the SQL. This is the same SQL as the VBA code. with a few points to
    note: [] is replaced by ' ', of course. Each text block is enclosed in single quotes.
    Where TableName is included, the preceding text block's quote is terminated, a
    comma is included, and then the TableName variable, another comma, and the next
    text block's opening quote are included. Spaces are included at the end of each line
    so that the command starting on the next line does not end up hard against the text
    and cause a SQL error. Build a string, and as we include the actual value of series, it is
    enclosed in double quotes. The code is illustrated in the following snippet:

```
'SELECT DISTINCT Country.'Country Code', Country.'Country
Name', ' , TableName , '.'Series Code' ',
'FROM Country INNER JOIN ' , TableName , ' ON
Country.'Country Code' = ' , TableName , '.'Country Code'
' ,
'WHERE ' , TableName , '.'Series Code' = "' , TheSeries ,
'" ',
'ORDER BY Country.'Country Name''
);
```

    This code is not all that different from the VBA method of joining strings with
    an ampersand.

6. We treat this a little differently. The following line prepares a statement using the
    text we just put together:

```
PREPARE stmt1 FROM @t1;
```

7. Execute the statement, clean things up, and finish as before, like so:

```
EXECUTE stmt1;
DEALLOCATE PREPARE stmt1;
```

8. Save and run the script. Test the new stored procedure with the following call to get
    the results shown:

```
call spCountryList_par("Jobstats","FP.CPI.TOTL")
```

This results in the following output:

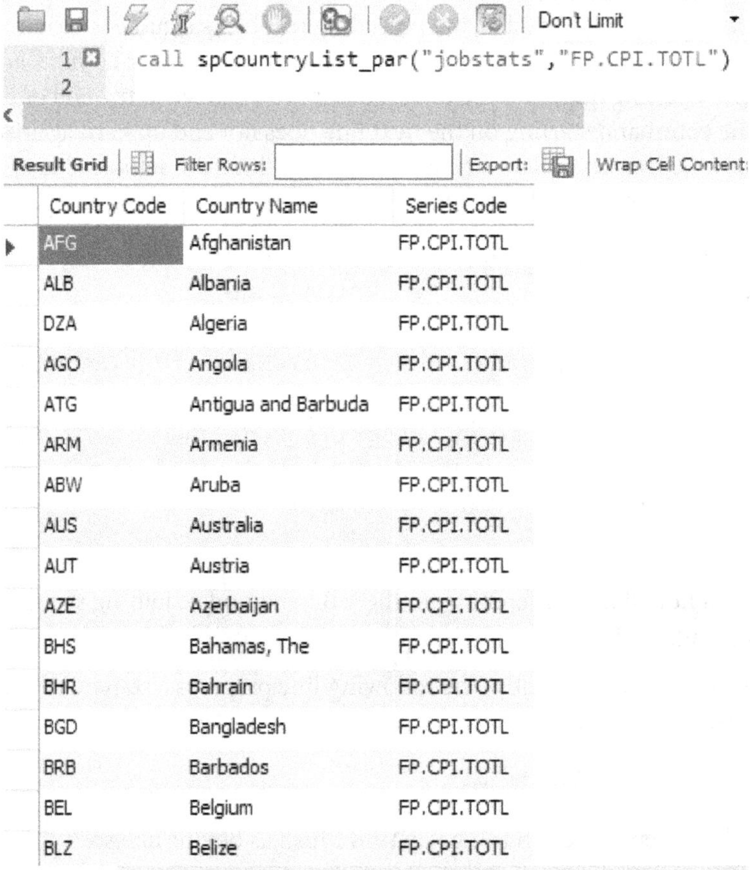

Figure 10.17 – Testing the stored procedure and the expected output

9. For the VBA code to run and assign the stored procedure, we are passing in the
    TableName variable and the value in cmbSeries, as illustrated in the following
    code snippet. When you change series, you will notice the country counter change.
    Since not all countries have statistics for the various series, you might get a zero.
    Try other combinations:

```
SQL = "Call spCountryList_par('" & TableName & "','" &
Me.cmbSeries & "')"
```
```
Call CreatePassThrough(SQL, "spCountryList_par", True,
False)
```
```
Me.cmbCountry.RowSource = "spCountryList_par"
```

Adding multiple parameters is no more difficult than adding parameters to a VBA function. Parameters make our stored procedures and functions very useful, and once they are created, they do not need to be changed until the logic built into them requires changing. We will constantly be changing the passthrough queries to get our different variables passed in, but with the `CreatePassThrough` function, this is simple.

> **Note**
>
> All the passthrough queries we have created were deleted and recreated for this book. In the real world, some of these will be created and remain fixed, such as counters with no parameters. You will figure out the best approach as you develop them and as your needs dictate.

Activity 10.05 – Multiple parameters stored procedure (date list)

Working through the migration project, you have identified two date controls used to filter data. The existing code uses VBA to generate SQL statements for the date dropdowns. The generated SQL is filtered by the users' selections to extract a specific date range for the statistics and put the date lists in the dropdowns. You want to convert these queries to parameterized stored procedures.

In this activity, you will create a stored procedure to determine dates, generate a passthrough query, and assign it to both date dropdowns.

The code tagged as `SQL 10` currently determines the range of dates and assigns them to both the **Start Year** and **End Year** dropdowns.

> **Note**
>
> Both the **Start Year** and **End Year** dropdowns will use the same passthrough, so it only needs to be generated once and assigned to both of them. Name the SQL file `Create Procedure spDateRange_par.sql`.

Perform the following steps to complete this activity:

1. Create a new SQL file named `Create Procedure spDateRange_par.sql` to generate a stored procedure.

2. Copy and paste the code from the `spCountryList_par.sql` SQL file you created in the previous exercise into the new file. You will modify this code.

3.   Refer to *Exercise 8.13* for the steps, if required.

4.   The parameters are the same. The `SELECT` statement will only return one value, `Year`; the **Order** field will be `Year`; everything else will remain the same.

After implementing the steps, the expected output should look like this:

Figure 10.18 – Both date comboboxes will change based on the series selected

There is no real difference between this activity and the previous activity, except that we are reading data with a single stored procedure and assigning it to two date comboboxes to provide a start and end date option for filtering the user's selections. The stored procedure returns a valid date range for the series selection, and VBA assigns it to the controls and then sets the default displayed date accordingly. A passthrough query, as with any normal MS Access query, can be assigned to multiple controls and used in code.

> **Note**
> The solution to this activity can be found in the Appendix.

Exercise 10.06 – Multiple parameters stored procedure (crosstab queries)

Up to now, we have been working with standard SQL statements; however, we have one more query to convert on the main form. You can find it behind the **View Data** button. It is designated as `SQL 11`, and it is a pivot query, commonly referred to in MS Access as a crosstab query. We have a problem; MySQL does not have a pivot function and cannot run such a query. You can imitate them in MySQL, but it is a complex process and well out of the scope of this book. This query is important in this demonstration application because it provides data for both the statistics table and the chart.

In this exercise, we will break this query down and get MySQL to retrieve the data, then get Access to do some of the work performing the final crosstab functions. Let's get started by first examining the query, as follows:

```
SQL = SQL & "TRANSFORM Sum(" & TableName & ".StatisticValue) AS SumOfStatisticValue "
SQL = SQL & "SELECT country.[Country Name], series.[Series Name] "
SQL = SQL & "FROM (" & TableName & " "
SQL = SQL & "INNER JOIN country ON " & TableName & ".[Country Code] = country.[Country Code]) "
SQL = SQL & "INNER JOIN series ON " & TableName & ".[Series Code] = series.[Series Code] "
SQL = SQL & "WHERE (((country.[Country Code]) = '" & Me.cmbCountry & "') "
SQL = SQL & "And ((series.[Series Code]) = '" & Me.cmbSeries & "')"
SQL = SQL & "And ((" & TableName & ".Year) >= " & Me.StartYear & " "
SQL = SQL & "And (" & TableName & ".Year) <= " & Me.EndYear & ") "
SQL = SQL & "And ((series.Group) = '" & Me.cmbGroups & "'))"
SQL = SQL & "GROUP BY country.[Country Name], series.[Series Name] "
SQL = SQL & "ORDER BY " & TableName & ".Year "
SQL = SQL & "PIVOT " & TableName & ".Year;"
```

Figure 10.19 – The existing crosstab query with the SELECT part between the highlighted lines

Notice the SQL between the `SELECT` and `ORDER BY` lines is one complete query. We can migrate this part to a stored procedure, but we need to make a couple of minor changes, which we will point out when needed. Let's start on the stored procedure, as follows:

1. Create a file and name it `Create Procedure spCTSource_par.sql`.

2. Add the following lines of code:

   ```
   USE ms_access_migration;
   DROP PROCEDURE IF EXISTS  spCTSource_par;
   DELIMITER //
   ```

3. This query requires a lot of parameters. Add the following lines. Each parameter is on its own line for readability:

   ```
   CREATE PROCEDURE spCTSource_par
   (
   IN TableName VARCHAR(25),
   IN TheSeries VARCHAR(25),
   IN TheGroup VARCHAR(25),
   IN TheCountry VARCHAR(100),
   IN StartYear VARCHAR(20),
   IN EndYear VARCHAR(20)
   )
   ```

4. Start the `BEGIN` process and set up `CONCAT`, as follows:

   ```
   BEGIN
   SET @t1 = CONCAT(
   ```

5. Here is a MySQL-formatted query to match the VBA statement shown in
    Figure 8.72:

```
'SELECT country.'Country Name', series.'Series Name', ',
TableName ,'.'Year', ', TableName ,'.'StatisticValue' '
'FROM (', TableName ,' '
'INNER JOIN country ON ', TableName ,'.'Country Code' =
country.'Country Code') '
'INNER JOIN series ON ', TableName ,'.'Series Code' =
series.'Series Code' '
'WHERE (((country.'Country Code') = "' , TheCountry , '")
'
    'And ((series.'Series Code') = "' , TheSeries , '") '
    'And ((', TableName ,'.Year) >= "' , StartYear , '" '
    'And (', TableName ,'.Year) <= "' , EndYear , '") '
    'And ((series.Group) = "' , TheGroup , '"))'
    'GROUP BY country.'Country Name', series.'Series
Name', ', TableName ,'.'Year' '
'ORDER BY ', TableName ,'.Year '
```

The `Year` and `StatisticValue` fields have been added to the `SELECT` statement,
and the `Year` field has been added to the `GROUP BY` statement. `TableName` is
included, as names will come from different tables depending on the parameters
passed in.

6. The rest is the same as we covered previously, as indicated in the following code
    snippet:

```
);
PREPARE stmt1 FROM @t1;
EXECUTE stmt1;
DEALLOCATE PREPARE stmt1;
END//
DELIMITER ;
```

7. Save the file, then load and run it in Workbench to create a stored procedure.

8. Now, for VBA, to make it work with Access, locate SQL 11 in VBA and comment out the entire SQL block of code.

9. As there are a lot of parameters to pass in, build a string of parameters only, as illustrated in the following code snippet. This will create a single string of parameters—that is, 'jobstats', 'FP.CPI.TOTL', 'Job Statistics', 'DZA', '2004', and '2016':

```
Dim txtPars As String
txtPars = "'" & TableName & "',"
txtPars = txtPars & "'" & Me.cmbSeries & "',"
txtPars = txtPars & "'" & Me.cmbGroups & "',"
txtPars = txtPars & "'" & Me.cmbCountry & "',"
txtPars = txtPars & "'" & Me.StartYear & "',"
txtPars = txtPars & "'" & Me.EndYear & "'"
```

10. Now, create a passthrough, as follows:

```
SQL = "Call spCTSource_par(" & txtPars & ")"
Call CreatePassThrough(SQL, "spCTSource_par", True,
False)
```

11. Recreate the crosstab query but use our new data source, like this:

```
SQL = "TRANSFORM Sum(spCTSource_par.StatisticValue) AS
SumOfStatisticValue "
SQL = SQL & "SELECT spCTSource_par.[Country Name],
spCTSource_par.[Series Name] "
SQL = SQL & "FROM spCTSource_par "
SQL = SQL & "GROUP BY spCTSource_par.[Country Name],
spCTSource_par.[Series Name] "
SQL = SQL & "PIVOT spCTSource_par.Year; "
```

Your final SQL 11 code should look like this:

```
'Dynamically build the SQL statement using the values selected
'SQL 11
'-------------------------------------------------------
'When converting, comment out the code lines below (between the lines)
'-------------------------------------------------------
'    SQL = ""
'    SQL = SQL & "TRANSFORM Sum(" & TableName & ".StatisticValue) AS SumOfStatisticValue "
'    SQL = SQL & "SELECT country.[Country Name], series.[Series Name] "
'    SQL = SQL & "FROM (" & TableName & " "
'    SQL = SQL & "INNER JOIN country ON " & TableName & ".[Country Code] = country.[Country Code]) "
'    SQL = SQL & "INNER JOIN series ON " & TableName & ".[Series Code] = series.[Series Code] "
'    SQL = SQL & "WHERE (((country.[Country Code]) = '" & Me.cmbCountry & "') "
'    SQL = SQL & "And ((series.[Series Code]) = '" & Me.cmbSeries & "')"
'    SQL = SQL & "And ((" & TableName & ".Year) >= " & Me.StartYear & " "
'    SQL = SQL & "And (" & TableName & ".Year) <= " & Me.EndYear & ") "
'    SQL = SQL & "And ((series.Group) = '" & Me.cmbGroups & "'))"
'    SQL = SQL & "GROUP BY country.[Country Name], series.[Series Name] "
'    SQL = SQL & "ORDER BY " & TableName & ".Year "
'    SQL = SQL & "PIVOT " & TableName & ".Year;"
'-------------------------------------------------------
'Enter your new code between the lines below
'-------------------------------------------------------

    Dim txtPars As String

    'Create formatted string of parameters
    txtPars = "'" & TableName & "','"
    txtPars = txtPars & "'" & Me.cmbSeries & "','"
    txtPars = txtPars & "'" & Me.cmbGroups & "','"
    txtPars = txtPars & "'" & Me.cmbCountry & "','"
    txtPars = txtPars & "'" & Me.StartYear & "','"
    txtPars = txtPars & "'" & Me.EndYear & "'"

    'Set new SQL and create the passthrough
    SQL = "Call spCTSource_par(" & txtPars & ")"
    Call CreatePassThrough(SQL, "spCTSource_par", True, False)

    'Create the new Transform SQL
    SQL = "TRANSFORM Sum(spCTSource_par.StatisticValue) AS SumOfStatisticValue "
    SQL = SQL & "SELECT spCTSource_par.[Country Name], spCTSource_par.[Series Name] "
    SQL = SQL & "FROM spCTSource_par "
    SQL = SQL & "GROUP BY spCTSource_par.[Country Name], spCTSource_par.[Series Name] "
    SQL = SQL & "PIVOT spCTSource_par.Year; "

'-------------------------------------------------------
```

Figure 10.20 – The new Transform code in SQL should look like this

12. We are done. The rest of the code uses the SQL variable to assign the recordset and also to show or hide columns and other housekeeping stuff. The data changes are complete. The main data collection is now moved to MySQL; however, the TRANSFORM processing remains in Access. The Transform SQL is much simpler.

13. Run the Access form and ensure both the display list and the chart work.

Crosstab queries are among the most difficult to work with, and MySQL's inability to process them natively can cause major problems when converting to MySQL. But as you can see, with a little thought, it can be done relatively easily. Here, we moved the main data collection to MySQL, and we got Access to finish off by creating a new crosstab based on the MySQL sourced data, and it is still fast.

In the samples in this training course, we had one function (**Populate Lists**) running multiple SQL statements to perform its task to update the screen values. We converted each one individually to various methods using passthrough queries, functions, stored procedures, and SQL statements. This was for training purposes, and as we were updating various form controls, each had to be a separate entity.

Often in business, you will have a function performing several tasks on the database— for instance, a sales system in filling an order will perform several tasks for each sale, such as creating a sales record, reducing the inventory for each item purchased, generating a reorder if stock levels get below a threshold, generating a picking list, generating a consignment record, and updating a customer's purchase history in a single function from VBA.

If you are running multiple SQL statements in your VBA code to update multiple tables (as in the preceding example), you can embed all the statements into a single stored procedure (such as `spProcessSale(OrderList, CustomerID)`), pass in the required parameters (Items purchased, `CustomerID`), and call it once. This approach could reduce your VBA code from hundreds of lines to only a few, reduce MS Access processing to almost nothing, and make your code much simpler.

The Bad Bits form

In the previous sections, we have not worked with bit fields except to set the default values. Bit fields are known as `Yes/No` fields in MS Access. When migrated to MySQL, they will become either a `Bit` or a `TinyInt` type, depending on how you migrated the table. Both types have some very peculiar properties when linked back to MS Access, which you need to know about. Let's have a further look here:

- `Bit`
  - Will only accept -1 or 0 (`True/False`)
  - Will accept `NULL` but then will no longer work with Access
- `TinyInt`
  - Will accept -128 to 127
  - Will accept `NULL` and continue to work with Access

A TinyInt type may be the best choice for an MS Access Yes/No field. It will take NULL, 0 is False, and any other value is True. Access will put -1 if selected in a CheckBox control. However, if you do have a bit field, you will get an interesting and baffling issue. If you are not aware of what to look for, this section is an informational exercise only; there will be no activity. Let's get started with the demonstration.

Exercise 10.07 – Bad Bits demonstration

As mentioned earlier, bit fields can cause unique and perplexing issues with MS Access and MySQL. The purpose of this demonstration is to show you these issues in a controlled manner so that if and when you do come across them in the real world, you will be able to identify the issue and fix it. Follow the next steps:

1. From the main form in the application, click the **The Bad Bits** button. The following form will open:

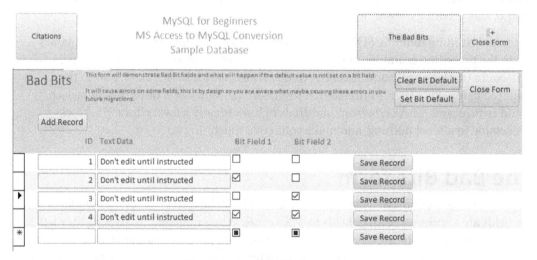

Figure 10.21 – The Bad Bits form

2. If you set the defaults earlier, click the **Clear Bit Default** button. This will set the defaults on the two bit fields back to NULL for this exercise. You have learned enough about passthrough queries, so we will not step through the code behind both of these buttons; they are commented.

3. You have four records on the screen. Edit the **Text Data** and **Bit Field** values on these records only. You can edit them as expected.

4. Now, add two more records, but only enter data in the **Text Data** field. A single character will do; you'll notice the records have been added.

5. Just to be sure, edit record 1 again; it still works.

6. Now, edit one of the new records and try to leave the field. You will receive the following pop-up message from Access:

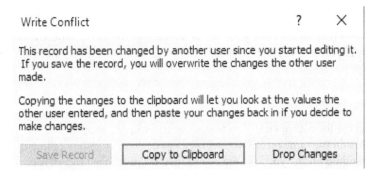

Figure 10.22 – Write conflict error

7. This is a most curious message, as it gives no clue as to what happened. There are no other users or code trying to modify the record. This is a perplexing issue and has had many programmers in despair. The issue is caused because there are NULL values in one or more of the bit fields in this record. You cannot fix this issue just by selecting the checkboxes or by running a query in Access to update them.

8. Click the **Set Bit Default** button, and this will set the defaults on the bit fields. You can check this in Workbench.

9. Now, try to edit one of the new records. It still didn't work. If there is NULL in the field of an existing record, you will have the same issue even after setting the defaults.

10. To properly fix this, it needs to be fixed in Workbench, as you cannot fix it in Access. Open a query tab in Workbench and run the following code. You can paste it all in and run it in one step. The code will update all records to the default of zero if the field value is NULL:

```
UPDATE ms_access_migration.badbits SET badbits.BitField1
= 0
WHERE badbits.BitField1 Is Null;

UPDATE ms_access_migration.badbits SET badbits.BitField2
= 0
WHERE badbits.BitField2 Is Null;
```

11. Now, try to edit one of the new records. It works! Problem solved. And that completes this exercise.

To avoid this situation, do the following:

- Always check for NULL values in MS Access Yes/No fields and set them to zero in Access before migrating a table.
- Always set the bit-field defaults in MySQL immediately after migrating a table.
- If you get this message, check your bit-field settings and values on the table as your first check. If you still get the message, then you may have a real write conflict.

> **Hint**
>
> Access considers a form or code module as a user. If you have an unsaved record on the form or open in a code module and you call another module and try to edit the same record, you will get a write conflict. Before calling the second module, ensure you save the record first. If the unsaved record is on a form, a simple Me.Dirty = False in your VBA code will force-save the record of the form, then call the second code module.

If you forget any of the preceding points, you will run into issues, but you can fix these with a simple query in MySQL.

Unbound forms

Unbound forms are in a class of their own. They are lightweight when it comes to data, and fast because they only ever display the values of one record. The record is not bound to the form. However, they do have one drawback: you need to program all the data handling. But once this is done, they are fast. The main reasons to use unbound forms are outlined here:

- Slow networks
- Remote users
- Large recordset and database
- Record selection is performed on the server, and only one record is transferred across the network to the application

The **Users** form in the sample database is an unbound form. Did you notice that when you opened it, the data was just there? This section is not an exercise or activity but a walk-through of the main points of setting up an unbound form, concentrating on two main functions: LoadForm and SaveData. All code is documented to help you work through it. Follow these next steps:

1. The form has no *record source*; however, when initially designing the form, assign a recordset temporarily so that you can get the fields on easier.

2. The fields on the form have no *data source*. Using the temporary record source, drag and drop the fields in place. When this is done, remove the form's record source and all field data sources. It is important that each field has the same name as the table field that will eventually provide data to it. Dragging them from the temporary record source will give them the right name.

3. Put on the buttons for record navigation and so on, as in the **Users** form. The code behind all the buttons on the **Users** form is similar to the code we have been working with in this section, so you will be familiar with it. The comments will explain what each bit of code is doing.

4. We will now step through the important area—the code in the `UnboundFormRoutines` module. There are two functions we will step through.

5. The top of the module has the following declaration. This is to store the form's original fields and data when we load it, and it will then be used when saving to check if anything on the form has changed:

```
Option Compare Database
Private OriginalData(20, 1) As Variant
```

6. The `LoadForm` declaration is shown here. It accepts a `Form`, an SQL statement to load the form, and the `TableName`. It will be called with `LoadForm(Me, "Select * FROM Users WHERE ID = 1", "Users")`:

```
Public Function LoadForm(TheForm As Form, SQL As String,
TableName As String) As Boolean
```

7. Next, we load the SQL and assign it to a recordset. We check the recordset to ensure we have data and only one record, and we message the user if there are no records or too many records. The code is illustrated in the following snippet:

```
Call CreatePassThrough(SQL, "tmpLoadForm", True, True)
Set RS = CurrentDb.OpenRecordset("tmpLoadForm",
dbReadOnly)
If RS.EOF And RS.BOF Then
    MsgBox "No record to load the form with", vbOKOnly +
vbCritical, "Cannot load form, no record"
    LoadForm = False
    GoTo ExitFunction
Else
    RS.MoveLast
    RS.MoveFirst
```

```
    If RS.RecordCount > 1 Then
        MsgBox "Too many records, There should only be
one, please check the filters", vbOKOnly + vbCritical,
"Cannot load form, too many records"
        LoadForm = False
        GoTo ExitFunction
    Else
    End If
End If
```

8. Next, we initiate the array and assign the SQL and table to the first element of the array (0); as this was declared at the top of the module, it will be available later when we need it. We also assign the Pos position counter to 1 where we will start storing the fields and data. The code is illustrated in the following snippet:

```
For Count1 = LBound(OriginalData) To UBound(OriginalData)
    OriginalData(Count1, 0) = Empty
    OriginalData(Count1, 1) = Empty
Next

OriginalData(0, 0) = SQL
OriginalData(0, 1) = TableName
Pos = 1
```

9. Next, we instruct to ignore errors, and if a field is not on the form then we can simply ignore it. The For/Next loop will cycle through all the recordset fields and store the value in the matching form control. It also records the field and data in the array and increments the position counter by 1. The code is illustrated in the following snippet:

```
On Error Resume Next
For Each Fld In RS.Fields
    TheForm.Controls(Fld.Name).Value = Fld.Value
    OriginalData(Pos, 0) = Fld.Name
    OriginalData(Pos, 1) = Fld.Value
    Pos = Pos + 1
Next
```

10. Finally, we remove the temporary passthrough and close the recordset, and we are done. Here's the code we execute:

```
CurrentDb.QueryDefs.Delete "tmpLoadForm"
ExitFunction:
    RS.Close
    Set RS = Nothing
```

The preceding code can be used on any form that is set up, so you can reuse this as you like.

Of course, we need to save the data later. The function for this is SaveFormData. The function simply accepts a form and is called with Call SaveFormData(Me) from behind the **Save Data** button. It can be called from anywhere appropriate in your code. Follow these next steps:

1. We start by clearing a string that will be used to build our insert data (if any) and initiating a loop through the array elements starting at position 1, as follows:

```
UpdateFields = ""
For Count1 = 1 To UBound(OriginalData)
```

2. On each loop-through, we check if the element is empty, which will indicate the end of the data loaded when the form was opened. If it is, we will exit the For loop; otherwise, we'll continue. The code is illustrated in the following snippet:

```
If IsEmpty(OriginalData(Count1, 1)) Then
        Exit For
    Else
```

3. Next, we compare the value in the array with the value in the matching form control. We use the array elements' field name and value, and if they are different, then we will move into the piece of code to add it to the string. The code is illustrated in the following snippet:

```
If Nz(OriginalData(Count1, 1), "") <> Nz(TheForm.
Controls(OriginalData(Count1, 0)).Value, "") Then
```

4. Next, check if anything has already been added to the update values. If it does have a value, then we put a comma at the end of the string, as illustrated in the following code snippet:

```
If UpdateFields <> "" Then
    UpdateFields = UpdateFields & ","
End If
```

5. Check if the value stored is actually a date (or can be converted to one). If it is, reformat it to ensure it is in the correct format for MySQL, like so:

```
If IsDate(TheForm.Controls(OriginalData(Count1, 0)).
Value) Then
TheForm.Controls(OriginalData(Count1, 0)).Value =
Format(TheForm.Controls(OriginalData(Count1, 0)).Value,
"YYYY-MM-DD")
End If
```

6. Now, we check if the value is a numeric value. If it is, we don't want quotes included when adding to the string. We also check if the control is a checkbox. If it is, we use the **Absolute** (**ABS**) function to ensure the value is positive and not -1, 'Active' = 1, as illustrated in the following code snippet:

```
If IsNumeric(TheForm.Controls(OriginalData(Count1, 0)).
Value) Then
If TheForm.Controls(OriginalData(Count1, 0)).ControlType
= 106 Then
UpdateFields = UpdateFields & "'" & OriginalData(Count1,
0) & "' = " & Abs(TheForm.Controls(OriginalData(Count1,
0)).Value)
Else
UpdateFields = UpdateFields & "'" & OriginalData(Count1,
0) & "' = " & TheForm.Controls(OriginalData(Count1, 0)).
Value
End If
```

7. If it was not a numeric value, then it will be added to the string with enclosing quotes—that is, 'Name' = 'Bob', as illustrated in the following code snippet:

```
UpdateFields = UpdateFields & "'" &
OriginalData(Count1, 0) & "' = " & "'" & Nz(TheForm.
Controls(OriginalData(Count1, 0)).Value, "") & "'"
End If
```

8. Now, we check if anything actually changed by checking the string. If not, we exit. The code is illustrated in the following snippet:

```
If UpdateFields = "" Then
    GoTo ExitFunction
Else
```

9. Here, we put the updated SQL together into one SQL statement. We have built the main part. First, we get the WHERE clause out of the original SQL, and then we create SQL from the parts, as follows:

```
startpos = InStr(1, OriginalData(0, 0), "WHERE")
endpos = InStr(1, OriginalData(0, 0), "ORDER BY")
If endpos = 0 Then
    Tmp = Right(OriginalData(0, 0), Len(OriginalData(0,
0)) - (startpos - 1))
Else
    Tmp = Mid(OriginalData(0, 0), startpos, endpos -
startpos)
End If
SQL = "UPDATE '" & OriginalData(0, 1) & "' SET " &
UpdateFields & " "
SQL = SQL & Tmp
```

10. Finally, we create a passthrough query and call it, and we then delete the passthrough, like so:

```
Call CreatePassThrough(SQL, "ptTemp", False, False)
DoCmd.SetWarnings False
DoCmd.OpenQuery ("pttemp")
DoCmd.SetWarnings True
CurrentDb.QueryDefs.Delete "ptTemp"
```

11. We then leave the function.

Using these two functions will help you make an unbound form work. These are basic operations, and we are sure that you will be able to expand and improve them.

Another way to unbind a form from a linked table

Unbound forms are great for supercharging forms that have a lot of records if linked, and are very good for remote-access users. However, if a form only ever has a few records, it may not be worth going to all that effort. Here is a nice trick:

1. Open the **Bad Bits** form in design mode. **Bad Bits** has a recordset using the linked table. It is the last object actually bound to a linked table in the application.

2. If you don't see **Record Source** under **Data**, check for **Form** in the dropdown.

3. Paste the following line into the **Record Source** property of the form, replacing the existing SQL statement:

```
SELECT * FROM [ODBC;DSN=ms_access_migration].BadBits;
```

This allows us to locate the record source, as indicated in the following screenshot:

Figure 10.23 – Bad Bits form properties with the record source

4. Run the form. It works and is fully editable, and is not attached to a linked table. This is a barely documented trick to connect directly to a database.

In the next section, we will solve an exercise wherein we will be removing linked tables.

Exercise 10.08 – Removing all linked tables

We have done a lot to the database—we have changed every single query to use passthrough queries and made a huge difference to the speed, even on this small database/application. We have also completely liberated the application from all linked tables, and they are no longer needed. In this exercise, we will remove all references to the linked table and then remove the linked table. Perform the following steps to implement this:

1. Open the main form in **Design View**.

2. Remove the *row source* for all five dropdowns. We were changing these in code and not saving the form, so the original row source is still there.

3. Ensure there is no *record source* in the main form, as illustrated in the following screenshot:

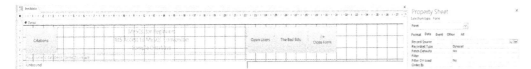

Figure 10.24 – Clicking the top-left square dot to view the form properties

4. Set the **Record Source** property in the lstDisplayData subform to spCTSource_par, as illustrated in the following screenshot:

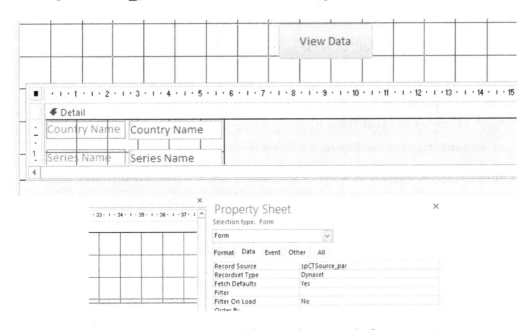

Figure 10.25 – Clicking the top-left square dot to view the form properties

5. Set the **Row Source** property for the graph to `spCTSource_par`, as illustrated in the following screenshot:

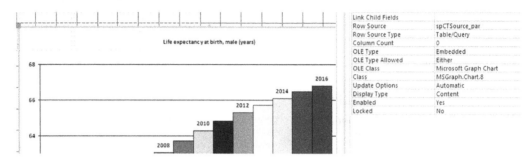

Figure 10.26 – Clicking chart to display the properties

6. Run the form. Does it still work? If so, then delete all linked tables from the Access database, except the `Lookups` table.

7. We need the `Lookups` table locally because it holds the connection information. This is easy to do. *Right-click* the `Lookups` table in the navigation bar and then select **Convert to Local Table**. Access will pull the table and its contents locally.

8. Close and open the form again; it still works.

The form and application are now free of all linked tables, and there are no linked tables in the application. Thus, we have removed all linked tables.

Summary

We converted several VBA SQL statements to statements to be run on the MySQL server and called them with passthrough queries. Some were simply passed to the server, while others were changed to functions or stored procedures, and the program still works. We looked at several possible issues that can arise when migrating a database to an application, and we worked through solutions; and finally, we removed the application's reliance on linked tables completely. In the next chapter, we will continue working with MS Access, deploying more advanced methods of using passthrough queries.

In *Chapter 11, MS Excel VBA and MySQL – Part 1*, we will be working with Excel and the MySQL database. Topics will include setting up connection functions with DSN and DSN-less capabilities; reading data from MySQL and setting ranges; populating data sheets, charts, and individual worksheet cells with MySQL data; and finally, we will be working with MySQL for Excel to create pivot tables and charts, and updating MySQL data directly from Excel.

11
MS Excel VBA and MySQL – Part 1

Setting up and properly demonstrating the use of Excel with MySQL is quite involved, so the topic is split over two chapters. In this chapter, we will begin by setting up a sample MySQL database using a `.sql` script file and learn how to activate the **Developer** tab and the **Visual Basic for Applications (VBA) integrated development environment (IDE)** so that you can develop in VBA. We will then connect to the MySQL server and retrieve data using Excel VBA, create a dashboard with the data from MySQL, and populate drop-down lists and individual cells with VBA and MySQL data. By the end of the chapter, you will be able to create pivot tables and charts from MySQL data and you will have learned how to use MySQL for Excel to load, modify, and update MySQL records directly in the database, and we'll finish off by learning how to push worksheets from Excel into a new MySQL table.

This chapter consists of the following topics:

- Introduction to Excel
- Exploring the **Open Database Connectivity (ODBC)** connection
- Exploring the Excel VBA structure
- Learning about VBA libraries

- Connecting to the MySQL database using VBA

- Reading data from MySQL using VBA

- Populating charts

- Activity 11.01—Creating a chart artist (artist track sales)

Introduction to Excel

Excel is the most popular data-consuming application in today's business world. It is used to analyze data and present it in a graphical manner, making it easy to understand at a glance. In fact, most businesses and personal users would not be able to function properly in today's data-centric world without Excel; so, people with advanced Excel skills are highly sought after.

In *Chapter 9, MS Access – Part 1*, and *Chapter 10, MS Access – Part 2*, you learned how to migrate an MS Access database to MySQL and retrieve and use the data using **VBA code** and **ODBC connections**.

For *Chapters 11* and *12*, the assumption is that you are now familiar with MySQL and Workbench, having worked through the previous chapters, so references to them will be high-level only. We also assume that you have basic Excel skills. We will be connecting to the MySQL server in three ways: through **Data Source Name (DSN)-less connections using VBA**, an **ODBC connection**, and **MySQL for Excel**. Using these different methods, we will be creating functions to connect to the database, creating functions to read from the database, converting the data to pivot tables, populating a data validation drop-down list, creating a permanent connection to the database where you can update the data directly, and then creating a dashboard, which is very popular now in business.

Excel is great for analyzing and displaying large amounts of data in order to obtain valuable information for business and personal purposes. However, only a small percentage of people have the skills to dynamically access data to provide real-time information. Often, the data is collated and copied into Excel on a periodic basis, whether weekly or monthly, and this is often a labor-intensive task. Integrating Excel directly with the data source or sources using VBA provides real-time data analysis with the latest available data for those important business decisions. Dynamic data access is perfect for Excel dashboards, which are all the rage among managers in today's business environment.

> **Note**
>
> Exercise and activity files and solutions for this chapter can be found here:
>
> ```
> https://github.com/PacktWorkshops/The-MySQL-
> Workshop/tree/master/Chapter11
> ```

To demonstrate the concepts in this chapter, we will use a simple sample MySQL database. We will need to first import this database into our MySQL instance so that it is accessible by our Excel file. The next exercise will demonstrate how this can be done.

Exercise 11.01 – Setting up a sample MySQL database

We will be using a database called `chinook`. The database is a sample database representing an online media sales site. This database was created by Luis Rocha, who has kindly made it freely available. This version has reduced data to save time. The full version is freely available on GitHub at `https://github.com/lerocha/chinook-database`. The `chinook` database includes the following tables:

Figure 11.1 – chinook database tables

As we progress through this chapter, you will be asked to run `.sql` script files that will add several prepared views and stored procedures that we will be using.

To install the database using MySQL Workbench, follow these steps:

1. Open **Workbench** and log in to your MySQL server.

2. From the top menu, select **Server** and then **Data Import**.

3. Select **Import from Self-Contained File**.

4. Click the ellipsis (three dots) and locate the file named `Chinook.sql` in the course resources folder.

> **Note**
>
> The Chinook.sql file can be found here:
>
> https://github.com/PacktWorkshops/The-MySQL-
> Workshop/tree/master/Chapter11

5. Click **Start Import** to import the database. The database will be created and populated with data. After the Chinook.sql script has finished running, the database will be visible in the **Workbench Schema** panel.

6. Open the **Workbench Schemas** panel.

7. Click **Refresh**.

8. The chinook database will be visible. Click on chinook and then Tables. The table list should look like this:

Figure 11.2 – chinook database tables

This chapter is concentrating entirely on working with data from MySQL and Excel. You will be creating **Structured Query Language (SQL)** in VBA to use database tables, views, and stored procedures.

> **Note**
>
> You are not expected to create views and stored procedure objects in this chapter as they have already been created for you.

Exploring the ODBC connection

Be sure you have the Excel sample database installed before attempting to create an ODBC DSN connection.

> **Important Note**
>
> We need to create an ODBC connection to the new database named `chinook` and name the connection `chinook` (the same name as the database). Several exercises in this chapter will require this connection. If you need to jog your memory about how to create an ODBC connection, refer to ODBC connections in *Chapter 6, Exercise 6.11*.

Now that the database is installed and an ODBC DSN has been created, we can start. As we will be working with VBA, we will start with the **Developer** menu.

The Developer menu

This section will introduce you to the **Developer** menu and the VBA IDE and explain how an Excel VBA program works within the Excel environment. It is very similar to the **Microsoft Access** (**MS Access**) environment, but of course, there are differences you need to be aware of to work in the environment. There are also some differences in the VBA language as well. Some MS Access commands are not available in Excel, and Excel has some commands that are not available in MS Access. This section is primarily an information section to get you started. We will only be covering options regarding the **Developer** menu that you will need to complete this chapter.

Exercise 11.02 – Activating the Developer tab and the VBA IDE

To get access to the VBA IDE and other developers' tools in Excel, you need to have the **Developer** tab visible. The **Developer** tab is not activated by default when you install MS Office; however, once you activate it, it will remain activated until you deactivate it.

To activate the **Developer** tab, proceed as follows:

1. Open a new Excel document. The **Developer** menu is located in the top menu bar of Excel, as illustrated in the following screenshot:

Figure 11.3 – Activated Developer tab

2. If you cannot see it, then you need to activate it. If you can see it, you can go directly to *Step 7*.

3. Click the **File** menu, as indicated in green in the previous screenshot, and then select **Options**. This will open the **Excel Options** window, as illustrated in the following screenshot:

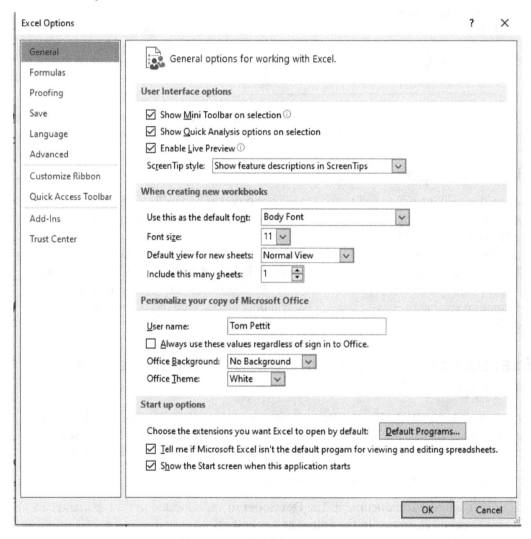

Figure 11.4 – Excel Options screen

4. Select **Customize Ribbon**. The screen will change to the following:

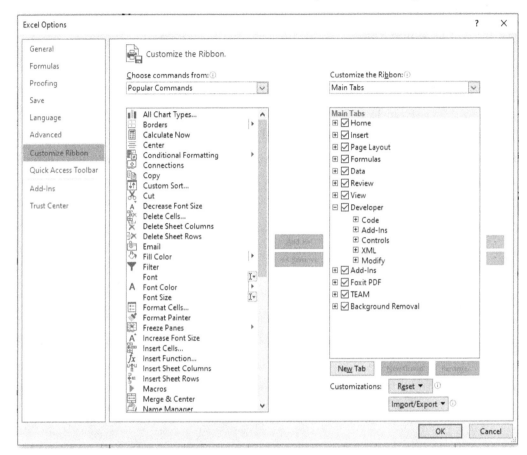

Figure 11.5 – Developer tab checked

5. Tick the **Developer** checkbox as shown in the preceding screenshot.

6. Click **OK**, and you will be returned to the Excel main screen.

7. The **Developer** menu will now be visible. Click it, and you will see the following options:

Figure 11.6 – The Developer tab is visible when activated

Congratulations—you have now activated the **Developer** menu!

8. Open the VBA IDE by clicking **Visual Basic** in the **Developer** menu, as shown here:

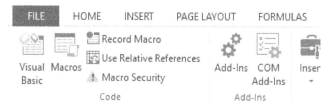

Figure 11.7 – Visual Basic button on the Developer tab

The IDE will open, as shown here:

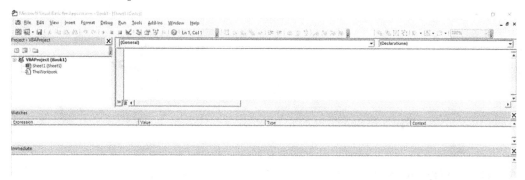

Figure 11.8 – Visual Basic IDE

The **Developer** tab offers a host of new options to assist you with your development. These include access to the **VBA IDE, macro development and recording**, and **data access**. We are mainly interested in the VBA IDE for this chapter.

The Excel VBA IDE looks and feels just like the **MS Access IDE**. The center of your screen may be gray. You need to double-click on Sheet1 to open the code window shown in the preceding screenshot. In the next section, we will explore the Excel VBA structure.

Exploring the Excel VBA structure

When you first create a workbook, there will be two entries in the **VBAProject Project** panel, and they will be `Sheet1 (Sheet1)` and `ThisWorkBook`, as explained in more detail here:

- **Worksheets**

  As you add more sheets to your workbook, they will appear in the panel. They can have **private subroutines (private subs)** that work within the worksheet only and **public subs** that can be called from other worksheets or functions. They can have `private` and `public` functions to return values. **Public routines** must be called with a fully qualified worksheet name, as illustrated here:

  ```
  myResult = worksheets("Sheet1").<function name>
  ```

  It has several events available relating to the worksheet. The name of the sheet in this example is `Sheet1`. This is the name inside the brackets and will change if you assign a new name to the tab. `Sheet1` outside the brackets is the Excel name for the sheet, and this will not change. You can refer to the sheet in your VBA code using either of these names.

- `ThisWorkBook`

  The Excel equivalent to the **AutoRun** macro in MS Access is the `Private Sub Workbook_Open()` event. Any code in this event will automatically run when the workbook is opened. You will usually use this to call functions to secure the workbook, run code or functions to set up your application and data connections, open user forms, create and test data connections, and for other tasks you may want to be done before the first worksheet opens for the user. Alternatively, the `Workbook_BeforeClose()` sub can be used to do tasks prior to the workbook being closed, such as saving data, saving the workbook, and closing data connections. There are several other events in `ThisWorkbook` that can be used as required.

In the next section, we will check how to prepare an Excel project.

Preparing your Excel project

When you start a new Excel project, you will have a good idea of some of the basic requirements you want to include in the project, such as VBA code, where you will be sourcing the data, approximately how many worksheets you need, and their names. If you plan ahead, you can set up the Excel project with your basic requirements from the start, which will help you stay focused and reduce distractions. In the next few exercises, we are going to work through what we need to do to prepare for our project.

Modules are where you place your VBA code. You create modules to group related functions together, and you should name them accordingly—for example, MySQLDatabase for MySQL-related functions such as connecting to the database, reading data, and writing data. You can then copy and paste the entire module from one project to another to avoid recoding them and to standardize your code.

Exercise 11.03 – Creating a code module

In this exercise, we will create our first code module. As we progress through this chapter, this will be where we will place the VBA functions and subs we will be developing.

To add a new code module, follow these steps:

1. Open the **VBA IDE** screen by clicking the **Developer** tab and then **Visual Basic**. The **VBA IDE** screen will open, as illustrated in the following screenshot:

Figure 11.9 – Opening the VBA IDE screen

2. Working in the **VBA IDE** screen, click **Insert** and then **Module**, as illustrated in the following screenshot. A module will be added to the list as Module#, starting at one. Functions in modules can be called from anywhere:

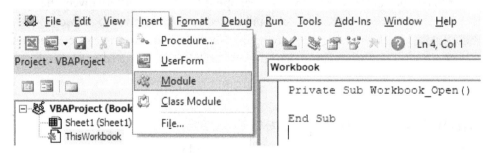

Figure 11.10 – Inserting a module

3. To rename the module, press *F4* to open the **Properties** panel, as illustrated in the following screenshot:

Figure 11.11 – The Properties and Project panels displaying Module1

4. Rename the module `MySQLDatabase`, replacing the default name with a descriptive name that will indicate which functions will be in the module, as illustrated in the following screenshot. As with Access, the module name has no relevance to the application, and the name is there so that you know which routines it contains. When you create other workbooks, you can copy the code to the new workbook and save yourself a lot of development time:

Figure 11.12 – Module1 renamed MySQLDatabase

We have learned how to create a new module. In the upcoming exercise, we will be focusing on saving the Excel file as a `.xlsm` file type.

When you create a new workbook in Excel, it will default to the `.xlsx` file type. This file type cannot run macros or VBA code. Since we will be using VBA code, we need to save the file as a `.xlsm` file by completing the following steps:

1. Select **File** from the top menu and then **Save As**, as illustrated in the following screenshot:

Figure 11.13 – File tab

You will see this menu when you click on the File option:

Figure 11.14 – Save As option

2. From the **Save As** menu, select **Computer**. Select **Browse** and locate your work folder.

3. Name the file MySQL Excel Training.

4. Select Excel Macro-Enabled Workbench (*.xlsm) in the **Save as type:** dropdown.

5. Click **Save**. You should see the following result:

Save As

Figure 11.15 – Browsing to location and options to save the file as a .xlsm file

Your file will now be saved in the selected folder and will be macro-enabled, ready for the upcoming exercises. Depending on your macro security settings and macro-enabled documents in MS Office, when you first open the file, you may be prompted to allow macros and trust the file. Be sure to answer **Yes**.

> **Note**
>
> When you download files from the internet from unknown sources, always be careful. If you do not trust the source 100%, answer **No** and check the VBA before allowing the macros to run. Unscrupulous people can insert malicious code into workbooks. If the VBA is locked and inaccessible, or if it is running or installing unknown **application programming interface** (**API**) references, be extra vigilant.

Excel offers many different file types. The default is .xlsx, which can hold formulas but not macros or VBA code. If you intend to include macros or VBA code, then you must use the .xlsm file type. Some of the more common file types are outlined here:

- A **comma-separated values** (**CSV**) file is a comma-delimited text file. It cannot contain macros or formulas. It is the most common file type for sharing data between other applications and Excel.

- **XLSB** is a binary file type. It can contain macros and VBA. Beware of running these files from unknown sources as they bypass MS Office security. You will not be prompted to allow the macros to run.

- There are many other file types (too many to discuss here), and each has specific features and uses. You may want to do some research on what they are and how they should be used.

Learning about VBA libraries

VBA in its native format in all of the MS Office suite of applications provides the most common functionality you will use in your applications, but it does not provide everything; otherwise, it would be large and unwieldy. For this reason, specific functionality is available in library files that can be shared and used by VBA. You add a reference when you require the library, which then makes its functions available to your code. These library packs are offered by Microsoft and third-party vendors, or you can create your own library files of functions you want to share with your application. Library files are created in languages such as C# and **Visual Basic 6** (**VB6**) and are compiled so that their code is not available to view or edit, thus protecting the logic built into them.

The most common library file extensions are listed here:

- **Dynamic-linked library (DLL)**
- **Object Linking and Embedding Type Library (OLE TLB)**
- **Active X controls (OCX)**

In the next exercise, we will learn how to reference a library.

Exercise 11.04 – Referencing a library

In this exercise, we will be setting our first reference to a DLL library. We will be using **ActiveX Data Objects** (**ADO**) to read the database, so we will need to reference the library before we can use its functionality.

To set a reference to the library, follow these steps:

1. You can only reference a library from a VBA code window. From the **VBA IDE** screen, open the `MySQLDatabase` module. *Note*: You can use *ANY* code module.

2. Select **Tools** and then select **References…**, as shown in the following screenshot:

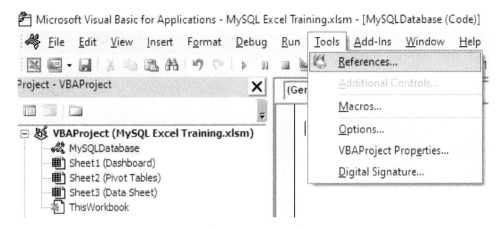

Figure 11.16 – Going to Tools and then References… to open References

3. The **References** window will open. There will be several libraries already selected; these are the default libraries used by Excel VBA. You can see an illustration of this in the following screenshot:

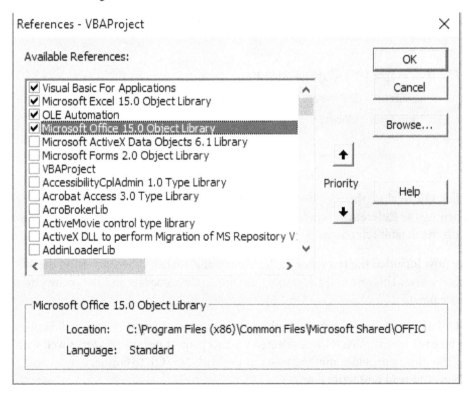

Figure 11.17 – References view for VBA

4.   Locate and select **Microsoft ActiveX Data Objects x.x Library**. The libraries are in alphabetical order. **x.x** refers to the version number. Select the latest, which will be **6.1**. The following screenshot provides an illustration of this:

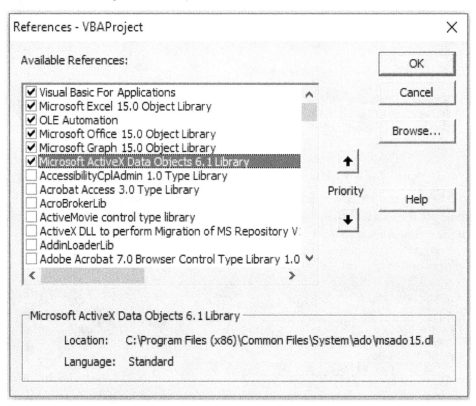

Figure 11.18 – ADO library selected

5.   Click **OK** to close the window. Quickly check that the reference was added by opening the **References** window again. You should see the new reference, along with the default references.

You have now included the reference to the library, and we can now start using the functions it offers. This one will allow us to use the ADO recordset and the connection features we need.

The ADO data type is one of several methods of communicating with external databases that can be used by Excel. ADO was selected for this project because it works well with MySQL. The library provides methods we can use with MySQL to connect, open and close tables, and read and write data.

Library references can also allow you to use other applications from within your code, and most business-related applications will provide a library to allow you to use the application from your VBA code. Here are just a few applications that provide libraries that can be used from Excel VBA; there are many more:

- **MS Outlook**—to send and receive emails and set calendar events
- **MS Word**—to create, open, and read Word documents
- **MS Visio**—to create and work with Visio files
- **Adobe Acrobat**—to open and modify **Portable Document Format** (**PDF**) files

Worksheets are the primary interface your application will have with its users. They will also store data for the application and can be hidden and locked to protect data. It is useful to know which worksheets you will require before starting development and to add these worksheets. Renaming them and placing them in the correct order will help you keep focused. Of course, you can add or remove them at any time.

In the upcoming exercise, we will learn how to insert worksheets.

Exercise 11.05 – Inserting worksheets

You can insert worksheets using the + button to the right of the tab names at the bottom of the screen. We are going to need several worksheets, so let's add and name them now.

We are going to need three initial worksheets within this project. Proceed as follows:

1. When you open the workbook, one worksheet has already been created. Click on Sheet1 to make it active.

2. You can insert new worksheets using the + button to the right of the tab names at the bottom of the screen. Click the + symbol twice to add another two worksheets to the book, as illustrated in the following screenshot:

Figure 11.19 – How to insert a new worksheet

3. *Right-click* on Sheet1 and select **Rename**, as illustrated in the following screenshot. The Sheet1 tab will go gray, and your cursor will flash:

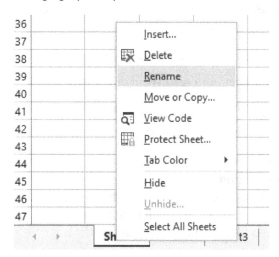

Figure 11.20 – Renaming a worksheet from the right-click menu

4. Rename Sheet1 Dashboard.

5. Repeat for Sheet2 and Sheet3. Rename them as follows:

    Sheet2 to Pivot Tables

    Sheet3 to Data Sheet

6. Select **Visual Basic** from the **Developer** menu and see the changes in the **Project** panel, as illustrated in the following screenshot:

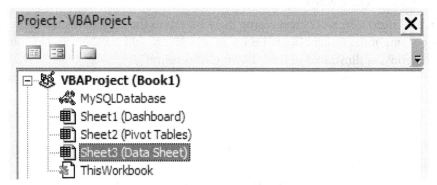

Figure 11.21 – Users' worksheet names

Some of your applications may have many worksheets in them to display charts and graphs, interact with the user, and store data. Follow these guidelines when working with worksheets:

- Be sure you rename each sheet as you create them with a meaningful name.
- Try not to keep the default `Sheet#` name provided by Excel.
- Refer to sheets in your code with the name you have provided.
- Do not rename a sheet after you have referenced it in your code or elsewhere. This will stop your application from working correctly.

In the next section, we will look at connecting to the MySQL database using VBA.

Connecting to the MySQL database using VBA

We are starting with VBA because, while there are tools to automate the retrieval of data, you won't necessarily learn the finer details of data handling as you will with VBA. Here is an analogy: if you learn how to drive a manual car, driving an automatic is easy, and you can always drive a manual again if required. However, if you only learn to drive in an automatic car, you will struggle with a manual.

All variable names in this section will be fully descriptive (therefore, long) to make it clear what their purpose is. During your own development, you may opt to shorten them.

Setting the scene

The manager of *Chinook Music Downloads* wants an Excel dashboard showing information about music sales and other information at a glance and has given you an Excel sheet with some blank charts and other information that you need to get operational. You need to extract the required information from the MySQL sales database and populate the dashboard. You have also been asked to provide a method in Excel to view and update customer and employee details directly in the database from Excel.

> **Note**
> The file you are to work with is located in the course resources and is named `MySQL Excel Training Template.xlsm`. It can be found here: `https://github.com/PacktWorkshops/The-MySQL-Workshop/tree/master/Chapter11`.

This file is already set up the same as the file you created in the previous exercise, with the addition of a dashboard layout and some preloaded data for upcoming exercises. All upcoming exercises will work with this file. You can copy the file you just created to another location if you wish; you will not need it.

We need to copy the `MySQL Excel Training Template.xlsm` file to your work folder and rename it as you wish. Do not work on the template file; you may need to reset it later.

In this section, we are going to create a global connection variable (a reusable function to connect to the database using a *DSN-less connection*), read some data, and place it on a worksheet.

When we connect to the database, we assign the connection to a variable so that we can use it. We can theoretically declare this variable in functions and subs. However, we can only use them within a function or sub. Most of the time, we will want to reuse the connection variable from many pieces of code. So, instead of declaring it each time, we will be declaring it as a *global* variable to make it available from all functions, subs, and event code. Global variables are declared in modules and must appear before any functions or subs—in other words, at the top of the module. They will be given a `Public` **identifier** (**ID**), which indicates that they should be accessible in a global context.

Our connection variable will be of type `ADODB.Connection`, which is a built-in variable type used to specify a database connection in VBA.

Exercise 11.06 – The connection variable

In this exercise, we will create a global variable to store the connection to the MySQL database so that when it's set, we can use the variable to work with the database from any code, function, or sub.

Follow these steps to complete this exercise:

1. From the **Developer** tab, select **Visual Basic**.
2. *Double-click* on `MySQLDatabase` to open the code window.
3. In the code window, enter the following text:

```
'Global Connection Variables
Public g_Conn_DSNless As ADODB.Connection
```

Line 1 is a comment, so don't forget the comma. g_ is to signify this is a *global variable*. Conn means connection, and DSNless signifies the connection will be a DSN-less connection. It is worth noting that devising a meaningful variable-naming system will help you immensely in your development, and it will also help you remember the names of your variables and what their intended use is. It will also be of great help to future developers maintaining your application, or even when you have to revisit the code at some future date. Try to stay away from single-letter variables.

4. That's it—our variable is declared. You can test if it is acceptable by compiling the code.

5. Select **Debug** from the VBA IDE's top menu and click on **Compile VBAProject**, as illustrated in the following screenshot:

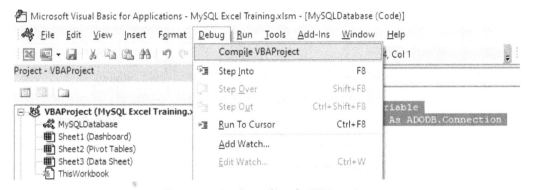

Figure 11.22 – Compiling the VBA project

The VBA language is an interpreted language, which means that as each line of code is processed when the application is run, the VBA system will check the syntax is correct, and then interpret it to machine language and execute it. This process takes a little time, but it is useful because you can run the code immediately after typing it or run it in the **Immediate** window for testing. Compiling the code regularly during development will check the syntax of the code is correct and ensure any libraries the code requires are correctly referenced. It will then store the code internally in a compiled state that is closer to the final machine language and speeds up program execution. Of course, as soon as you modify any of the code, you need to compile the application again. Compiling will not check for logical errors. It is recommended you compile the application regularly as you enter or change VBA code to identify possible problem code.

If all is good with the compile, nothing will happen; however, if there is a problem, a message will be displayed. If you get an error at this point, refer to *library references* in the previous section and check the library has been referenced (ticked), as shown in *Figure 11.18*.

That's it—too easy. We cannot test this connection at this stage, but if you have compiled without errors, then we are off to a good start. As we work through the upcoming exercises, you will see the advantages of using a global variable in action.

The advantages of using a global variable are listed here:

- We only need to declare it once.

- It can be opened and closed and then reused as required.

- We can assign different data sources to the same variable as required.

- We ensure consistency throughout the code.

- The value of the variable (in this example, the connection) can be accessed or set from any code throughout the application.

> **Note**
> We will require another global connection variable for the upcoming exercises. Following the preceding steps, please create another global connection variable named g_Conn_ODBC.

Now that we have connection variables declared and verified the syntax is correct, we need to create a function to make an actual connection to the MySQL database. We will start with the DSNless connection.

Connection functions in VBA

To create a connection with our `ADODB.Connection` variable, we will need to provide a few important pieces of information. This information primarily corresponds to the ODBC connection information. This information is specified in a String variable, in the following format:

```
'Prepare the connection string
str = "DRIVER={<ODBC Driver>};"
str = str & "SERVER=<Server IP Address>;"
str = str & "PORT=3306;"
str = str & "DATABASE=<Database name>;"
str = str & "UID=<User ID or Account Name>;"
str = str & "PWD=<Password>;"
str = str & "Option=3"
```

`Option=3` is a special portion of ADODB that configures our MySQL connection. It sets it so that column width is not optimized, and only found rows are returned by MySQL. These are simply modifications that allow VBA to interact more consistently with MySQL. This option is not required; however, it does help to create a more stable experience for our book examples. To see a full list of options, you can visit the following web page: `http://web.archive.org/web/20120120203736/http://dev.mysql.com/doc/refman/5.0/en/connector-odbc-configuration-connection-parameters.html`.

Once our string is specified, we can open it as a connection using the Open method of the `ADODB.Connection` object. If this is successful, we have a fully functional MySQL connection. The next exercise shows how to apply this code to our database.

Exercise 11.07 – Creating a connection function

We need to be able to make a connection to the MySQL server when required. In this exercise, we will create a routine that you can call to make a MySQL database connection when required.

A function can return a value to the calling code, whereas a sub cannot; for this reason, we will create a function to make a MySQL connection so that we can tell the calling code if we successfully created a connection or not, for the calling code to be able to deal with any errors appropriately. Proceed as follows:

1. Continuing in the `MySQLDatabase` module, enter the following line of code after the variable declarations:

    ```
    Public Function ConnectDB_DSNless(oConn As ADODB.
    Connection) As Boolean
    ```

    VBA will add the `End Function` statement. Be sure to add all of the following commands between the function's declaration and `End Function`. It will also add a line immediately under variable declarations. These lines will be added between functions and subs; they provide an easy way to see where the functions begin and end—especially when you scroll up and down the code.

2. Add some comments on what the function does. This is to jog your memory and for other developers who may need to modify the code at a later stage. You can see some example comments in the following screenshot:

    ```
    'This Function will create a DSNless connection and
    assign it to the
    'input variable.
    '
    ```

```
'Input: oConn, ADODB.Connection variable to assign the
connection to
'Output: Boolean, Success (True) or Failure (False)
```

3. Add an error-handler instruction and declare some variables to use, as follows:

```
On Error GoTo HandleError

    'Declare the variable we will use
    Dim Msg As String
    Dim str As String
```

4. Set the connection variable to a new variable. Up to now, it has just been a declaration; the following code will set it as an actual connection type:

```
    'Set the passed in connection variable to a new
connection
    Set oConn = New ADODB.Connection

    'Use the Client cursor so we can read the number of
records returned
    oConn.CursorLocation = adUseClient
```

5. Enter the following lines of code. Fill in your specific details for ODBC Driver, Server IP, User, and Password. Be sure to remove the < > arrowheads as well. Note that each of the parameters is separated by semicolons and that there are no spaces in the final string:

```
'Prepare the connection string
str = "DRIVER={<ODBC Driver>};"
str = str & "SERVER=<Server IP Address>;"
str = str & "PORT=3306;"
str = str & "DATABASE=<Database name>;"
str = str & "UID=<User ID or Account Name>;"
str = str & "PWD=<Password>;"
str = str & "Option=3"
```

6. Now, open the connection. If there is an error, the code will jump to the error handler, as defined earlier in the code; otherwise, it will continue with the next statement, as follows:

```
'Open the connection, if there is a problem, it will
happen here
    oConn.Open str
```

7. If the connection to MySQL was successful, then the program execution will continue with the next line of code. We then pass back `True` to indicate the connection was successful, as follows:

```
'No problem, good, pass back a True to signify
connection was successful
    ConnectDB_DSNless = True
```

8. Declare a sub to exit the function, and the error handler will then have a point to resume to exit the function, immediately followed by an actual `Exit Function` statement. At this point, the function will terminate. The code is illustrated in the following snippet:

```
LeaveFunction:
    'and leave
    Exit Function
```

9. Declare an `errorHandler` sub to handle any errors. Here, we are displaying a message to the user with the error number and description included:

```
HandleError:

    'There was a problem, tell the user and include the
error number and message
    Msg = "There was an error - " & Err & " - " &
Error(Err)
    MsgBox Msg, vbOKOnly + vbCritical, "Problem
Connecting to server"
```

10. After the user has clicked **OK**, pass back `False` to the calling routine to indicate failure to connect, then resume the code at the `LeaveFunction` sub to exit. The `End Function` statement will already be there; don't put it in twice. The code is illustrated in the following snippet:

```
'Pass back a False to signify there was an issue to
the calling code
```

```
        ConnectDB_DSNless = False

        'Leave the function
        Resume LeaveFunction

    End Function
```

To test the function, type the following in the **Immediate** window. You should receive `True` to indicate a successful connection was made, as illustrated in the following screenshot:

```
Immediate

    ? ConnectDB_DSNless(g_Conn_DSNless)
    True
```

Figure 11.23 – Testing function and result from the Immediate window

> **Note**
>
> *Ctrl* + *G* will open the **Immediate** window, or you can select it from the **View** menu. Once you open it, it will open by default until you hide it.

If you receive a message and a `False` value, check your connection values and try again. The message should indicate what the problem was.

11. Close the connection by typing `g_Conn_DSNless.Close` into the **Immediate** window, and press *Enter*. There will be no response; however, to test that it was in fact opened and then closed, try to close it again. You will get the following error message:

Figure 11.24 – Error when trying to close an already closed connection

This also indicates that the global connection is available to all functions and subs as well as the **Immediate** window; a private function cannot be accessed from the **Immediate** window.

Creating a reusable function to make a connection will reduce the amount of coding throughout the application. By allowing the connection variable to be passed in, we can use the same function for many connections, if required. We leave it up to the calling routines to close them when they are no longer required.

To make the function even more flexible, we could have included connection parameters as input to the function, passed them in, and built the connection string using the parameters.

While you can leave a connection open all the time, good practice dictates connections should only be opened when required and closed when they are no longer required. This will reduce the load on the server.

A DSN-less connection has its pros and cons, as outlined here:

Pros

- Does not require an ODBC connection to be set up on the user's machine. This is useful if a lot of people will be using the workbook.
- Slightly faster when there are many concurrent connections.

Cons

- Connection details are stored in the workbook—either in a worksheet or directly in the VBA.

The correct ODBC drivers will still need to be installed on the user's machine regardless of the type of connection you choose to use.

> **Note**
> The VBA file for this exercise can be located here:
> ```
> https://github.com/PacktWorkshops/The-MySQL-
> Workshop/tree/master/Chapter11/Exercise11.08
> ```

Let's move on and read some data using our new connection and VBA.

Reading data from MySQL using VBA

To read data, we will need a database query, as well as to store the results. To create a database query, we simply write it in as a string variable in VBA. To execute the query, we use the open method of a special object called `Recordset`, specifying the query, as well as the connection we wish to execute it against. This `Recordset` object can store the results of a query and make each field accessible by name. For example, suppose we run the following query:

```
Dim SQL as String
Dim RS as Recordset

SQL = "SELECT username, password FROM Login"
Set RS = New ADODB.Recordset

RS.Open SQL, g_Conn_DSNless
```

If our query is successful, the RS variable will contain all of the username and password fields from the `Login` table. To access these fields, we use the `RS.Fields` method. The next exercise shows a full example of a query that retrieves data through a `Recordset` object.

Exercise 11.08 – ReadGenreSales

In this exercise, we are going to read data from two database tables and place it in the worksheet named `Data Sheet`. There are a lot of comments included in the code to explain each step. Be sure to enter comments as well, as comments will assist you and other developers later. Follow these steps to complete the exercise:

1. Continue with the `MySQLDatabase` function.

2. Declare a `ReadGenreSales` function and enter comments about what the function is doing, as follows:

```
Public Function ReadGenreSales () as Boolean
'This function will read Genre Sales data from the MySQL
database
'It will place the data in the worksheet named 'Data
Sheet'
'It will cycle through the Field headings and use them
for column headings in Row 1
'It will then place the data starting at Row 2
```

3. Declare variables and set up error handling, like so:

```
'Declare the variables to use
Dim SQL As String          'To store the SQL statement
Dim RS As Recordset        'The Recordset variable
Dim Msg As String          'To display messages
Dim Counter As Integer     'A counter
Dim MyNamedRng As Range    ' A range variable


'Setup error handling
On Error GoTo HandleError
```

4. Build a SQL statement to read the data, as follows:

```
        'Build the SQL statement to read from the two
databases
    SQL = ""
    SQL = SQL & "SELECT "
    SQL = SQL & "genre.Name, "
    SQL = SQL & "Sum(invoiceline.Quantity) AS 'Units
Sold' "
    SQL = SQL & "FROM "
    SQL = SQL & "genre "
    SQL = SQL & "INNER JOIN track ON track.GenreId =
genre.GenreId "
    SQL = SQL & "LEFT JOIN invoiceline ON invoiceline.
TrackId = track.TrackId "
    SQL = SQL & "Group BY "
    SQL = SQL & "genre.Name "
    SQL = SQL & "Order BY "
    SQL = SQL & "genre.Name"
```

5. Connect to the server and test if it is working, as follows:

```
    'Make the connection to the server, test if it was
successful
    If ConnectDB_DSNless(g_Conn_DSNless) = True Then
        'Connection succeeded so we can continue
processing
```

6. Once the connection works, set up the `recordset` variable, like so:

```
'Set the recordset variable
Set RS = New ADODB.Recordset
```

7. Load the `recordset` variable using the SQL and the connection, like so:

```
'Load the recordset, pass in the SQL and the
connection to use
RS.Open SQL, g_Conn_DSNless
```

8. Test if there are records to work with, as follows:

```
'Test there are records.
'A recordset can only be at End Of File and
Beginning Of File at the same time when the recordset is
empty
If RS.EOF And RS.BOF Then
```

9. If execution gets in here, then there are no records. Tell the user, close the `recordset`, and leave the function, as follows:

```
'No data, close the recordset
RS.Close
Set RS = Nothing


'tell user and then leave the function
Msg = "There is no data"
MsgBox Msg, vbOKOnly + vbInformation, "No
data to display"
GoTo leavefunction
```

10. Once we get to this part, we have got the data. Now, process it, as follows:

```
Else
'We have data
```

11. Add headings for data on the worksheet. Use the field names for this purpose, as illustrated in the following code snippet:

```
'Insert Field headings for column headings
'We cycle through the field collection
For Counter = 0 To RS.Fields.Count - 1
```

```
                    'Put the fieldname in the cell on row 1
                    'When cycling through objects or data,
it is easier to refer to the worksheet cells by their
numeric values
                    Cells(1, 1 + Counter) =
RS.Fields(Counter).Name
            Next
```

12. We are done adding the headings. Now, we need to add the data in Excel. Use the `CopyFromRecordset` command to copy the entire dataset with one command, as follows:

```
                'Starting at cell in Row 2, Column 1, copy
the entire recordset into the worksheet
                Worksheets("Data Sheet").Cells(2,
1).CopyFromRecordset RS
```

13. Set a named range for the data. We use the `RS.RecordCount` value to calculate how many rows the range should cover, as illustrated in the following code snippet:

```
                'Set and create a named range covering the
column with the Genre name, data only
                Set MyNamedRng = Worksheets("Data Sheet").
Range("A2:A" & RS.RecordCount + 1)
                ActiveWorkbook.Names.Add Name:="Genre",
RefersTo:=MyNamedRng
```

14. Now, we have our data in place. Start closing everything down, like this:

```
                'Close the recordset
                RS.Close
                Set RS = Nothing

                'Close the connection
                g_Conn_DSNless.Close
                Set g_Conn_DSNless = Nothing
```

15. Pass back `success`, as follows:

```
                'Pass back success
                ReadGenreSales = True
            End If
```

16. If we get in here after the connection test, then the connection failed. So, leave the function, as illustrated in the following code snippet:

```
Else
        'Connection failed if gets in here, just drop
through to leave
        'The connection routine will have displayed a
message so nothing to do but leave
        ReadGenreSales = False
        GoTo leavefunction
    End If
```

17. Leave the function, like so:

```
LeaveFunction:
    'Leave the function
    Exit Function
```

18. Add error-handling code to deal with any errors that may occur, as follows:

```
HandleError:
    'In this sample we are just going to display the
error and leave the function
    'you may want to log the error or do something else
    'depending on your requirements
    MsgBox Err & "-" & Error(Err), vbOKOnly + vbCritical,
"There was an error"
```

19. Pass back a fail and leave the function, as follows:

```
    'Pass back Failed
    ReadGenreSales = False
    Resume LeaveFunction
```

20. An `End Function` statement will have already been created when you declared the function. Ensure that it is present as the very last statement of the function code block, as shown here:

```
End Function
```

21. Test the function by typing ? `ReadGenreSales` into the **Immediate** panel. You should get `True` returned to indicate it was successful, as illustrated in the following screenshot:

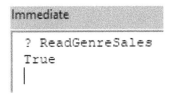

Figure 11.25 – Testing the function and seeing the result in the Immediate window

And the data sheet should be displaying the data, as we can see here:

	A	B
1	**Name**	**Units Sold**
2	Alternative	14
3	Alternative & Punk	244
4	Blues	61
5	Bossa Nova	15
6	Classical	41
7	Comedy	9
8	Drama	29
9	Easy Listening	10
10	Electronica/Dance	12
11	Heavy Metal	12
12	Hip Hop/Rap	17
13	Jazz	80
14	Latin	386
15	Metal	264
16	Opera	
17	Pop	28
18	R&B/Soul	41
19	Reggae	30
20	Rock	835
21	Rock And Roll	6
22	Sci Fi & Fantasy	20
23	Science Fiction	6
24	Soundtrack	20
25	TV Shows	47
26	World	13

Figure 11.26 – Output from the ReadGenreSales function test

22. Select the new named range, `Genre`, from the dropdown, as indicated in the following screenshot. The **Name** genre (column A) should be selected:

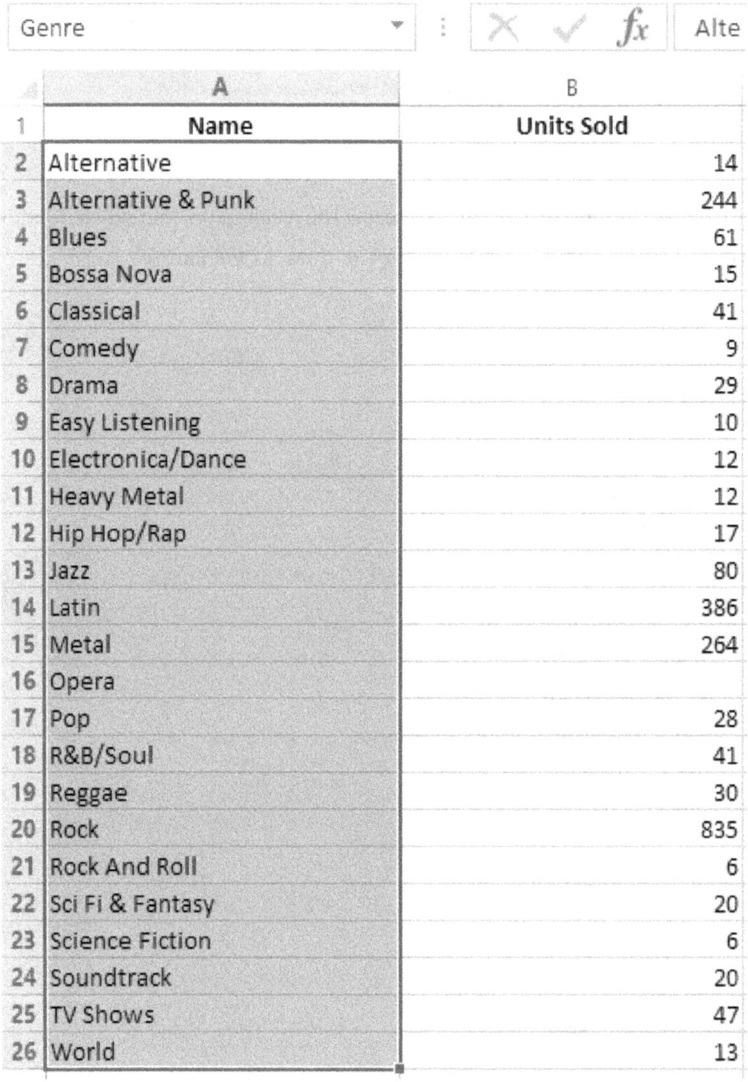

	A	B
	Genre ▾ ⋮ ✕ ✓ *fx*	Alte
1	**Name**	**Units Sold**
2	Alternative	14
3	Alternative & Punk	244
4	Blues	61
5	Bossa Nova	15
6	Classical	41
7	Comedy	9
8	Drama	29
9	Easy Listening	10
10	Electronica/Dance	12
11	Heavy Metal	12
12	Hip Hop/Rap	17
13	Jazz	80
14	Latin	386
15	Metal	264
16	Opera	
17	Pop	28
18	R&B/Soul	41
19	Reggae	30
20	Rock	835
21	Rock And Roll	6
22	Sci Fi & Fantasy	20
23	Science Fiction	6
24	Soundtrack	20
25	TV Shows	47
26	World	13

Figure 11.27 – Shaded area indicating the cells the named range refers to

The majority of the code you will write will be to set up reading data, handling errors, and—of course—commenting. The actual retrieval and displaying of data in a worksheet is very simple. How you use the data after it is displayed is up to your specific requirements for the task at hand.

> **Note**
>
> The VBA file for this exercise can be located here:
>
> `https://github.com/PacktWorkshops/The-MySQL-Workshop/tree/master/Chapter11/Exercise11.09`

When a list of data is required, it needs to be stored in a worksheet, which is usually hidden from the user. When using VBA, we can calculate the coordinates of data and assign a name to the coordinates. Using a named range simplifies assigning data to graphs, charts, and other controls by allowing us to use the name rather than the actual cell coordinates. In the next section, we will do an exercise based on the `Genre` dropdown.

Exercise 11.09 – Genre dropdown

The first chart on the dashboard is the **Genre Sales** chart. We need to provide a drop-down list of genres for the user to make a selection so that we can populate the chart. In *Exercise 11.08 – ReadGenreSales*, we read a list from the MySQL database, placed the data on the data sheet, and named it `Genre`. We are now going to assign the genre data to a drop-down list.

In this exercise, we will assign a `Genre`-named range to the dashboard cell `B5`. We will populate a list using the **Data Validation** method, which will create a dropdown for us. By using the **Data Validation** method for the dropdown, we stop the user from entering invalid data and provide a list of appropriate values for the user to select from. By setting this up, we can then confidently use the selection in VBA code to display the correct data and display it in charts, which we will be doing later.

To create a **Data Validation** drop-down list, follow these steps:

1. Start by testing that the Genre-named range is pointing to the correct data by selecting the named range. The worksheet should change to Data Sheet with the cells A2 to A26 selected, as illustrated in the following screenshot:

	A	B
	Genre ▾	⋮ ✕ ✓ ƒx Alte
1	**Name**	**Units Sold**
2	Alternative	14
3	Alternative & Punk	244
4	Blues	61
5	Bossa Nova	15
6	Classical	41
7	Comedy	9
8	Drama	29
9	Easy Listening	10
10	Electronica/Dance	12
11	Heavy Metal	12
12	Hip Hop/Rap	17
13	Jazz	80
14	Latin	386
15	Metal	264
16	Opera	
17	Pop	28
18	R&B/Soul	41
19	Reggae	30
20	Rock	835
21	Rock And Roll	6
22	Sci Fi & Fantasy	20
23	Science Fiction	6
24	Soundtrack	20
25	TV Shows	47
26	World	13

Figure 11.28 – Genre list

2. Return to the Dashboard worksheet and click on cell B5.

3. From the top menu, click **DATA**, select the **Data Validation** option, and select **Data Validation…** from the small menu, as illustrated in the following screenshot:

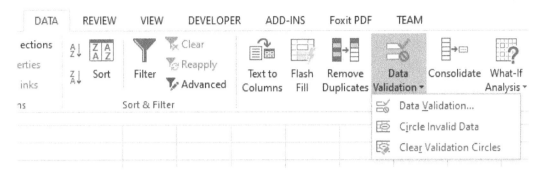

Figure 11.29 – Location of the Data Validation button

4. The **Data Validation** form will open. Select `List` in the **Allow:** dropdown and type `=Genre` for **Source:**, as illustrated in the following screenshot, then click **OK**:

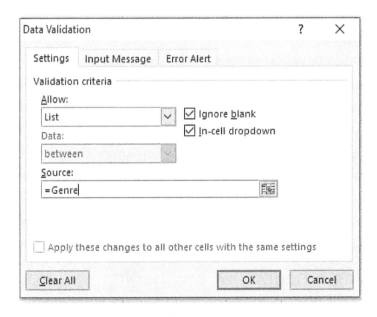

Figure 11.30 – Data Validation properties window

5. The **Data Validation** form will close. Return to the `Dashboard` worksheet. Cell `B5` should now have a small down arrowhead. When this is selected, the list should open, as illustrated in the following screenshot:

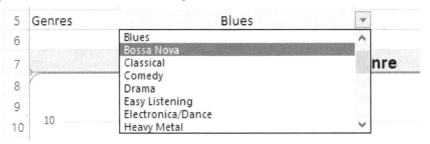

Figure 11.31 – Genre drop-down list

The `Genre` list is now ready. When we loaded the `Genre` list data onto the data sheet, we knew which row and column the data started at by determining the number of records returned from SQL when we loaded it into the `recordset` variable. We could then calculate in VBA exactly which rows the `Genre` list covered and assign this range to a named range we called `Genre`. By calculating the coordinates and setting the named range in VBA, if `Genre` categories are added or removed later, the named range will be set accordingly, and the **Data Validation** list will always show the correct values.

Placing your data on other sheets is a good practice; you can hide sheets and even lock them away from prying eyes. When placing data on data sheets, keep in mind that the data may grow either vertically or horizontally, so ensure you leave space between groups of data to make room for potential expansion. You can also use as many sheets as you like, but be sure to name them appropriately. Don't rely on the `Sheet1`, `Sheet2`, `Sheet3`... names assigned by Excel; your data may be difficult to find.

In addition to populating data, we may want to keep data updated from the source. One way to achieve this is by using auto-running functions. In the next section, we will learn how these work in VBA.

Auto-running functions when opening a workbook

In many cases, we want to load data immediately when a workbook is opened. This allows us to refresh data when the workbook is viewed, allowing us to ensure data is always up to date. To auto-run functions in VBA, we can use a special function called `ThisWorkbook_Open()`, which executes code when a workbook has been opened.

In the next exercise, we will prepare the `ReadArtistSales` function to run when a workbook is opened.

Exercise 11.10 – Auto-running functions when opening a workbook

During the last few exercises and activities, we created two new functions to load and store data, we created named ranges to identify data (or parts of it), and then we used the named ranges to populate data validation drop-down lists. We will now prepare calls to the functions so that they will run when a workbook is opened.

In this exercise, we will use the `ThisWorkbook` object to achieve this. `ThisWorkbook` is located in the VBA/IDE window and provides several useful events related to the workbook. We will use the `Open` event. Any VBA code in the `ThisWorkbook_Open()` event will be executed when a workbook is opened, before any other code is run, effectively providing an auto-run feature.

To auto-run functions, proceed as follows:

1. Open the VBA IDE, as illustrated in the following screenshot:

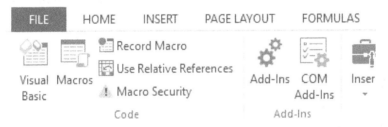

Figure 11.32 – Opening the VBA IDE screen

2. Locate the `ThisWorkbook` module and *double-click* on it to open the VBA window, as illustrated in the following screenshot:

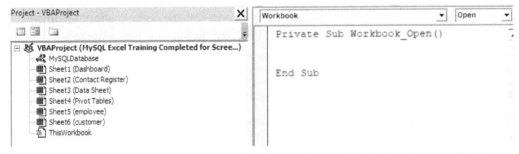

Figure 11.33 – The ThisWorkbook module showing the Workbook_Open() event sub

3. Select `Workbook` and `Open` from two dropdowns. You will then see the `Private Sub Workbook_Open()` sub. It will be empty at this stage.

4. Enter the following code:

```
Private Sub Workbook_Open()
    'Load the intial data
    ReadGenreSales
    ReadArtistSales
End Sub
```

5. Click the **Save** button.

6. Close and restart the workbook to test that both the Genre and Artist dropdowns will be correctly populated with their lists.

The Private Sub Workbook_Open() event is Excel's auto-run method. By placing calls to the ReadGenreSales and ReadArtistSales functions in this event, we ensure they will be executed when an Excel file is opened. You can enter any amount of code in this event; however, it is usually better to keep it simple and limit its code to call functions designed to perform any complex operations. In the next section, we will populate our first chart.

Populating charts

Dashboards are very popular, and managers love them, but why? Because they provide important information at a glance about the current status of the business without the need to sift through large sets of numbers to retrieve statistics, and help managers make important business decisions. Charts play an important role in business by displaying comparisons or changes in data over time in a simple-to-understand graphical way. Managers use charts in meetings with upper management, stakeholders, and external clients when presenting information. These people know little or nothing about the finer details of the business and will not relate to numbers alone, but they will relate to a well-presented chart.

Populating a chart – Genre sales

When you were assigned to create a dashboard, your manager said that the information was to be displayed graphically so that they could use the charts in their meetings with the **board of directors (BOD)**.

The **Genre Sales** chart will display all sales for the selected genre from the dropdown in `Dashboard` cell B5. To do this, we need to be able to determine when a value has changed in cell B5. We cannot directly determine when a cell has changed, but we can detect when a cell has changed in a worksheet because any change will fire a `Worksheet_Change(byVal Target as Range)` event. We can read the `Target` parameter, which will tell us which cell caused the event to fire—that is, which cell was changed.

We need to do this task in two parts: first, we need a function to load the required data from MySQL, and then we need a function to determine if the changed cell is in fact B5 and call the load routine. The next two exercises will deal with each part individually.

Exercise 11.11 – Loading Genre Sales chart data

In this exercise, we will prepare a sub that will accept a single string parameter (the selected genre), load the filtered data from MySQL, and create a named range for the data. This sub will be called when the user selects a genre and will be the subject of the next exercise. Proceed as follows:

1. Open the VBA IDE, as illustrated in the following screenshot:

Figure 11.34 – Opening the VBA IDE screen

2. In the **Project** panel, locate and *double-click* on the `Dashboard` worksheet to open the worksheet's VBA window, as illustrated in the following screenshot:

Figure 11.35 – The Dashboard VBA IDE

3. Declare a private sub. We are going to pass in the selected genre name as a string, so declare a parameter as well. This is a sub, so it will not be returning any values. The code is illustrated in the following snippet:

```
Private Sub GenreSales(ByVal pGenre As String)
```

4. Declare the variables we will be using, as follows:

```
Dim RS As Recordset
Dim SQL As String
Dim MyNamedRng As Range
```

5. Start by clearing the existing range data from the location in which we are going to insert the data. The first time this sub is run, the range does not exist and will cause an error, so before attempting to clear the range, we ignore errors. After the range is cleared, we start checking for errors. The code is illustrated in the following snippet:

```
On Error Resume Next
Worksheets("Data Sheet").Range("GenreSales").
ClearContents
On Error GoTo HandleError
```

6. Connect to the database. If the connection was successful, start processing, as follows:

```
If ConnectDB_DSNless(g_Conn_DSNless) = True
Then
```

7. Prepare the SQL statement. We have a view in the database compiling the data, so we only need to filter to the genre passed in and select the fields we want to display. The code is illustrated in the following snippet:

```
SQL = ""
SQL = SQL & "SELECT SaleMonth, 'Units
Sold' "
SQL = SQL & "FROM vw_genresales "
SQL = SQL & "WHERE Name = '" & pGenre &
"' "
SQL = SQL & "ORDER BY SaleMonth ASC"
```

8. Set the `recordset` variable and open the recordset with the connection, as follows:

```
Set RS = New ADODB.Recordset
RS.Open SQL, g_Conn_DSNless

'Test there are records.
If RS.EOF And RS.BOF Then
    'No data
    GoTo Leavesub
Else
```

9. Load the data into the `Data Sheet` worksheet, starting at row 2, column 5, as follows:

```
Worksheets("Data Sheet").Cells(2,
5).CopyFromRecordset RS
```

10. Define and set a named range and add it to the `Names` collection, as follows:

```
'Set and create a named range covering new data
Set MyNamedRng = Worksheets("Data
Sheet").Range("E2:F" & RS.RecordCount + 1)
ActiveWorkbook.Names.Add
Name:="GenreSales", RefersTo:=MyNamedRng
```

11. The rest of the code finalizes the routine, including error handling. You can view it in the following snippet:

```
    End If
  Else
  End If

Leavesub:
    'Close recordset
    RS.Close
    Set RS = Nothing
    Exit Sub

HandleError:
```

```
MsgBox Err & " " & Error(Err)
Resume Leavesub
End Sub
```

12. Click **Save**.

And we are done. This is a sub, and we cannot call it from the **Immediate** panel to test it, so we will move on to the next exercise and test it when it is called from the next routine.

Running code on changes to a document

As with running at **Open**, we can also run code when a worksheet changes. This is useful for situations where you want to recalculate or reconstruct objects in your worksheet. For example, you can have a cell that allows a user to input different scenarios. When the value changes, you can generate new charts based on the scenario selected.

To run code on change, you can use the `Worksheet_Change` function. This function runs if anything changes in a worksheet. The following exercise demonstrates an example of using this function.

Exercise 11.12 – Detecting and working with worksheet changes

To detect when a user has changed a drop-down value by selecting a genre from the list, we need to use a `Worksheet_Change` event on the `Dashboard` worksheet. The code will need to be able to detect which cell was changed and then direct the program flow accordingly. Proceed as follows:

1. Continuing in the `Dashboard` worksheet's VBA window, select `Worksheet` and then select the `Change` event from the top dropdowns. You will be presented with a `Worksheet_Change` code construct, as shown in the following screenshot. Enter the following code inside this construct:

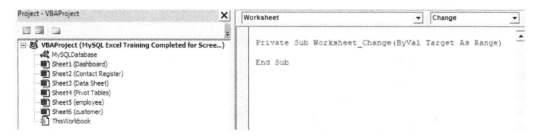

Figure 11.36 – The Dashboard worksheet's Worksheet_Change event sub

2. To test what cell was changed, we test the `Target` parameter. This is a range object and, among other things, contains the address and the value of `Target`. We are interested in both. Test the address of `Target` using a `Select Case` statement to test if the changed cell is `B5`, as follows:

```
'Test the active cell (the one that changed)
    Select Case Target.Address

        Case "$B$5"
```

3. If `Target` is referring to `B5`, then we need to process it. Start by calling the data load routine we created in the previous exercise. `Target` will contain the selected genre's text. This is very convenient, as we only need to pass in `Target`. The code is illustrated in the following snippet:

```
            'The change was in the dropdown, target has
    the value
            Call GenreSales(Target)
```

4. Activate the `Dashboard` worksheet so that we can use the `With/End` construct, as follows:

```
            'Set the chart details Population
            Worksheets("Dashboard").
    ChartObjects("chrtPopulation").Activate
```

5. Set the parameters for the chart, including `datasource`, `title`, and `series name`, like so:

```
            With ActiveChart
                .SetSourceData Source:=Sheets("Data
    Sheet").Range("GenreSales"), PlotBy:=xlColumns
                .HasTitle = True
                .ChartTitle.Text = "Genre Sales - " &
    Target
                .SeriesCollection(1).Name = "Sales"
            End With
```

6. We are finished with B5. Include `Else` to ignore cell changes we are not interested in, as illustrated in the following code snippet. If you want to include other cells, just add a case test and code (remember that for the upcoming activity):

```
        Case Else
            'Nothing to work with so leave
            GoTo Leavesub
    End Select


Leavesub:
        Exit Sub
```

Test the code by selecting a genre from the `Dashboard's Genre` dropdown. The data will be in the `Data Sheet` tab, and the chart should change with your selections, as illustrated in the following screenshot:

E	F
2015 01 01	1.00
2015 04 01	1.00
2015 09 01	2.00
2016 04 01	2.00
2016 07 01	2.00
2016 12 01	3.00
2017 07 01	1.00
2017 10 01	1.00
2018 03 01	2.00
2018 10 01	2.00
2019 01 01	1.00
2019 07 01	2.00

Figure 11.37 – Data output in the Data Sheet tab of the selected genre

The data values indicate the year/month and the total sales in the month for the selected genre. The view in MySQL compiling the data calculates the sales and also modifies the output date to show the first of the month because the chart as defined in this exercise expects a valid date format in order to it sort correctly.

The following two charts are displaying data for two separate selections graphically. Here's the first one, showing data for the **Rock** genre:

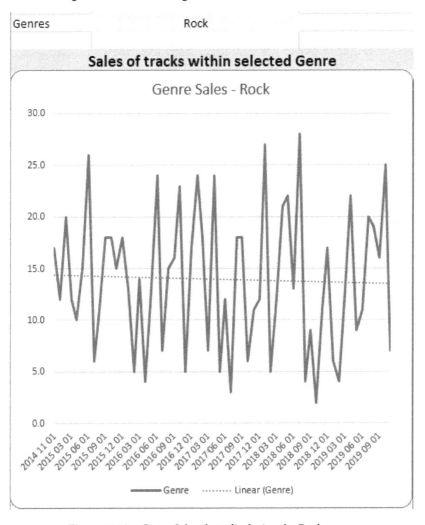

Figure 11.38 – Genre Sales chart displaying the Rock genre

The following chart shows data for the **Alternative & Punk** genre:

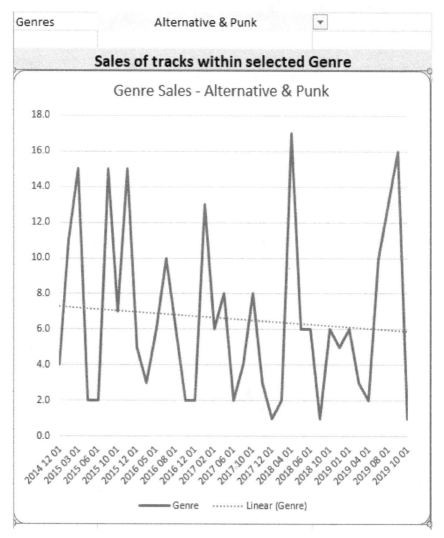

Figure 11.39 – Genre Sales chart displaying the Alternative & Punk genre

Reading MySQL data and populating charts requires only a few basic steps: query the database, place the data somewhere, assign the data to a named range, and assign the named range to a chart. Most of the coding will relate to error checking and formatting the chart. The order in which you read and place columns of data on a worksheet is important. If the data is not in the order the chart expects, it will not be displayed correctly.

Activity 11.01 – Creating a chart (artist track sales)

The manager has changed their requirements for artists' tracklists. They now want a bar chart with artists' tracks and their sales. Fortunately, we can still use the dropdown to select an artist, and there is a view in the `chinook` database that will provide data, named `vw_artist_track_sales`.

Your task is to do the following:

1. Create a new function and name it `ArtistTrackSales`.

2. The function is to read the data from the MySQL view named `vw_artist_track_sales`, filter the data to the selected artist in the dropdown in `P5`, and place the data in the workbook named `Data Sheet` in columns `L` and `M`.

3. You need to then modify the existing `worksheet_change` event to call your new function to load the data.

4. Then, create a bar chart and place it beneath the dropdown, display the data generated from the function, name the chart `chrtArtistTrackSales`, and format the area around the chart on the dashboard to fit in with the rest of the sheet.

> **Hint**
> This will be very similar to the genre dropdown/chart process.

After following the steps, the data in `Data Sheet` in columns `L` and `M` should look like this:

L	M
Save The Children	2
Abraham, Martin And John	2
Seek And You Shall Find	1
Heavy Love Affair	1
You Sure Love To Ball	1
Praise	1
You've Been A Long Time Coming	1
When I Had Your Love	1

Figure 11.40 – ArtistTrackSales output in the Data Sheet tab

You should have a new named range, as illustrated in the following screenshot:

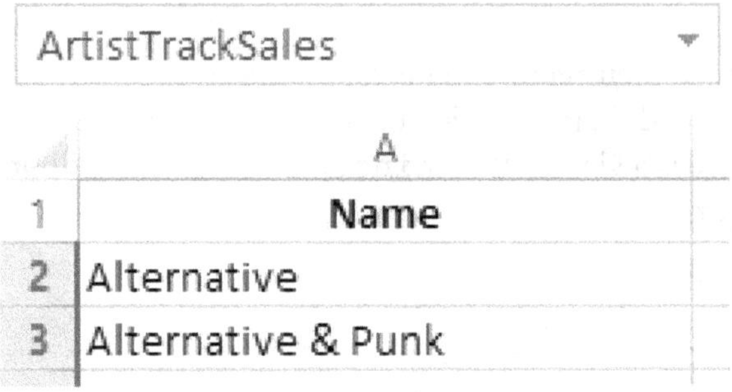

Figure 11.41 – ArtistTrackSales named range in range list

This is what happens when you select the named range:

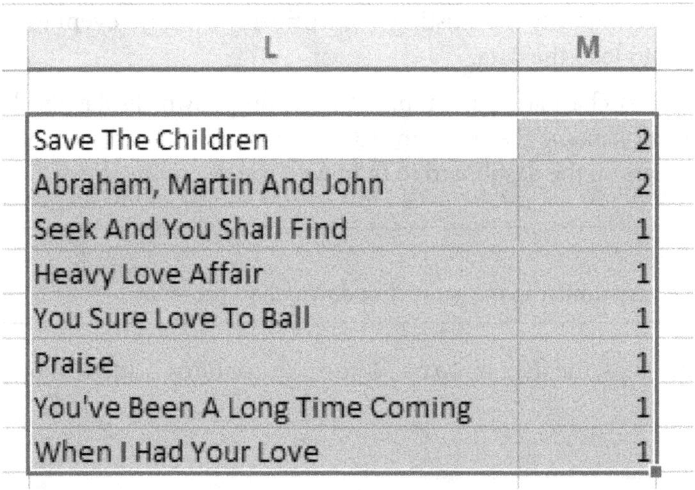

Figure 11.42 – ArtistTrackSales highlighted when the named range is selected

And finally, a new chart and formatting appear on the dashboard, as follows:

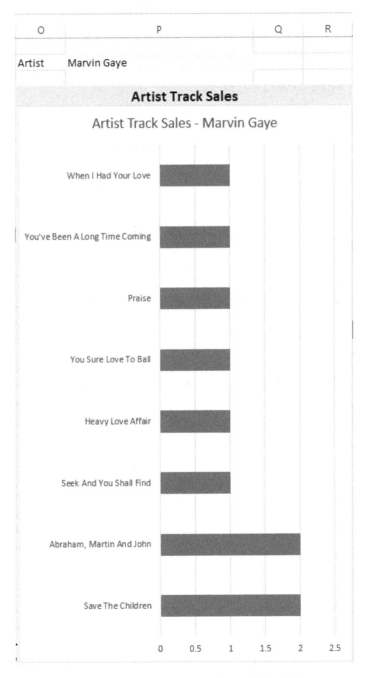

Figure 11.43 – New chart to display ArtistTrackSales

Once you have developed data load functions and charts, implementing new requirements is often simply duplicating what you have already developed and modifying it to suit the new requirements. The first charts and data reads are often the longest to create, with subsequent charts and functions being quicker to implement.

Up to now, we have been using a DSN-less connection to the database. We will now move on to using a DSN we have called `chinook`.

> **Note**
> The solution to this activity can be found in the Appendix.

Summary

In this chapter, you learned how to create a **named DSN** to the `chinook` database, add and use the **DEVELOPER** tab, and start the VBA IDE. We then created reusable functions to connect to MySQL using two different types of ODBC connections. We also learned how to read and import data from MySQL into a worksheet, define named ranges for the imported data, and assign the named ranges to charts using VBA. We set chart labels and categories using VBA and created and used drop-down lists to load filtered data, display data, and use it in charts. We then set data collections from MySQL that will run when a workbook is opened.

In *Chapter 12*, we will continue working with MySQL and Excel.

12
Working With Microsoft Excel VBA – Part 2

To set up and properly demonstrate using Excel with MySQL is quite involved, so the topic is split over two chapters. In this chapter, we will begin by setting up a sample MySQL database using a `.sql` script file and learn how to activate the **Developers** tab and the VBA IDE so that you can develop in VBA. We will then connect to a MySQL server and retrieve data using Excel VBA, create a dashboard with the data from MySQL, and populate drop-down lists and individual cells with VBA and MySQL data. By the end of the chapter, you will be able to create pivot tables and charts from MySQL data, and you'll know how to use MySQL for Excel to load, modify, and update MySQL records directly in the database. You'll finish off by learning how to push worksheets from Excel into a new MySQL table.

This chapter covers the following concepts:

- An introduction to MySQL connections
- Connecting to the MySQL database using ODBC
- Exploring generic data read functions
- Creating connections to MySQL in Excel

- Inserting data using MySQL for Excel

- Updating data using MySQL for Excel

- Pushing data from Excel

- Pivot tables

- Activity 12.01 – building a MySQL-based Excel document

An introduction to MySQL connections

You will now continue working with Excel and MySQL by generating several VBA functions to read from MySQL using DSN and DSN-less connections to display the results. You will generate graphs and charts to analyze the data and learn how to autorun a macro so that the spreadsheet will automatically read the latest data and update the information, chart, and graphs. Sometimes, you just want to read specific data from a database but not create MySQL functions, procedures, or views to do it, so you will also be creating a generic data reader. This is a VBA function that you can pass in a SQL statement, which will be executed on the server and the results passed back; you will also learn how to use the returned data. These tasks will move you on to an advanced level of Excel programming.

> **Note**
>
> All exercise and activity solution files for this chapter can be located here:
> `https://github.com/PacktWorkshops/The-MySQL-Workshop/tree/master/Chapter12`.

Connecting to the MySQL database using ODBC

DSN-less connections are great for portability; however, as mentioned earlier, the connection's login details are in the Excel workbook. This can be a security risk if your data is sensitive. Another issue is that they will work for anyone who happens to get hold of the spreadsheet. As long as the relevant driver is on their computer, the driver details are also in the connection routine, so someone with a little ODBC and VBA knowledge will figure that out quickly.

In the next exercise, we will create a new function to connect to the database using a DSN.

Exercise 12.01 – creating a DSN connection function

A DSN offers more security; you need to set up the connection on the user's computer before they can use the spreadsheet. When using a DSN, the login details are not visible in the workbook. Also, as a bonus for you, the developer, it requires less coding to use in your application.

In this exercise, we will create a DSN connection based on a named ODBC connection:

1. Double-click the **MySQLDatabase** module in the **VBAProject** panel to open the VBA code window:

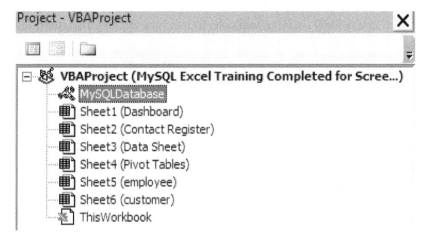

Figure 12.1 – The VBA code window

2. Move to the end of the code window after the code that you created in previous exercises and activities.

3. Start a new function with two parameters (a connection variable and a string to pass in the name of the named ODBC). The function will return a Boolean value:

```
Public Function ConnectDB_ODBC(oConn As ADODB.Connection,
ODBCName As String) As Boolean
```

4. Set up error handling and declare the variables that we will use:

```
On Error GoTo HandleError


    Dim Msg As String
    Dim str As String
```

5. Set up the connection variable. With the named ODBC, it is important to tell the connection which cursor to use. We need to use the client cursor; otherwise, we will not get the data back as we expect to:

```
'Set the passed in connection variable to a new
connection
    Set oConn = New ADODB.Connection
    oConn.CursorLocation = adUseClient
```

6. Now, prepare the connection string. This is simple – we use the string we passed in that represents the name of the named DSN. An advantage of this method is that we can pass in any DSN we have available, so it is very flexible. This method is very simple compared to a DSN-less connection:

```
'Prepare the connection string
    str = "DSN=" & ODBCName & ";"
```

7. Now, open the connection, passing in the connection string that we just built:

```
'Open the connection, if there is a problem, it will
happen here
    oConn.Open str
```

8. Pass back True to indicate success and leave the function:

```
    ConnectDB_ODBC = True


LeaveFunction:
    'and leave
    Exit Function
```

9. Include error handling as follows:

```
HandleError:
    'There was a problem, tell the user and include
the error number and message
    Msg = "There was an error - " & Err & " - " &
Error(Err)
    MsgBox Msg, vbOKOnly + vbCritical, "Problem
Connecting to server"
```

```
        'Pass back a False to signify there was an issue to
    the calling code
        ConnectDB_ODBC = False


        'Leave the function
        Resume LeaveFunction
```

10. Finally, close off the function block:

```
    End Function
```

And we are done. We cannot test this until we create a routine to use it; we will do that in the next exercise. Using the DSN is much simpler to set up from a programming perspective; there are no sensitive connection details in the code, just the DSN. We do not need to use login names or passwords because they are set up and stored in the DSN.

> **Note**
>
> The VBA code for this exercise can be found here: `https://github.com/PacktWorkshops/The-MySQL-Workshop/tree/master/Chapter12/Exercise12.01`.

To use the `ConnectDB_ODBC` function, we will create a new function to read data to populate the `Genre Tracks Sales vs Tracks with No Sales` chart, named `TracksStats`.

The *Genre Tracks Sales vs Tracks with No Sales* chart, for each genre, will display how many tracks in the genre have at least one sale and how many have no sales. It will also display the number of tracks in each category.

To create a function, we can use the following steps:

1. In the **MySQLDatabase** module, create a new function. The function will return a Boolean value to indicate success or failure:

```
    Public Function GenreTrackSalesStats() As Boolean
```

2. Declare the variables we will use, as well as error handling:

```
        Dim SQL As String        'To store the SQL statement
        Dim RS As Recordset      'The Recordset variable
        Dim Msg As String        'To display messages
```

```
Dim Counter As Integer   'A counter
Dim MyNamedRng As Range ' A range variable


On Error GoTo HandleError
```

3. Prepare the SQL statement. The SQL combines two views to retrieve our data. The MySQL view, `vw_genre_count`, returns the total number of tracks in each genre, and `vw_genre_count_no_sales` returns the count of racks with no sales. Both have a common field name called `count`. By combining the two views in SQL and subtracting the no sales count from the total count, we can determine the number of sales in each genre:

```
SQL = ""
SQL = SQL & "SELECT "
SQL = SQL & "vw_genre_count.Count   - vw_genre_count_
no_sales.Count AS Sales, "
SQL = SQL & " vw_genre_count_no_sales.Count AS
NoSales, "
SQL = SQL & "vw_genre_count.Genre "
SQL = SQL & "FROM "
SQL = SQL & "vw_genre_count "
SQL = SQL & "LEFT JOIN vw_genre_count_no_sales ON vw_
genre_count_no_sales.Genre = vw_genre_count.Genre "
SQL = SQL & "Order BY "
SQL = SQL & "vw_genre_count.Genre"
```

4. Now, call the new `ConnectDB_ODBC` function. The difference between this and the DSN-less connection is that here, we are using the global ODBC variable and also passing in the DSN that we want to use. You can use this function with any DSN:

```
If ConnectDB_ODBC(g_Conn_ODBC, "chinook") = True Then
```

5. Set the `recordset` variable and open `recordset`, passing in the SQL and the global ODBC variable:

```
Set RS = New ADODB.Recordset
RS.Open SQL, g_Conn_ODBC
```

6. Test whether we have data and deal with a situation in which we have no data:

```
If RS.EOF And RS.BOF Then
        RS.Close
        Set RS = Nothing
        Msg = "There is no data"
        MsgBox Msg, vbOKOnly + vbInformation, "No
 data to display"
        GoTo LeaveFunction
```

7. Include `Else` to handle when we have data and enter code to process the data. Start by setting the column headings. Place the data in the **H**, **I**, and **J** columns:

```
Else
        For Counter = 0 To RS.Fields.Count - 1
                Worksheets("Data Sheet").Cells(1, 8 +
Counter) = RS.Fields(Counter).Name
        Next
    Worksheets("Data Sheet").Cells(2,8).
CopyFromRecordset RS
```

8. Create the named range for the data. The **H** and **I** columns contain the data for the two series in the chart:

```
        Set MyNamedRng = Worksheets("Data Sheet").
Range("H2:I" & RS.RecordCount + 1)
                ActiveWorkbook.Names.Add Name:="Sales",
RefersTo:=MyNamedRng


        Set MyNamedRng = Worksheets("Data Sheet").
Range("J2:J" & RS.RecordCount + 1)
                ActiveWorkbook.Names.Add
Name:="TrackStatGenre", RefersTo:=MyNamedRng
```

9. Close the recordset and connection:

```
        RS.Close
        Set RS = Nothing
        g_Conn_ODBC.Close
        Set g_Conn_ODBC = Nothing
```

10. Now, set some options in the chart and activate it:

```
Worksheets("Dashboard").
ChartObjects("TrackStats").Activate
        With ActiveChart
            .HasTitle = True
            .ChartTitle.Text = "Genre Tracks Sales
vs. No Sales"
```

11. Set the chart's data source and category to the named ranges we defined and also set the series names:

```
            .SetSourceData Source:=Worksheets("Data
Sheet").Range("Sales"), PlotBy:=xlColumns
            .Axes(xlCategory).CategoryNames =
Worksheets("Data Sheet").Range("TrackStatGenre")

            .SeriesCollection(1).Name = "Sales"
            .SeriesCollection(2).Name = "No Sales"
        End With
        GenreTrackSalesStats = True
    End If
```

12. From here, handle failed connections and exit the function:

```
    Else
        'Connection failed if gets in here, just drop
through to leave
    End If


LeaveFunction:
    'Leave the function
    Exit Function


HandleError:
    GenreTrackSalesStats = False
    Resume LeaveFunction


End Function
```

13. Test the function by typing the function name in the **Immediate** window, and your result should be as follows:

```
Immediate

 ? GenreTrackSalesStats
 True
 |
```

Figure 12.2 – The test and result for GenreTrackSalesStat in the Immediate panel

Typing ? `GenreTrackSalesStats` and pressing *Enter* will run the function to retrieve the data. If it was successful, it will return `True`; if not, it will return `False`. The returned value can be tested by your code to decide on the next action.

The output is as follows:

Sales	NoSales	Genre
14	26	Alternative
203	129	Alternative & Punk
53	28	Blues
14	1	Bossa Nova
36	38	Classical
8	9	Comedy
27	37	Drama
10	14	Easy Listening
11	19	Electronica/Dance
12	16	Heavy Metal
15	20	Hip Hop/Rap
68	62	Jazz
340	239	Latin
231	143	Metal
0	1	Opera
26	22	Pop
37	24	R&B/Soul
28	30	Reggae
745	552	Rock
6	6	Rock And Roll
20	6	Sci Fi & Fantasy
5	8	Science Fiction
19	24	Soundtrack
43	50	TV Shows
13	15	World

Figure 12.3 – The GenreTrackSalesStats named range in the data sheet workbook

The data is presented in the data sheet. **Sales (H)** indicates how many tracks had sales in **Genre (J)**, and **NoSales (I)** indicates how many tracks had no sales in the genre. The data is presented in a specific order (**Sales**, **NoSales**, and **Genre**) to meet the requirements of the chart that will display it:

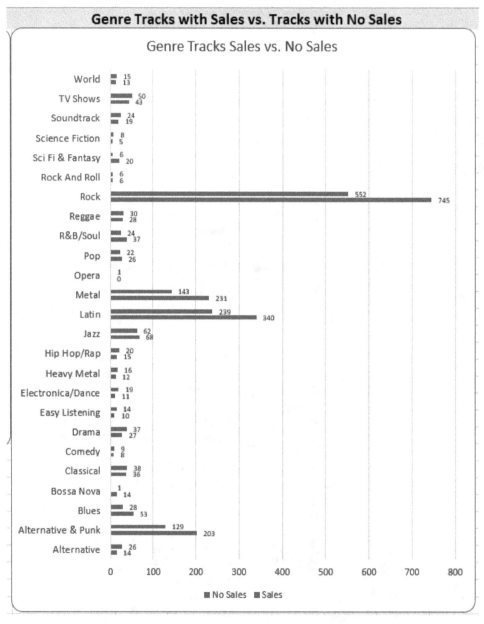

Figure 12.4 – The chart displaying Sales and No Sales for GenreTrackSalesStats

Using a DSN connection in your code is no different from using a DSN-less connection, but as you can see, there are no login details stored in the workbook. We can have both connection types in the same workbook (although we will usually settle on one type and use it). By setting up the connection function for DSN with a parameter to pass in the name of the DSN, any database can be accessed that we have set up a DSN for. This means that we can extract data from any number of databases – even on different servers and in different locations – in one workbook.

> **Note**
>
> The VBA for this exercise can be found here: `https://github.com/PacktWorkshops/The-MySQL-Workshop/tree/master/Chapter12/Exercise12.02`.

In this section, we learned how to create a function to use a DSN connection, called and used the function, and explored the differences between a DSN-less and a named ODBC connection. We also prepared a third function to autorun on startup to update data.

In the next section, we will create a generic data reader function to which we can pass a SQL statement to be executed on the MySQL server and have the result passed back from the function to use in our application.

Exploring generic data read functions

Throughout the execution of your programs, you will often want to query the database and return a single value only. Writing individual functions for this can result in a lot of code, making the application bloated and causing future maintenance to be problematic due to duplicated code. As a developer, your main goal is to create small and efficient code to generate accurate results. Developing small and flexible functions that are callable from any other code will help you to achieve this goal. For example, we can create a `read` function that is designed to read data from our MySQL database. This will allow you to call a single function every time you need to read data, reducing code repetition and improving development efficiency. This process can be applied with other functions, such as `read`, `update`, and `insert`, allowing us to complete these operations in multiple areas without repeating code.

Exercise 12.02 – a generic data reader

In this exercise, you will create a single function that you can use to query the database and return a single value only. You will pass into the function a SQL statement to be executed, and the result will be returned, which you can then use in your VBA code. Learning how to create this type of function will enable you to modularize your VBA code to simplify development and improve the readability of your logic.

We will then demonstrate its use by populating the **N8**, **N9**, and **N12** cells in the dashboard worksheet:

1. In the **MySQLDatabase** module, add a new function. The function will have one parameter, **SQL**, as a string, and it will return a variant-type value. We do not know what type of value will be returned by the SQL statement. It can be numeric, a date, a long string, or any of the data types that MySQL can return, so a variant data type will allow any data type to be returned, and it will be the responsibility of the VBA code to handle the data type appropriately:

    ```
    Public Function runSQL_SingleResult(SQL As String) As
    Variant
    ```

2. Declare the variables to use and set up error handling:

    ```
    Dim RS As Recordset       'The Recordset variable
    Dim Msg As String         'To display messages


    On Error GoTo HandleError
    ```

3. Make the connection to the database. We will be using the DSN connector and passing in the DSN name to use:

    ```
    If ConnectDB_ODBC(g_Conn_ODBC, "chinook") = True Then
    ```

4. Prepare and open the recordset variable, RS. Pass in the SQL statement that was passed into the function using the ODBC connection:

    ```
    Set RS = New ADODB.Recordset
    RS.Open SQL, g_Conn_ODBC
    ```

5. Test whether a record was returned. If not, then close the recordset and connection, and return 0 before leaving the function:

```
If RS.EOF And RS.BOF Then
     RS.Close
     Set RS = Nothing

     Msg = "There is no data"
     MsgBox Msg, vbOKOnly + vbInformation, "No
data to display"
     runSQL_SingleResult = 0
     GoTo LeaveFunction
```

6. If there was a record returned, set the cursor position to the first record, read the value, and pass it back by assigning it to the function. This function will accept any SQL statement that you care to pass into it. We have no way of knowing in advance what the name of the return field is. As we are expecting only a single value to be returned, we can simply read the first (and only) field's value by referring to it, using its numeric value of 0:

```
Else
     RS.MoveFirst
     runSQL_SingleResult = RS.Fields(0)
```

7. Close the recordset connection and exit the function. In the error routine, we pass back 0:

```
     RS.Close
     Set RS = Nothing
End If

g_Conn_ODBC.Close
Set g_Conn_ODBC = Nothing

Else
End If

LeaveFunction:
  Exit Function
```

```
HandleError:
    runSQL_SingleResult = 0
    Resume LeaveFunction
End Function
```

8. Save the function.

9. Test it by typing the following in the **Immediate** panel:

```
? runSQL_SingleResult("SELECT Count(`TrackID`) FROM
Track")
```

The returned value will be 3503.

Now, let's populate the customer purchase details in **N8, N9**, and **N12**. For this exercise, we will do this in the Workbook_Open subroutine so that they are read and populated when the workbook is opened.

10. Add the following code in the Workbook_Open() subroutine:

- Call the runSQL_SingleResult function, pass in a query to read data from the vw_customer_count view, and assign the returned value to **N8**:

```
'Populate Customer Purchase Details
Worksheets("Dashboard").Cells(8, 14) = runSQL_
SingleResult("SELECT * FROM vw_customer_count")
```

- Call the runSQL_SingleResult function, pass in a query to call the spTotalSales stored procedure, and assign the returned value to **N9**:

```
Worksheets("Dashboard").Cells(9, 14) = runSQL_
SingleResult("call spTotalSales()")
```

- Call the runSQL_SingleResult function, pass in a query to find the last invoice date, and assign the returned value to **N12**:

```
Worksheets("Dashboard").Cells(12, 14) = runSQL_
SingleResult("SELECT MAX(InvoiceDate) FROM Invoice")
```

11. Test it by placing your cursor anywhere in the `WorkbookOpen()` function and pressing *F5* to run it. The results on the dashboard will be as follows:

M	N
Customer Purchase Details	
Total Customers	59
Total Sales	$ 2,238.62
Highest Customer Sales	
Avg Customer Sales	
Last Sale date	12-11-19

Figure 12.5 – The dashboard results from three single-result queries

In this exercise, we created a single function to run a SQL statement and pass back a single result of any data type. The code in *step 9* called it three times by running a standard SQL statement and returning an integer value, calling a MySQL stored procedure and returning a currency value, and running an aggregate query to return a date value. These three types were chosen to demonstrate the flexibility of this type of function and the variant data type that it returns.

> **Note**
>
> The VBA for this exercise can be found here: `https://github.com/PacktWorkshops/The-MySQL-Workshop/tree/master/Chapter12/Exercise12.02.`

In the next activity, you will add two lines of code to complete the **Customer Purchase Details** section of the dashboard.

In this section, we learned how to create a flexible function to read data and return results and how to use the function to retrieve varied results, based on the SQL statement passed into it.

In the next section, we will learn how to work with MySQL for Excel to load data, create pivot tables, and update data in the MySQL database directly from an Excel worksheet.

Creating connections to MySQL in Excel

Oracle has released a plugin for Excel called **MySQL for Excel**. This plugin gives you a simple-to-use window in the MySQL database using a DSN connection, which makes reading the data from tables simple and even provides a direct connection to the tables, allowing you to edit and update, delete records, and add new records. In this section, we will briefly look at these features to finish off our dashboard and to permit direct data editing on key tables. MySQL for Excel should have been installed when you installed MySQL; if not, return to the MySQL installation pages in the *Preface* for installation instructions. We will be concentrating only on importing data for display and editing. There are many more interesting features in MySQL for Excel that are not addressed in this book that may be worth following up.

Exercise 12.03 – creating a connection to MySQL

In this exercise, we are going to start MySQL for Excel and describe the opening panel.

To start MySQL for Excel, follow these steps:

1. Click the **DATA** tab on the ribbon. If the MySQL for Excel plugin is installed, there will be a button for it on the right of the tab:

Figure 12.6 – The location of the MySQL button in the ribbon

2. Click the **MySQL for Excel** button, and the MySQL panel will open on the right side of the screen. The screen will display any local and remote connections you may have set up. It also has two options, **New Connection** for setting up new connections and **Manage Connections** for managing any existing ones:

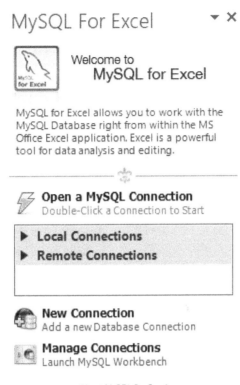

Figure 12.7 – The MySQL panel

Once installed, MySQL for Excel is easy to start and use. If other people are going to use your workbook, they will also need to have MySQL for Excel installed on their computer to use the plugin. The plugin can be installed independently of MySQL, and it is free to use.

Next, we are going to create a connection to the MySQL server using the methods provided by MySQL for Excel.

To create a new connection, do the following:

1. Click **New Connection**. The **MySQL Server Connection** window will open. Enter your connection details. Name the connection `Chinook` and set **Default Schema** as the `chinook` database:

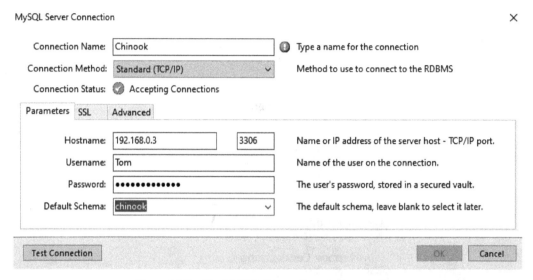

Figure 12.8 – The MySQL Server Connection window

2. Click **Test Connection** to test that the connection was successful. If it was, then you will see the confirmation screen. If not, check your connection details and try again. If successful, click **OK**:

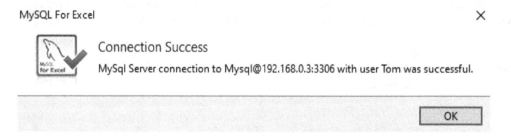

Figure 12.9 – The successful connection notification

3. Click **OK** on the **MySQL Server Connection** screen. The new connection will appear in either **Local Connections** or **Remote Connections** on the panel, depending on your specific setup:

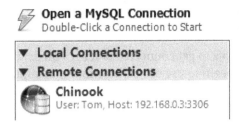

Figure 12.10 – The connection panel displaying the new connection

4. To ensure that the password is set on the server as well, click **Manage Connections**. Workbench will open, as well as a connection screen:

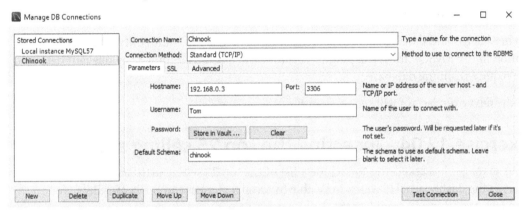

Figure 12.11 – Adding a password to the Workbench vault

5. Select the new connection in the **Stored Connections** panel and click **Store in Vault**. Another password window will open. Enter the password and click **OK**:

Figure 12.12 – The password entry screen

6. Click **Test Connection** to make sure all is okay, and if successful, click **Close** to close the window. This step ensures that the password is also stored on the server, so you will not need to enter it again later.

We have now made our connection to the server. You can create as many as you need for different databases and servers.

In this section, we learned how to get connected to a MySQL database through Excel. In the next section, we will begin to use our connection to manipulate data in our MySQL database.

Inserting data using MySQL for Excel

Using the connection that we have created, it is now possible to work with data between Excel and MySQL. One of the first things we will look at is how to get data from MySQL into an Excel workbook.

To be able to send data from MySQL to Excel, we will use the **Import MySQL Data** option in the MySQL for Excel plugin. This tool will allow us to move any relevant data we require from the database.

In the next exercise, we will see an example of inserting data.

Exercise 12.04 – inserting the top 25 selling artists

Your manager has asked you to include a list of the top 25 selling artists and their total sales in the dashboard. He wants to be able to see at a glance who the best-sellers are and feels that this will complete the dashboard.

In this exercise, we are going to include this list:

1. In the dashboard worksheet, click on the **M15** cell to make it active. This will be the insert point of the data.

2. *Double-click* on the **chinook** connection in the **MySQL For Excel** panel. A list of available database schemas will be displayed:

Figure 12.13 – The chinook connection in MySQL For Excel

This shows the following display:

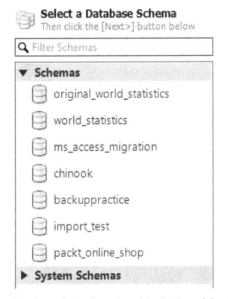

Figure 12.14 – The chinook database listed in the list of database schemas

3. *Double-click* on the **chinook** schema. The panel will then display a list of tables, views, and procedures that are available:

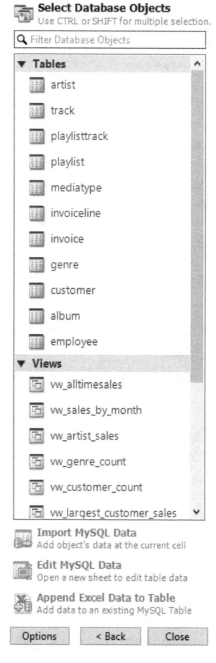

Figure 12.15 – Views and inactive options

4. Click on the **vw_artist_sales** view. This view provides the top 25 selling artists data that we want to insert into the dashboard. This view consists of two columns of data. Note that when you clicked, the top option, **Import MySQL Data**, was activated.

5. Click the **Import MySQL Data** option. A window will open, displaying 10 sample records and several options. For now, just click **Import**. Then, 2 columns of data consisting of headings and 25 data rows will appear, starting at the **M16** cell. There will also be a new named range added to the range list:

Top 25 Selling Artists	
Artist_Group ▼	Total Sales ▼
Iron Maiden	$ 139.81
U2	$ 106.60
Metallica	$ 91.32
Led Zeppelin	$ 87.01
Os Paralamas Do Sucesso	$ 44.75
Deep Purple	$ 43.83
Faith No More	$ 42.19
Lost	$ 41.04
Eric Clapton	$ 40.36
R.E.M.	$ 38.94
Queen	$ 37.06
Creedence Clearwater Revival	$ 37.05
Guns N' Roses	$ 36.11
Titãs	$ 33.73
Green Day	$ 32.80
Pearl Jam	$ 31.79
Kiss	$ 30.80
Van Halen	$ 29.26
Various Artists	$ 28.90
Chico Buarque	$ 27.30
Red Hot Chili Peppers	$ 26.81
Lenny Kravitz	$ 26.10
Chico Science & Nação Zumbi	$ 25.08
The Office	$ 24.96
Tim Maia	$ 24.11
	$ 1,127.71

Figure 12.16 – The top 25 selling artists with the total

6. When displaying a list of values, it is often desirable to have a sum of the values below the list so that the user can see at a glance what the total value is; add the following cell-based formula to the **N42** cell, =SUM(N16:N41), if not already there.

This data will be updated when you select **DATA** and **Refresh All** on the ribbon:

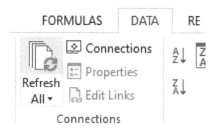

Figure 12.17 – Location of Refresh All on the ribbon

The first time you refresh the data, a new column may be inserted, and everything to the right of the insert point moves by one column. This appears to be a bug in the plugin. If this does happen, select the data and move it to the proper position. It does not do this on subsequent refreshes.

Once you have MySQL for Excel set up, inserting data is very easy. Once the data is on the sheet, you can access and refer to it as you would with any other data. The cell-based sum() function you included in the **N42** cell demonstrates this.

In this section, we learned how to insert data into MySQL using Excel. In the next section, we will learn how to update data in MySQL using Excel.

Updating data using MySQL for Excel

Once data is inserted into an Excel workbook, we may want to update the MySQL database based on changes made to the data. To help with this, MySQL for Excel implements functionality for editing data. This allows us to edit data in Excel and save the changes back to MySQL.

In the next exercise, we will bring in data from MySQL and place it on a new worksheet. This data can be edited, and the updated data can be written back to the MySQL database. The ability to edit MySQL data and save it back to the database helps you to maintain your data without the need to develop complicated forms.

Exercise 12.05 – updating MySQL data – employees

MySQL for Excel will insert the data into a new worksheet if you select the editing option. In this exercise, we will add the employee data:

1. Activate the MySQL panel by clicking **DATA** and **MySQL for Excel**:

Figure 12.18: Location of the MySQL for Excel button in the ribbon

2. Select the **Chinook** connection:

Figure 12.19 – The available connections

3. Select the **chinook** database:

Figure 12.20 – The available databases

4. Click on the **employee** table:

Figure 12.21 – The available tables in the selected database

5. Click the **Edit MySQL Data** option. The preview window will open:

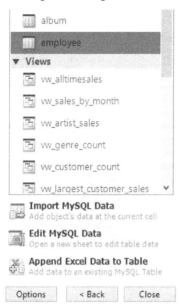

Figure 12.22 – The available options when a table is selected

6. We can preview the data in the employee table as seen here:

Preview MySQL Data

Table Name   employee
Row Count: 10

This is a subset of the data for preview purposes only.

Employee...	LastName	FirstName	Title	ReportsTo	BirthDate	HireDate	Address
1	Adams	Andrew	General Manager	NULL	18-Feb-62	14-Aug-02	11120 Jasper Ave NW
2	Edwards	Nancy	Sales Manager	1	08-Dec-58	01-May-02	825 8 Ave SW
3	Peacock	Jane	Sales Support Agent	2	29-Aug-73	01-Apr-02	1111 6 Ave SW
4	Park	Margaret	Sales Support Agent	2	19-Sep-47	03-May-03	683 10 Street SW
5	Johnson	Steve	Sales Support Agent	2	03-Mar-65	17-Oct-03	7727B 41 Ave
6	Mitchell	Michael	IT Manager	1	01-Jul-73	17-Oct-03	5827 Bowness Road NW
7	King	Robert	IT Staff	6	29-May-70	02-Jan-04	590 Columbia Boulevard West
8	Johnston	Laura	IT Staff	6	09-Jan-68	04-Mar-04	923 7 ST NW
9	Bloggs	Fred	IT Staff	6	NULL	NULL	NULL
10	zzz	zzz	zzz	6	NULL	NULL	NULL

Preview 10 rows.   Refresh

OK   Cancel

Figure 12.23 – The employee table preview window

7. Click **OK**. A new worksheet will be added and populated with the contents of the `employee` table:

EmployeeId	LastName	FirstName	Title	ReportsTo	BirthDate	HireDate	Address	City	State	Country	PostalCode	Phone	Fax	Email
1	Adams	Andrew	General Manager		18-02-62 0:00	14-08-02 0:00	11120 Jasper Ave NW	Edmonton	AB	Canada	T5K 2N1	+1 (780) 428-9482	+1 (780) 428-3457	andrew@chinookcorp.com
2	Edwards	Nancy	Sales Manager	1	08-12-58 0:00	01-05-02 0:00	825 8 Ave SW	Calgary	AB	Canada	T2P 2T3	+1 (403) 262-3443	+1 (403) 262-3322	nancy@chinookcorp.com
3	Peacock	Jane	Sales Support Agent	2	29-08-73 0:00	01-04-02 0:00	1111 6 Ave SW	Calgary	AB	Canada	T2P 5M5	+1 (403) 262-3443	+1 (403) 262-6712	jane@chinookcorp.com
4	Park	Margaret	Sales Support Agent	2	19-09-47 0:00	03-05-03 0:00	683 10 Street SW	Calgary	AB	Canada	T2P 5G3	+1 (403) 263-4423	+1 (403) 263-4289	margaret@chinookcorp.com
5	Johnson	Steve	Sales Support Agent	2	03-03-65 0:00	17-10-03 0:00	7727B 41 Ave	Calgary	AB	Canada	T3B 1Y7	1 (780) 836-9987	1 (780) 836-9543	steve@chinookcorp.com
6	Mitchell	Michael	IT Manager	1	01-07-73 0:00	17-10-03 0:00	5827 Bowness Road NW	Calgary	AB	Canada	T3B 0C5	+1 (403) 246-9887	+1 (403) 246-9899	michael@chinookcorp.com
7	King	Robert	IT Staff	6	29-05-70 0:00	02-01-04 0:00	590 Columbia Boulevard West	Lethbridge	AB	Canada	T1K 5N8	+1 (403) 456-9986	+1 (403) 456-8485	robert@chinookcorp.com
8	Callahan	Laura	IT Staff	6	09-01-68 0:00	04-03-04 0:00	923 7 ST NW	Lethbridge	AB	Canada	T1H 1Y8	+1 (403) 467-3351	+1 (403) 467-8772	laura@chinookcorp.com

Figure 12.24 – The new tab with data and insert record line

Note the following on screen:

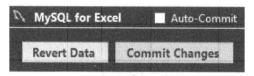

Figure 12.25 – The commit and revert edit options

> **Note**
>
> The **Options** window will be displayed on the tab whenever you have the **MySQL for Excel** panel activated and you click in the data area.

8. Select **Auto-Commit** to write changes to the data back to the database as soon as a change is made.

9. Clicking **Revert Data** will undo any changes you have made that are not committed.

10. Clicking **Commit Changes** will commit any changes to the database.

11. Make a change to any of the displayed data. The cell will turn blue.

12. The yellow line is where you can add a new line of data. Add some new data in the yellow line. When you're done, it will turn blue, and a new yellow line will appear below it:

	A	B	C	D	E	F	G	H	I	J	K	L	M	N	O
	EmployeeId	LastName	FirstName	Title	ReportsTo	BirthDate	HireDate	Address	City	State	Country	PostalCode	Phone	Fax	Email
	1 Adams	Andrew	General Manager			18-02-62 0:00	14-08-02 0:00	11120 Jasper Ave NW	Edmonton	AB	Canada	T5K 2N1	+1 (780) 428-9482	+1 (780) 428-3457	andrew@chinookcorp.com
	2 Edwards	Nancy	Sales Manager	1		08-12-58 0:00	01-05-02 0:00	825 8 Ave SW	Calgary	AB	Canada	T2P 2T3	+1 (403) 262-3443	+1 (403) 262-3322	nancy@chinookcorp.com
	3 Peacock	Jane	Sales Support Agent	2		29-08-73 0:00	01-04-02 0:00	1111 6 Ave SW	Calgary	AB	Canada	T2P 5M5	+1 (403) 262-3443	+1 (403) 262-6712	jane@chinookcorp.com
	4 Park	Margaret	Sales Support Agent	2		19-09-47 0:00	03-05-03 0:00	683 10 Street SW	Calgary	AB	Canada	T2P 5G3	+1 (403) 263-4423	+1 (403) 263-4289	margaret@chinookcorp.com
	5 Johnson	Steve	Sales Support Agent	2		03-03-65 0:00	17-10-03 0:00	7727B 41 Ave	Calgary	AB	Canada	T3B 1Y7	1 (780) 836-9987	1 (780) 836-9543	steve@chinookcorp.com
	6 Mitchell	Michael	IT Manager	1		01-07-73 0:00	17-10-03 0:00	5827 Bowness Road NW	Calgary	AB	Canada	T3B 0C5	+1 (403) 246-9887	+1 (403) 246-9899	michael@chinookcorp.com
	7 King	Robert	IT Staff	6		29-05-70 0:00	02-01-04 0:00	590 Columbia Boulevard West	Lethbridge	AB	Canada	T1K 5N8	+1 (403) 456-9986	+1 (403) 456-8485	robert@chinookcorp.com
	8 Johnston	Laura	IT Staff	6		09-01-68 0:00	04-03-04 0:00	923 7 ST NW	Lethbridge	AB	Canada	T1H 1Y8	+1 (403) 467-3351	+1 (403) 467-8772	laura@chinookcorp.com
	Bloggs	Fred	IT Staff												

Figure 12.26 – The edited data and new records are displayed in blue

13. Click **Revert**. This will undo any blue cells. You will be given the option to reload from the database or undo the changes.

14. Make some more changes to the data. The cells will again turn blue.

15. Click **Commit**. This will commit the changes. This time, the changes will be written to the database and the cells will turn green:

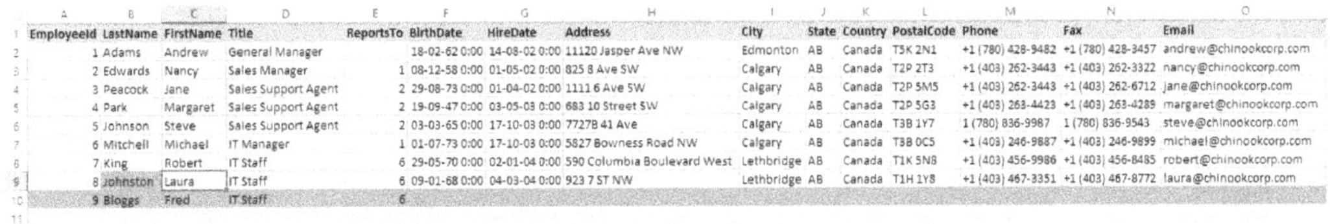

Figure 12.27 – The committed changes are shown in green

If **Primary Key** is set to **Auto Increment** in the table, then you do not need to enter the **EmployeeID** value. This will be added automatically when the update is committed.

The ability to update or add to the table data directly from Excel is very convenient, but with this convenience comes responsibility. Updating the wrong data can break the linking of records between tables. If other people use this feature, ensure that they have proper training and try to limit what they can update. Having said that, this is very easy and quick to implement.

You cannot add formulas for other objects to the worksheet where editable data has been placed in this method; the worksheet is protected. You can, however, refer to the data from another worksheet, using VBA or a cell formula – for example, type the following formula into **Dashboard A40**:

```
=SUM(employee!E3:E9)
```

The result will be 20, assuming that you did not change the existing values during the exercise. Delete the formula when you are done. This is a demonstration only.

In the next activity, you will create a new worksheet with editable data using the Customers table.

When you have mastered the process of importing MySQL data into worksheets using this method, you will be able to edit your MySQL data very quickly and efficiently without the need to develop an application, which is very useful for those quick edits that are often required when you are maintaining a database. Because this method opens up an entire table of data to free and unchecked editing, don't provide this to untrained users. It is not recommended to allow other people unfettered access to the data. You can save the sheets with the workbook, and you will be prompted to refresh the data when you open the workbook.

In the next exercise, we will be pushing some data from Excel into a new MySQL table.

Pushing data from Excel

Often, data is first stored in Excel before it is transferred to a database. This can happen for many reasons; most commonly, it is because the data is exported from a tool that can interface with Excel but not directly with a database.

In situations where you want to move data from Excel to MySQL, you can utilize the Export Excel Data to New Table functionality. This will create a new table in MySQL and load the Excel data into it. This functionality allows for the quick and easy loading of data.

Exercise 12.06 – pushing data from Excel to a new MySQL table

Excel is often used in business as a database and can hold a lot of organized data. When the data requirements have outgrown Excel, often the decision is made to migrate the data to a proper RDBMS such as MySQL. Without a doubt, at some time in your career, you will need to migrate data from Excel to a new table in MySQL. This exercise will show you how to do that with ease using MySQL for Excel.

Chinook Music Downloads has been maintaining an Excel contacts register. Management would like this data to be transferred to the primary *chinook* database, and as the primary developer, you have been assigned the task:

1. Open the **MySQL For Excel** panel using **DATA** and **MySQL for Excel**.
2. Open the **Contact Register** worksheet where the data is located.
3. Select the **Chinook** connection and the **chinook** database.
4. Click and open the **Contact Register** tab in Excel.
5. Select the data to be migrated. Select all cells in the **A3–G12** range:

Contact Date	Contact Name	Contact Comments	Action	Followup Required	Status	Closed
		Contact Register				
01-06-19	Anonymous	I am using a 2400bd dial up modem and I find your downloads speeds are too long, can you speed them up please	Issue is with users technology, there is nothing that can be done and we cannot contact the user	No	Closed	Yes
02-06-19	Mary Meenow	There are not enough wedding songs on your site	songs about weddings and asked if she could suggest any songs. Handed over to purchasing for further action.	Yes	Open	No
03-06-19	Fred Bloggs	I am having problems downloading since my credit card ran out of credit	Emailed customer suggesting that he should contact his credit card provider	No	Closed	Yes
04-06-19	Johny Doe	credit card was charged twice, can you please investigate and refund the overcharge. Invoice # 123456	refund has been issued along with a letter of apology and the issue has been handed to IT to investigate and fix	Yes, in one week to ensure customer is happy	Under Investigation	No
05-06-19	Anonymous	People are quick to complain, I just wanted to say your site is great, keep up the good work	Can't respond, due to anonimity. I would be nice to say thank you	No	Closed	Yes
06-06-19	Constance Complainer	It won't stop raining	has been issue with a note stating how it is beyond our capabilities at this time to control the weather	No	Closed	Yes
07-06-19	Ima Complainer	Can you supply snacks in the Listening Lounge page?	to supply snacks, the current technology is not capable of transmitting physical items and that we are eagerly awaiting on the development of input.	No	Closed	Yes
08-06-19	Anonymous	Some of your music has too much swearing	Asked for a list of specific tracks and will investigate putting ratings and warnings on individual tracks and albums	Yes	Under Investigation	No
09-06-19	Tom Pettit	Great list of song titles	issued a thank you response	No	Closed	Yes

Figure 12.28 – Selected Excel data for exporting to MySQL

6.   When the data is selected, the **Export Excel Data to New Table** option will activate:

Figure 12.29 – The option to export data to a MySQL database

7.   Click **Export Excel Data to New Table** to open the **Export Data** window:

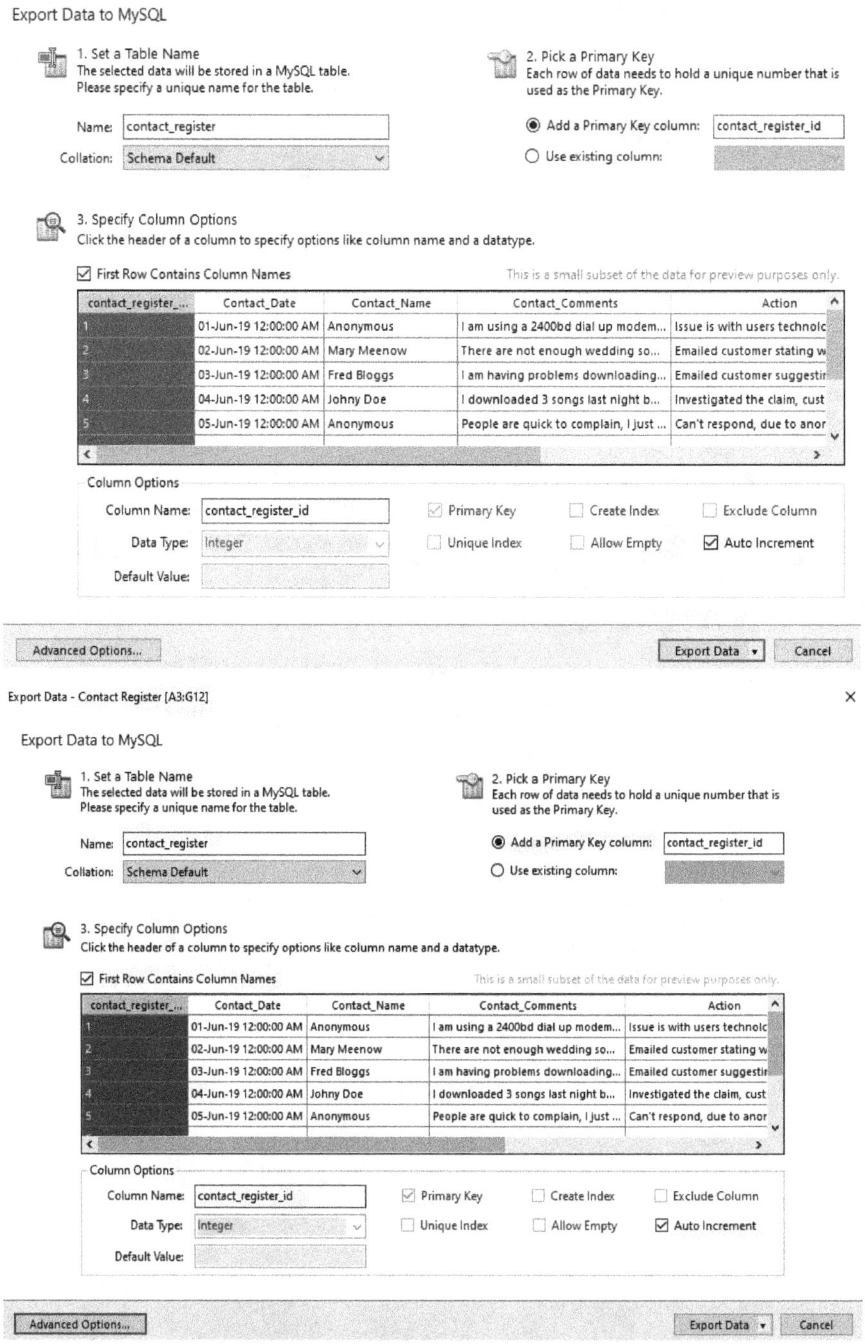

Figure 12.30 – The Excel Export Data screen

The default table name is taken from the worksheet name, but you can change this. It was detected that there is no primary key in the data; one is added and set to `Auto Increment`. You can change individual column details by clicking on the column in the display, and you can set default values for each column if required. The data type for each column is set, based on the data. You can change this if required for each column.

8. Examine each column and check the data types and values, and set any default values. **Closed** is set as a `Boolean` data type. This is appropriate for the column, but be sure to set a default value of `0`.

9. Click **Export Data**. After a few seconds, you should get a notification on the successful status of the operation:

Figure 12.31 – Export to table confirmation

10. Click **OK**. The notification window will close.

11. Open Workbench and refresh the **chinook** database. Check that the new table is there and view the data:

Figure 12.32 – The Workbench view of the new table, fields, and data

The ability to push Excel data into a new MySQL table will make data migrations very easy and quick. There are a lot of options to handle most of the situations that you may encounter. Some will migrate easily; others will require data validation and manipulation to ensure that the data is in a suitable state before you get a successful export.

In this section, we learned how to update data in MySQL through Excel. In the next section, we will look at how we can visualize and analyze data using pivot tables.

Pivot tables

Finally, we have reached the final topic of MySQL for Excel – pivot tables.

Pivot tables allow you to analyze your data by providing options to add different fields and values to a table, perform a multitude of mathematical operations on the data, and filter the data, both horizontally and vertically. Pivot tables are a powerful tool for analyzing data and preparing charts and graphs.

In this exercise, we will be importing data related to album sales and preparing it for display in a chart.

Exercise 12.07 – album sales

In this exercise, we will import some data from MySQL and create a pivot table with it. We will then add a chart to visualize the data:

1. Click on the **Pivot Tables** worksheet tab. A blank sheet will be displayed:

Figure 12.33 – Click on the Pivot Tables worksheet tab

2. Click on the **A1** cell and type `Album Sales - Pivot Data`, change the font size for the cell to **18** points, and make the text **bold** to identify the data. You can do this by *right-clicking* on the **A1** cell and selecting the **Format Cells** option, which will open the **Format Cells** window. Select **Font** to change the font styling:

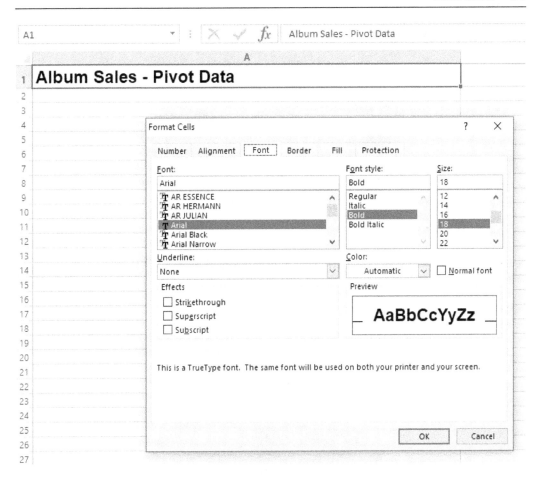

Figure 12.34 – The Format Cells window with Font selected

3. Click on the **A2** cell. This is where we are going to insert our data.

4. Open the MySQL for Excel panel via **DATA | MySQL for Excel:**

Figure 12.35 – Opening the MySQL for Excel panel

5.   Select the **Chinook** connection:

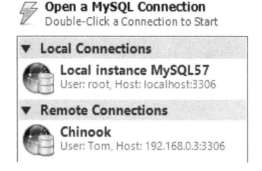

Figure 12.36 – Double-click the Chinook connection

6.   Once connected, select the **chinook** database from the schema list:

Figure 12.37 – Double-click the chinook database

7. We have a view in the database named **vw_albumsales**. Select the view and click **Import MySQL Data**:

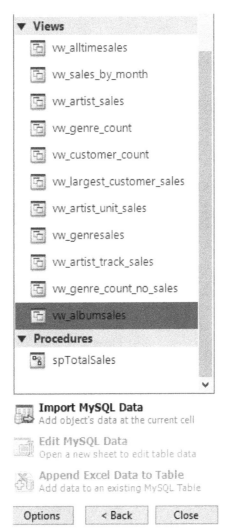

Figure 12.38 – Select vw_album sales and click Import MySQL Data

The **Import Data - Pivot Tables** screen will open, displaying **10** rows of sample data:

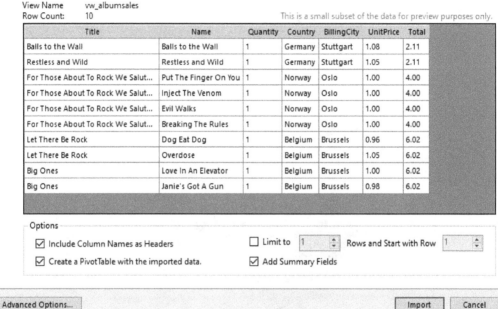

Import Data from MySQL

Choose Columns to Import
Click on column headers to exclude/include them when importing the MySQL view data in Excel.

View Name   vw_albumsales
Row Count: 10   This is a small subset of the data for preview purposes only.

Title	Name	Quantity	Country	BillingCity	UnitPrice	Total
Balls to the Wall	Balls to the Wall	1	Germany	Stuttgart	1.08	2.11
Restless and Wild	Restless and Wild	1	Germany	Stuttgart	1.05	2.11
For Those About To Rock We Salut...	Put The Finger On You	1	Norway	Oslo	1.00	4.00
For Those About To Rock We Salut...	Inject The Venom	1	Norway	Oslo	1.00	4.00
For Those About To Rock We Salut...	Evil Walks	1	Norway	Oslo	1.00	4.00
For Those About To Rock We Salut...	Breaking The Rules	1	Norway	Oslo	1.00	4.00
Let There Be Rock	Dog Eat Dog	1	Belgium	Brussels	0.96	6.02
Let There Be Rock	Overdose	1	Belgium	Brussels	1.05	6.02
Big Ones	Love In An Elevator	1	Belgium	Brussels	1.00	6.02
Big Ones	Janie's Got A Gun	1	Belgium	Brussels	0.98	6.02

Options

☑ Include Column Names as Headers   ☐ Limit to 1 ⬍ Rows and Start with Row 1 ⬍

☑ Create a PivotTable with the imported data.   ☑ Add Summary Fields

Advanced Options...   Import   Cancel

Figure 12.39 – The Import Data options screen

8. Tick both **Create a PivotTable with the imported data** and **Add Summary Fields**.

9. Click **Import**. The data will be imported, starting at the **A2** cell:

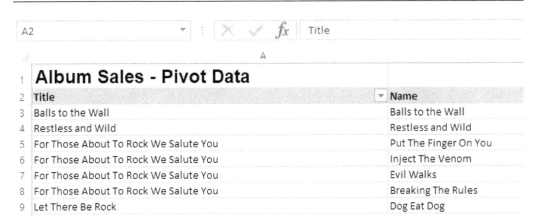

Figure 12.40 – Album sales data imported from MySQL

10. At first glance, it doesn't look all that different from previous imports; however, scroll to the right of the data and you will see the following box:

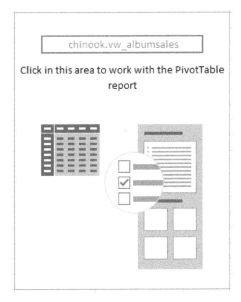

Figure 12.41 – The PivotTable placeholder

Note that it has the name of the view you imported. This is the pivot table placeholder and will be changed to display the table data when you have set it up. This box will provide a screen where you can adjust and manipulate the data, and once you have made the changes you want, it will display the pivot table from its location.

11. Click in the box; a new panel will open next to the MySQL panel. Note that the panel displays the field names. Select the fields that you want to include by ticking them:

Figure 12.42 – The PivotTable panel

12. Let's experiment with this a bit, starting with all the field boxes unticked. Position the screen so that the PivotTable box is in view.

13. Tick **Country**. Note that the box has now changed to display the data – specifically, the countries. This is the actual pivot table. Also, note that **Country** is displayed in the **ROWS** box at the bottom of the **PivotTable Fields** panel.

14. Tick **Title**. The titles are now grouped under the countries where they were sold. The title is also shown in the **ROWS** box.

15. Now, drag and drop the **Quantity** field into the **VALUES** box. The data will now have a new column and **Count of Quantity**, and the country line will have the total. These are the number of sales of the album in the country.

> **Note**
> Depending on your settings in Excel, you may get some other aggregate function other than Count – for example, you may get Sum. You can change the aggregate function to use on the field by clicking the down arrowhead to the right of the field name in the **Value** box and selecting the **Value Field** setting to change the option to perform on the field.

16. In the **ROWS** box, drag **Title** so that it is above **Country**. The pivot table now shows how many albums were sold and the countries in which they were sold.

17. Drag **Country** to the **COLUMNS** box. Now, the albums are listed on the left, the countries are listed across the top, and the units sold are totaled where the columns and rows intersect. You can see how useful a pivot table is at analyzing data.

18. We want to create a chart, so before we move on, untick **Title**.

19. Drag **Country** back to the **ROWS** box. You will now have two columns of data on the screen:

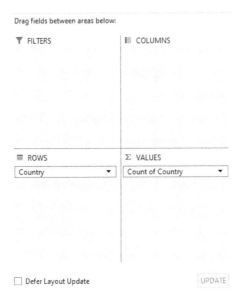

Figure 12.43 – The fields set up for the next step

After dragging `country` back to the **ROWS** box, you will see that the countries are displayed in the rows of the table:

Row Labels	Count of Quantity
Argentina	38
Australia	38
Austria	38
Belgium	38
Brazil	190
Canada	304
Chile	38
Czech Republic	76
Denmark	38
Finland	38
France	190
Germany	152
Hungary	38
India	74
Ireland	38
Italy	38
Netherlands	38
Norway	38
Poland	38
Portugal	76
Spain	38
Sweden	38
United Kingdom	114
USA	494
(blank)	
Grand Total	**2240**

Figure 12.44 – The corresponding data

We are currently showing a count of sales; it just so happens in this database that each sale is for one unit only, so count and sum return the same value. You can change how the data is summarized by performing the following steps.

20. Right-click on the **Value** column in the pivot table.

21. Select **Summarize Values By**. You will be presented with several options to summarize the values, such as **Count**, **Sum**, and **Average**.

22. To create a chart, click anywhere in the pivot table data area, and then click the **INSERT** tab and **Recommended Charts**:

Figure 12.45 – The Recommended Charts location in the ribbon

23. The **Insert Chart** window will open. It will be displaying the data in a chart. You can select the various charts. For now, select **Pie** and then **3-D Pie**:

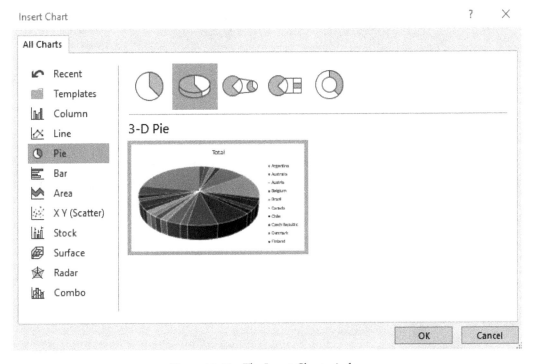

Figure 12.46 – The Insert Chart window

24. Click **OK**, and the chart will be placed on the worksheet:

Row Labels	Count of Quantity
Argentina	38
Australia	38
Austria	38
Belgium	38
Brazil	190
Canada	304
Chile	38
Czech Republic	76
Denmark	38
Finland	38
France	190
Germany	152
Hungary	38
India	74
Ireland	38
Italy	38
Netherlands	38
Norway	38
Poland	38
Portugal	76
Spain	38
Sweden	38
United Kingdom	114
USA	494
(blank)	
Grand Total	**2240**

Figure 12.47 – The new chart placed on the worksheet

25. We want the chart on the dashboard. *Right-click* the white area of the chart near the border and select the **Move Chart** option, located about halfway down the options. The following box will open:

Figure 12.48 – The options for moving the chart

26. Select **Dashboard** from the dropdown shown in the preceding screenshot and click **OK**. The chart will then be moved to the **Dashboard** worksheet. Excel will place the chart in a blank area on the **Dashboard** worksheet. This is likely to be to the right of the existing dashboard objects, but it could be below; you will need to find it:

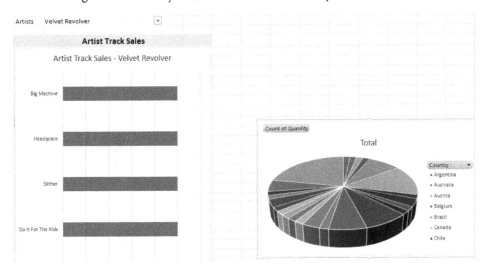

Figure 12.49 – Excel places the chart to the right of the existing objects on the target worksheet

27. Locate the chart on the **Dashboard** worksheet and drag it to where you want it placed:

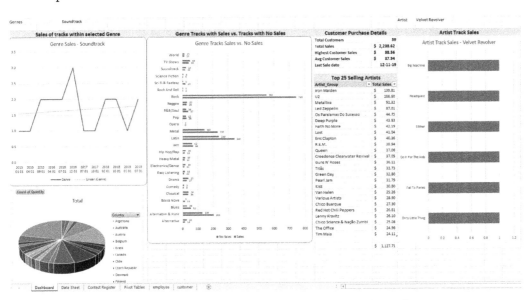

Figure 12.50 – The chart placed in position in Dashboard

28. Charts created with a pivot table have a dropdown on the category (in this chart, this is **Country**). This allows you to filter and change the chart. You also have options to format the chart, add labels, and so on:

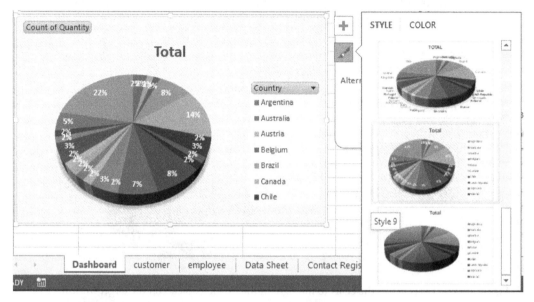

Figure 12.51 – The filtered chart with formatting

Pivot tables are amazing, and the power they provide to analyze your data in different ways is incredible. When used with MySQL for Excel, you can make your data tell a story. You can generate pivot tables from any suitable data in Excel; they are not limited to MySQL for Excel.

In this section, we learned how to start MySQL for Excel, create a connection to a MySQL database, import data from a view, import data in a form that can be edited or added to with the changes directly updating a database, export data to a new table in a MySQL database, and create pivot tables and generate flexible charts from them.

MySQL for Excel offers a unique, no-nonsense, and no-programming interface to a database. As you learn more about this plugin and its features, it will most likely become a staple and an important tool for creating impressive and easy-to-use applications. The ability to easily update data directly in MySQL is invaluable. However, great care should be taken if an application is to be released to other non-technical users who do not have the required knowledge or understanding of the effects of changing data – especially in foreign key fields that can adversely affect the integrity of a database.

Activity 12.01 – building a MySQL-based Excel document

> **Note**
>
> The Excel document used in this project, `CoffeeProducts.xlsx`, can be found at `https://github.com/PacktWorkshops/The-MySQL-Workshop/tree/master/Chapter12/Activity01`.

You are working for a coffee shop, and currently, all the data for the shop products are stored in an Excel file. Your manager would like you to create a MySQL database that contains the Excel data, as well as setting up Excel connections so that they can continue to manage the data through Excel. Perform the following steps to implement this activity:

1. Create a new MySQL database named `coffee_data`.

2. Push the current data in `CoffeeProducts.xlsx` to the MySQL database.

3. Add a new product named `Americano` with a price of `$3.50` and a size of `medium`.

> **Note**
>
> The solution for this activity can be found in the *Appendix*.

With this completed, you now have a fully functional Excel sheet that interacts with a MySQL database!

Summary

In this chapter, we created reusable functions to connect to MySQL using two different types of ODBC connections. We also learned how to read and import data from MySQL into a worksheet, define named ranges for the imported data, and assign the named ranges to charts using VBA. We set chart labels and categories using VBA and created and used drop-down lists to load filtered data, display data, and use it in charts. We then set some data collections from MySQL that will run when a workbook is opened.

We learned about the advantages of creating generic data readers that can run various SQL statements and return results for use in Excel. We imported data for editing or adding records and wrote the changes back to MySQL. We exported Excel data to MySQL as a new table and created pivot tables with attached charts.

The purpose of including *Chapter 9* and *Chapter 10* was to introduce you to several methods of using MySQL with Excel by using DSN connections, VBA, and MySQL for Excel. You have worked through several techniques and have developed a basic knowledge of these techniques. With Excel being so popular and a lot of companies looking for people with advanced knowledge of Excel programming and data manipulation, practicing and improving your skills with the techniques covered in this book will improve your employability profile. Consider undertaking a more advanced training course in Excel programming to further expand your skillset.

In the next chapter, we are going to cover different ways to get data into MySQL and export data from MySQL. Using various tools and processes, we will be able to efficiently manipulate MySQL data, allowing us to easily load external data sources.

Section 4: Protecting Your Database

This section covers the methods of backing up and securing your database data. You will learn how these methods can help to keep your data secure and safe.

This section consists of the following chapters:

13
Getting Data into MySQL

In this chapter, we will cover different ways to get data into MySQL and export it from inside the MySQL server to various formats. We will begin with adding data to tables and collections, and then move on to exporting data from MySQL to CSV files, and importing data from CSV, SQL, and JSON files. By the end of this chapter, you will be able to utilize the CSV storage engine to export and import data.

This chapter will cover the following topics:

- An introduction to data preparation
- Working with the X DevAPI
- Inserting documents
- Loading data from a SQL file
- Loading data from a CSV file
- Loading data from a JSON file
- Using the CSV storage engine to export data
- Using the CSV storage engine to import data
- Searching and filtering JSON documents

- Using JSON functions and operators to query JSON data

- Using generated columns to query and index JSON data

- Activity 13.01 – exporting report data to CSV for Excel

An introduction to data preparation

In the previous chapters, we covered using MySQL Shell in JavaScript mode, which we will use again in this chapter. We also covered connecting Microsoft Excel and Microsoft Access to the MySQL database directly. In this chapter, we will use CSV files to import and export data. Excel can be used to create and read these files.

When working with databases, it is essential to be able to import and export data. This can be when you are working on a new application and need some test data that you manually insert, when you are exporting data to a spreadsheet application that can be sent via email, or when importing data that is collected by something else such as a hardware device to then allow you to create reports. When working with multiple database instances, it may be necessary to copy data from a development setup to a production instance or vice versa.

In this chapter, we will see how to insert a single record, multiple records, and documents into a database. We will also learn how to load data from various file formats such as SQL, CSV, and JSON. Finally, we will make use of the CSV storage engine to import and export data.

We are going to use MySQL Workbench to demonstrate the examples unless mentioned otherwise.

Working with the X DevAPI

The X DevAPI is available in MySQL Shell and the official MySQL connectors for various programming languages. The X DevAPI is what provides the NoSQL interface for MySQL. A NoSQL interface is a way to work with a database without having to use SQL. This allows you to work with a database without using or learning SQL.

> **Note**
> NoSQL is a term used to classify database interfaces that are not SQL; in most cases, the interface that is provided is based on JSON and tries to let a database behave in a way that's more similar to the APIs that most web services provide. Also, most NoSQL interfaces don't require you to have a pre-defined format of your data (known as a schema); this is called **schema-less**.

The main benefit of this is that it feels more natural to developers. The drawback is that a lot of the flexibility that SQL provides is not available in the X DevAPI, making it more difficult to do things such as reporting. On the other hand, having to have a table definition for every table and then getting all the rows to conform to this is sometimes seen as a pain point when working with a SQL database. One of the selling points of MySQL is that it works with both SQL and NoSQL and allows you to use tables and collections of JSON-formatted documents. The X DevAPI uses the **X protocol**, which was introduced in MySQL in the 5.7 release. Besides the X DevAPI, there is also an X AdminAPI that is used for administrative operations.

The X DevAPI allows you to work with tables and with collections. You will need to have Python 3.6 or any older version installed. The reason for this is that `mysql-connector` is a dependency module that allows us to talk to the `mysql` instance. In Python 3.7, the `verbose` argument was removed and will throw an error if it is imported into the script.

First, you need to use the `pip` module to install the `mysql-connector` module:

```
pip install mysql-connector
```

The script for this can be found at `ch08_02_X_DevApi.py`.

The following example illustrates how the script looks in SQL and again in the X DevAPI:

```python
#!/usr/bin/python3
import mysql.connector
import mysqlx


def sql_example():
    con = mysql.connector.connect(
        host='localhost',
        user='msandbox',
        password='msandbox',
        database='test',
    )
    c = con.cursor()
    c.execute("SELECT name FROM animals")
    output = "Animals in the animals table\n"
    for row in c:
        output += "SQL Animal: {0}\n".format(row[0])
    c.close()
    con.close()
```

```python
    return output


def nosql_example():
    session = mysqlx.get_session(
      host='localhost',
      user='msandbox',
      password='msandbox',
    )
    schema = session.get_schema('test')
    animals = schema.get_collection('animals_collection')
    output = "Animals in the animals collection\n"
    for doc in animals.find().fields('name').execute().fetch_
all():
        output += "NoSQL Animal: {0}\n".format(doc['name'])
    session.close()
    return output


def test_sql_example():
    assert sql_example() == """Animals in the animals table
SQL Animal: dog
SQL Animal: Camel
SQL Animal: None
"""


def test_nosql_example():
    assert nosql_example() == """Animals in the animals
collection
NoSQL Animal: monkey
NoSQL Animal: zebra
NoSQL Animal: lion
"""


if __name__ == "__main__":
    print(sql_example())
```

```
print(nosql_example())
print(test_sql_example())
print(test_nosql_example())
```

In order to demonstrate the script, you will create a table called `animals` and a collection called `animals_collection`:

1. Connect to the MySQL client with Workbench and the appropriate user.

2. Select the `test` database for execution:

   ```
   USE test;
   ```

3. Since you already have a table called `animals`, you will drop it and recreate it:

   ```
   DROP TABLE animals;
   ```

4. Create the table called `animals`:

   ```
   CREATE TABLE animals (
       id int(11) NOT NULL,
       name varchar(255) DEFAULT NULL,
       PRIMARY KEY (id)
   );
   ```

5. Create the collection called `animals_collection`:

   ```
   CREATE TABLE animals_collection (
       doc json DEFAULT NULL,
       _id varbinary(32) GENERATED ALWAYS AS (json_
   unquote(json_extract('doc',_utf8mb4'$._id'))) STORED NOT
   NULL,
       PRIMARY KEY ('_id')
   );
   ```

6. Insert data into `animals`:

   ```
   INSERT INTO animals VALUES (1,'dog'),(2,'Camel'),(3,NULL);
   ```

7. Insert documents into `animals_collection`:

   ```
   INSERT INTO animals_collection ('doc') VALUES ('{\"_id\":
   1, \"name\": \"monkey\"}'),('{\"_id\": 2, \"name\":
   \"zebra\"}'),('{\"_id\": 3, \"name\": \"lion\"}');
   ```

8. Execute the script using the command line, making sure you navigate to the right folder:

```
python ch08_02_X_DevApi.py
```

The output should be as follows:

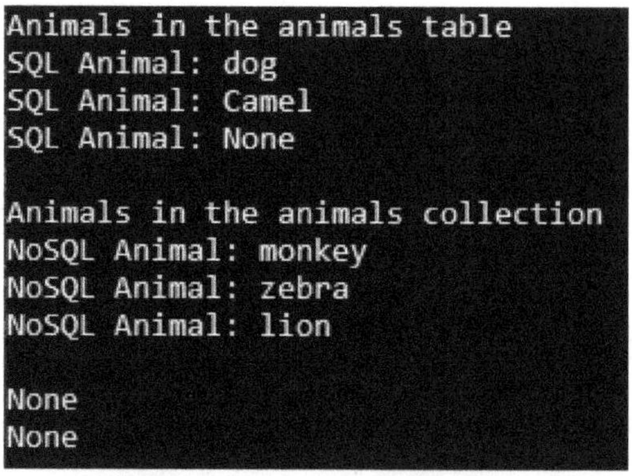

```
Animals in the animals table
SQL Animal: dog
SQL Animal: Camel
SQL Animal: None

Animals in the animals collection
NoSQL Animal: monkey
NoSQL Animal: zebra
NoSQL Animal: lion

None
None
```

Figure 13.01 – Output of the script

The two None means that the assertion was successful and the records in the database match our expectations.

An example of the X DevAPI

You are in a company that rents out electric scooters. Every time a scooter is picked up by an employee of the company, some data is downloaded from the scooter. This data is in JSON format and depends on the firmware version and the brand of the scooter.

If you store this in tables, then you will have to have a column for every piece of information that may be in every JSON file. And for every new brand and firmware version, you have to change the table definition to allow for new pieces of information to be stored.

Using a NoSQL interface to store the information as JSON documents allows you to store data for every piece of firmware and every model without making any changes to the database for new models and/or firmware versions.

Using MySQL Shell with the X DevAPI

We will use MySQL Shell to use both the SQL and X DevAPI interfaces. Let's look at a few of the commands that will help us in the upcoming exercises and activities.

To connect to the server, use the following command:

```
\connect mysqlx://root@localhost:33060
```

To create a classic session, use the following command:

```
\connect mysql://root@localhost:3306
```

To connect to the database, use the following command:

```
\use (Database Name)
```

There is no equivalent for the DESCRIBE command in the X DevAPI. So, you can either switch to SQL mode with \sql, run the DESCRIBE command and switch back to JS mode with \js, or you can use \sql <command> to run the DESCRIBE SQL command but stay in JS mode. To describe any table in the database, use the following command:

```
\sql DESCRIBE (Table Name)
```

To insert any values to the table, use the following command:

```
db.<Table Name>.insert().values(<values>)
```

To check what was inserted into the table, use the following command:

```
db.<Table Name>.select()
```

In the upcoming exercise, you will be using MySQL Shell in JS mode to insert values to the table.

Exercise 13.01 – inserting values with MySQL Shell in JS mode

In this exercise, you will insert values to a table using MySQL Shell in JS mode. It also has Python mode and SQL mode. The mode is shown in the prompt by default. Follow the following steps to accomplish this:

1. Open MySQL Shell.

2. Connect to the MySQL server using the \connect command. Provide the appropriate localhost and port number to connect. In this case, the localhost is 127.0.0.1 or localhost, and the port is 33060.

An example of a connection to the server looks like this:

```
MySQL  JS > \connect root@localhost:33060
Creating a session to 'root@localhost:33060'
Please provide the password for 'root@localhost:33060': *********
Save password for 'root@localhost:33060'? [Y]es/[N]o/Ne[v]er (default No): Y
Fetching schema names for autocompletion... Press ^C to stop.
Your MySQL connection id is 52 (X protocol)
Server version: 8.0.21 MySQL Community Server - GPL
No default schema selected; type \use <schema> to set one.
```

Figure 13.02 – A connection to the server

> **Note**
>
> Refer to the commands before the exercise to make a connection to the server.

3. Switch to the right database if you were not connected to it yet. Connect to the test database with the following command:

```
\use test
```

This produces the following message:

```
Default schema `test` accessible through db.
```

Figure 13.03 – MySQL Shell connected to the test database

4. Describe the animals table using the following command:

```
\sql DESCRIBE animals
```

This produces the following output:

```
Fetching table and column names from `test` for auto-completion... Press ^C to stop.
+--------+--------------+------+-----+---------+-------+
| Field  | Type         | Null | Key | Default | Extra |
+--------+--------------+------+-----+---------+-------+
| id     | int          | NO   | PRI | NULL    |       |
| name   | varchar(255) | YES  |     | NULL    |       |
+--------+--------------+------+-----+---------+-------+
2 rows in set (0.0012 sec)
```

Figure 13.04 – MySQL Shell – the DESCRIBE output

5. Insert data to a table called animals using the following commands:

```
db.animals.insert().values(4, 'Cheetah')
db.animals.insert().values(5, 'Leopard')
```

6. Check what was inserted into the table and return the data:

```
db.animals.select()
```

This produces the following output:

```
+----+---------+
| id | name    |
+----+---------+
|  1 | dog     |
|  2 | Camel   |
|  3 | NULL    |
|  4 | Cheetah |
|  5 | Leopard |
+----+---------+
```

Figure 13.05 – MySQL Shell – using SELECT()

In this exercise, the `insert()` function is used with one or two `values()` functions to add records to a table called `animals`. You can also use `\sql DESCRIBE animals` or `\sql` to switch to SQL mode and then later user `\js` to switch back to JavaScript mode. Here, we are already connected to the `test` database. If you need to switch to a different database, you can use `\use <database>` to switch. In the next section, we will explore inserting documents.

Inserting documents

You may want to add some test data into a database while developing an application. Inserting documents can be done with MySQL Shell. A document is a JSON data structure that's similar to a record in SQL.

JSON stands for **JavaScript Object Notation**. It is a method to describe a data structure in text format. As most programming languages and databases support this, it is a good format for data exchange.

A collection of documents is similar to a table with records. A table has a structure to describe what each record should look like. But this is not the case for documents. You can define some additional requirements for collections, but this is not the default. A document is also more flexible, as it can have nested data.

Here is an example of nested data in a JSON document:

```
{
  "type": "book",
  "title": "Harry Potter and the Philosopher's Stone",
```

```
    "translations": {
        "Afrikaans": "Harry Potter en die Towenaar se Steen",
        "German": "Harry Potter und der Stein der Weisen",
    }
}
```

Here, "type", "title", and "translations" are all top-level elements, and "Afrikaans" and "German" are nested. A good example of when documents are useful is web shops, where there are different kinds of properties to be stored for each item – a T-shirt has a size and a color, a TV has a number of HDMI connections, an RC car has a battery type, and so on. Some of the properties might be nested.

Not having a strict structure and having everything in a single document instead of data spread out across multiple related tables can also be problematic. The good thing, however, is that MySQL allows you to combine NoSQL and SQL. You can use SQL to query a collection, and then the X DevAPI interface allows you to query tables in a similar way to how you would query collections.

You can either use an existing collection, or you can create a new one with the following command:

```
db.createCollection(<collection>)
```

To add the documents, use the following command:

```
db.<collection>.add()
```

To search the documents stored in the table, use the following command:

```
db.<collection>.find()
```

To search any particular document stored in the table, use the following command:

```
db.<collection>.find('<collection attribute> = "<value>"')
```

An index in a database is similar to an index in a book. Whereas a book index helps you to find the right page without having to check every one, a database index allows the server to quickly find the right records. This won't affect the output of any queries, but it does speed them up, especially when larger quantities of data are stored. Use the following command to create an index for a collection:

```
db.<collection>.createIndex(name, IndexDefinition)
```

Let's say you want to provide an index to the `code` field in the `countries` collection. Your query to create it should look like the following:

```
db.countries.createIndex('code',
  {"fields":
    {"field": "$.code"}
  }
)
```

The way to specify a part of the document is done with JSONPath, which will be explained in the next chapter in more detail.

In the next exercise, you will insert documents into a table.

Exercise 13.02 – inserting documents into a table

In this exercise, you will create a collection, insert records into it, and display the information. Ensure that you are connected to the test database. Follow the following steps to accomplish the exercise:

1. Create a collection named `countries` by writing the following command:

   ```
   db.createCollection('countries')
   ```

2. Add three records to the collection:

   ```
   db.countries.add({"code": "FR", "name": "France"})
   db.countries.add({"code": "DE", "name": "Germany"})
   db.countries.add({"code": "IT", "name": "Italy"})
   ```

3. Find the inserted records in the collection:

   ```
   db.countries.find()
   ```

This produces the following output:

```
{
    "_id": "00005e3bf4f20000000000000001",
    "code": "FR",
    "name": "France"
}
{
    "_id": "00005e3bf4f20000000000000002",
    "code": "DE",
    "name": "Germany"
}
{
    "_id": "00005e3bf4f20000000000000003",
    "code": "IT",
    "name": "Italy"
}
3 documents in set (0.0006 sec)
```

Figure 13.06 – MySQL Shell – using find()

> **Note**
>
> It is also possible to insert multiple records at once by specifying a list of JSON documents to the add() function.

4. Add two documents for `Belgium` and `Poland` in a single statement:

```
db.countries.add(
    {"code": "BE", "name": "Belgium"},
    {"code": "PL", "name": "Poland"}
)
```

5. Find a record in the collection where `code` is equal to `PL`. Use the `find` method to implement this:

```
db.countries.find('code = "PL"')
```

This produces the following output:

```
{
    "_id": "00005e3bf4f20000000000000005",
    "code": "PL",
    "name": "Poland"
}
1 document in set (0.0079 sec)
```

Figure 13.07 – MySQL – using find() with a filter

6. Create an index on the newly created collection using the `createIndex()` method:

```
db.countries.createIndex('code',
  {"fields":
    {"field": "$.code"}
  }
)
```

The statement here creates an index called `code`, and the index indexes one single field, the code in the document.

In this exercise, you created a new collection, inserted a few records into it, viewed the records, and finally, created the index. In the next section, you will explore loading data from a SQL file.

Loading data from a SQL file

A SQL file is usually generated with the `mysqldump` command so that we can export from one database system and later import into a new one. The `mysqldump` utility comes with MySQL. One file can hold data and definitions for multiple tables and schemas. Another source of the SQL file is when installing or upgrading third-party software. It is a file that contains all the changes needed to make the database ready for the new version.

Let's say you want to load the `world.sql` file. First, create a database using the following command:

```
CREATE DATABASE world;
```

Ensure that you are using the `world` database by writing the following command:

```
USE world;
```

Use the following query to load the `world.sql` file:

```
source /path/to/world.sql
```

You will be able to access the content inside the `world` database. Ensure that you have given the correct path of the `world.sql` file (which is stored in your local system).

In order to access all the tables present in the loaded database, write the following command:

```
SHOW TABLES FROM world;
```

In the next section, you will complete an exercise where you will load data from a SQL file and view its details.

Exercise 13.03 – loading data from a SQL file and viewing tables

In this exercise, you will load the data from the `world` database and access table information present inside it. Follow the following steps to accomplish this:

> **Note**
>
> The `world` database used in this exercise can be found here: `https://github.com/PacktWorkshops/The-MySQL-Workshop/tree/master/Chapter08`.
>
> This database is provided by Oracle MySQL as a sample database. The data comes from Statistics Finland. More details can be found at `https://dev.mysql.com/doc/world-setup/en/`. This database contains three tables – a table with cities, a table with countries, and a table with languages. A country can have multiple cities and multiple languages.

1. Open the MySQL client and import the `world.sql` file with the following command:

    ```
    source /path/to/world.sql
    ```

    This should result in many lines, like these:

    ```
    Query OK, 1 row affected (0.00 sec)
    ```

    ```
    Query OK, 1 row affected (0.00 sec)
    ```

```
Query OK, 1 row affected (0.00 sec)


Query OK, 1 row affected (0.00 sec)
```

> **Note**
> Please make sure that you provide the correct path to the `world.sql` file.

Now, you should be able to list the tables you just imported.

2. Write the following command to view all the tables present in the `world` database:

```
SHOW TABLES FROM world;
```

This produces the following output:

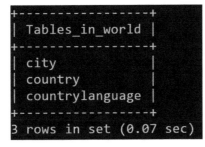

```
+-----------------+
| Tables_in_world |
+-----------------+
| city            |
| country         |
| countrylanguage |
+-----------------+
3 rows in set (0.07 sec)
```

Figure 13.08 – Tables in the world database after importing

3. View the details of the `city` table using the `describe` command:

```
DESCRIBE world.city;
```

This produces the following output:

```
+-------------+----------+------+-----+---------+----------------+
| Field       | Type     | Null | Key | Default | Extra          |
+-------------+----------+------+-----+---------+----------------+
| ID          | int      | NO   | PRI | NULL    | auto_increment |
| Name        | char(35) | NO   |     |         |                |
| CountryCode | char(3)  | NO   | MUL |         |                |
| District    | char(20) | NO   |     |         |                |
| Population  | int      | NO   |     | 0       |                |
+-------------+----------+------+-----+---------+----------------+
5 rows in set (0.00 sec)
```

Figure 13.09 – The city table definition

Thus, you have imported a SQL file and viewed its information. Note that depending on how a SQL file is created, it may have only data or the definition of tables. The `world.sql` file had both. You should only import SQL files from sources you trust, especially when importing as the root user. This is because any kind of statement can be put in such a file, including statements to change a configuration, add users, or change passwords.

4. Another way of importing a SQL file is by using the following command:

```
mysql < /path/to/world.sql
```

The result is similar, but there are a few slight differences. There is less output by default and the process will halt when there is an error in the file. To get more output about what statements are being run, you can invoke `mysql` with the verbose option.

There is another way of importing a SQL file using MySQL Workbench. In the next exercise, you will see how you can use Workbench to import the SQL file.

Exercise 13.04 – importing a SQL file using MySQL Workbench

In this exercise, you will make use of MySQL Workbench to import the `world.sql` file. In order for this example to work, you need to get rid of the previously imported world database. Perform the following steps to accomplish this:

> **Note**
>
> The `world` database used in this exercise can be found here: `https://github.com/PacktWorkshops/The-MySQL-Workshop/tree/master/Chapter13`.

1. Start **MySQL Workbench**.
2. Connect to your database.

3. Issue the following command:

```
DROP DATABASE world;
```

4. Now, select **Data Import/Restore** in the **Administration** tab.

5. Specify the `world.sql` file and click **Start Import**. The following screenshot shows the output that is generated after implementing the preceding steps:

Figure 13.10 – MySQL Workbench – Data Import

6. Once the import is complete, you will see the **Import Completed** status in the
    Administration window:

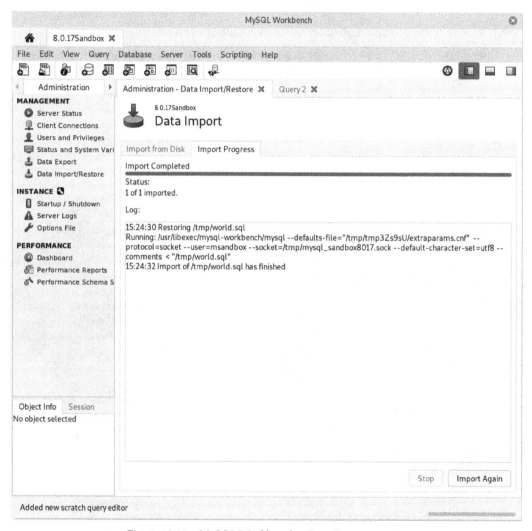

Figure 13.11 – MySQL Workbench – Data Import output

As you can see in the preceding screenshot, we are basically doing the same thing as we
did before. The `world.sql` file was successfully imported. Now, let's explore loading
data from a CSV file.

Loading data from a CSV file

CSV files are often used to exchange data between different systems. This can be between database systems from different vendors or between a database and a spreadsheet application such as Microsoft Excel. The biggest problem with **CSV** (which stands for **Comma Separated Values**) is that there is not a single standard. This leads to a plethora of sub-formats. The differences are mostly in which character is really used for separating values. It is not always a comma; it can be a semicolon or something else entirely. And the other points on which these files differ are how the values, which contain the separator character, escape and how newlines are handled. You might also come across TSV files, which are very similar to CSV but separated by tabs. These files can mostly be handled like CSV files and are quite common with MySQL.

In order to read and save these files, you need to first check the directory where they can be saved. In order to do that, we use the following command:

```
SHOW VARIABLES LIKE 'secure_file_priv';
```

This variable contains the value (path) where you can save your CSV file. The preceding code when used in a Windows OS will return the following path:

```
C:\ProgramData\MySQL\MySQL Server 8.0\Uploads
```

The SELECT...INTO OUTFILE Format

This format allows you to write selected rows of a table to a file. Let's say you want to export the data of the `city` table to a `city.csv` file and store it in the preceding path. Use the following query:

```
SELECT * FROM city INTO OUTFILE 'C:/ProgramData/MySQL/MySQL
Server 8.0/Uploads/city.csv'
FIELDS TERMINATED BY ',' OPTIONALLY ENCLOSED BY '"'
LINES TERMINATED BY '\n';
```

In the preceding query, the terms FIELDS TERMINATED BY ',', OPTIONALLY ENCLOSED BY '"', and LINES TERMINATED BY '\n' are column and line terminators.

They can be specified in the command to produce a certain format of output.

Once you have exported the table data to a CSV file, view it using the following command:

```
\! type "C:\ProgramData\MySQL\MySQL Server 8.0\Uploads\city.
csv" | more
```

The \! command on the MySQL prompt allows you to call system commands. type helps to display all the data of the file. Please note that type is used on the Windows OS. more helps to display the contents of this file one screen at a time.

The LOAD DATA INFILE...INTO format

In order to load data from a CSV file into a table, we use this format. Let's say we have created a copy of the city table named copy_of_city and want to load the data of city.csv into this table. In this case, we will use the following command:

```
LOAD DATA INFILE 'C:/ProgramData/MySQL/MySQL Server 8.0/
Uploads/city.csv' INTO TABLE copy_of_city CHARACTER SET latin1
FIELDS TERMINATED BY ',' OPTIONALLY ENCLOSED BY '"'
LINES TERMINATED BY '\n';
```

Now, you have learned how to write data from a table to a CSV file and read data from a CSV file to a table. In the next exercise, you will load data from a CSV file.

Exercise 13.05 – loading data from a CSV file

In this exercise, you will first export the city table of the world database into a CSV file, import it, and check whether everything works fine. Follow the following steps to accomplish this:

1. Open the MySQL client.

2. Connect to the world database:

   ```
   USE world;
   ```

3. To check which directories you can save files in, write the following command:

   ```
   SHOW VARIABLES LIKE 'secure_file_priv';
   ```

   This produces the following output:

```
+-------------------+--------------------------------------------------+
| Variable_name     | Value                                            |
+-------------------+--------------------------------------------------+
| secure_file_priv  | C:\ProgramData\MySQL\MySQL Server 8.0\Uploads\   |
+-------------------+--------------------------------------------------+
```

Figure 13.12 – Checking directories

4. Now, you can use the preceding path in order to export the data of the `city` table and save it in a CSV file.

5. Export the `city` table into the `city.csv` file by writing the following command:

```
SELECT * FROM city INTO OUTFILE 'C:/ProgramData/MySQL/
MySQL Server 8.0/Uploads/city.csv'
FIELDS TERMINATED BY ',' OPTIONALLY ENCLOSED BY '"'
LINES TERMINATED BY '\n';
```

This produces the following output:

```
Query OK, 4079 rows affected (0.01 sec)
```

Figure 13.13 – Data exported to the CSV file

6. To check the results of the `city.csv` file, write the following command:

```
\! type "C:\ProgramData\MySQL\MySQL Server 8.0\Uploads\
city.csv" | more
```

The output will look like the following:

```
1,"Kabul","AFG","Kabol",1780000
2,"Qandahar","AFG","Qandahar",237500
3,"Herat","AFG","Herat",186800
4,"Mazar-e-Sharif","AFG","Balkh",127800
5,"Amsterdam","NLD","Noord-Holland",731200
6,"Rotterdam","NLD","Zuid-Holland",593321
7,"Haag","NLD","Zuid-Holland",440900
8,"Utrecht","NLD","Utrecht",234323
9,"Eindhoven","NLD","Noord-Brabant",201843
10,"Tilburg","NLD","Noord-Brabant",193238
11,"Groningen","NLD","Groningen",172701
```

Figure 13.14 – Inspecting the contents of the city.csv file

This is done with the `INTO OUTFILE` part of a `SELECT` statement.

> **Note**
>
> Depending on how MySQL is configured, there may be limitations to where a file can be placed. You can run `SELECT @@global.secure_file_priv`, which should give you the path where you can write to. If this returns NULL, then you first need to change the configuration of the MySQL server.

7. Now, create a table called `copy_of_city` with the same structure as `city` using the following command:

```
CREATE TABLE copy_of_city LIKE city;
```

8. To import, use the `LOAD DATA` statement, as follows:

```
LOAD DATA INFILE 'C:/ProgramData/MySQL/MySQL Server 8.0/
Uploads/city.csv' INTO TABLE copy_of_city
FIELDS TERMINATED BY ',' OPTIONALLY ENCLOSED BY '"' LINES
TERMINATED BY '\n';
```

This produces the following output:

```
Query OK, 4079 rows affected (1.79 sec)
Records: 4079  Deleted: 0  Skipped: 0  Warnings: 0
```

Figure 13.15 – Loading the contents of city.csv into the copy_of_city table with LOAD DATA INFILE

In this exercise, you loaded data from the CSV file. In case of any errors, you can troubleshoot. You need to verify whether the table and the CSV file have the same number of columns and in the same order. You also need to verify whether the character set of the files is what you expect it to be. When in doubt, you can convert the file to UTF-8 first.

Besides the `LOAD DATA` options used to describe the format of the CSV, you can also specify which values go into which columns and even use expressions for this (for example, to add and to multiply two fields). To see the built-in help from the server, you can use `\help LOAD DATA`.

The `mysqlimport` utility that comes with MySQL can also help you to load data. In the end, this uses the same `LOAD DATA` command and has most of the same options.

In the next section, we will explore loading data from a JSON file.

Loading data from a JSON file

MySQL uses a format of a file with one JSON document per line to allow quick and easy import of documents into a database. To import JSON documents into collections in MySQL, we can use MySQL Shell with the `util.importJson()` function.

For example, if you need to import a JSON file named `languages.json`, then you write the following command:

```
util.importJson('/path/to/languages.json')
```

Ensure that you give the correct path to the JSON file.

In order to update the schema names, use the \rehash command. Once you have updated the schema names, you can view the data of the collection using the following command:

```
db.languages.find()
```

You can also sort the details of a collection using the sort() method. Here, you need to specify the column name inside the method. Consider a scenario where you need to sort the passenger details in the airports file, and then find the airport having the highest number of passengers. In this case, you write the following command:

```
db.airports.find().sort('passengers DESC')
```

You can also use JSONPath syntax to extract fields from the JSON stored in the doc column and make it available as a column. The ->> operator is a shorthand for extracting and unquoting. The ->> operator is also available to only do the extraction but not the unquote operation. Let's say you want to fetch column1 and column2 from mytable. You can write the following command to achieve this:

```
SELECT
    doc->>'$.column1' AS column1,
    doc->>'$.column2' AS column2,
FROM mytable;
```

Thus, you have learned to import a JSON file and view its details. Now, solve an exercise based on it to practice them.

Exercise 13.06 – loading data from a JSON file

In this exercise, you will create the_beatles.json file, input values, import the file using the importJson() function, and view the details in MySQL Shell. Follow the following steps to implement this exercise:

1. Create a JSON file named the_beatles.json and add the following contents within it:

```
{"name": "Rubber Soul"}
{"name": "Revolver"}
{"name": "Sgt. Pepper's Lonely Hearts Club Band"}
{"name": "Magical Mystery Tour"}
{"name": "Yellow Submarine"}
```

2. Open MySQL shell and connect to the test database with the help of the following command:

```
\use test
```

3. Use the `importJson()` function to import the preceding JSON file:

```
util.importJson('/path/to/the_beatles.json')
```

> **Note**
>
> Ensure that you provide the correct path of the JSON file in the preceding command.

The JSON file gets imported, and you can see the progress in the shell:

```
.. 5
Processed 156 bytes in 5 documents in 0.2406 sec (20.78 documents/s)
Total successfully imported documents 5 (20.78 documents/s)
```

Figure 13.16 – Loading documents into a collection

4. Call `rehash`:

```
\rehash
```

This produces the following output:

```
Fetching schema names for autocompletion... Press ^C to stop.
```

Figure 13.17 – Fetching schema names for autocompletion

5. Show the contents of the collection in the shell:

```
db.the_beatles.find()
```

This produces the following output:

```
{
    "_id": "00005e3bf4f20000000000000006",
    "name": "Rubber Soul"
}
{
    "_id": "00005e3bf4f20000000000000007",
    "name": "Revolver"
}
{
    "_id": "00005e3bf4f20000000000000008",
    "name": "Sgt. Pepper's Lonely Hearts Club Band"
}
{
    "_id": "00005e3bf4f20000000000000009",
    "name": "Magical Mystery Tour"
}
{
    "_id": "00005e3bf4f2000000000000000a",
    "name": "Yellow Submarine"
}
5 documents in set (0.0213 sec)
```

Figure 13.18 – Using find() to inspect a collection

As you can see in the preceding screenshot, _id is automatically generated. As \h importJson will tell you, there are many options – for example, to set the name and schema of the collection to which the data will be imported.

In the next sections, we will explore the CSV storage engine and how it can be used to export as well as import data.

Using the CSV storage engine to export data

MySQL supports multiple storage engines. The default storage engine is InnoDB, but there are a few more shipped with the server. There are also third-party storage engines available, an example of which is the MyRocks storage engine from Facebook, which allows MySQL to use RocksDB to store data. The job of a storage engine is to store and retrieve data, while the server knows how to parse and execute SQL queries. The CSV storage engine that comes with MySQL allows you to store data in a CSV file and query it via SQL. This can be used to export and import data. The server knows how to copy or move data from one storage engine to another, so this also works for data that is stored in any other storage engine. The main limitation of this is that you require direct access to the filesystem of the server.

You can also make your tables use a particular storage engine. Consider the following code where you modify your `languages` table to use the CSV storage engine:

```
ALTER TABLE languages ENGINE=CSV;
```

Let's say you want to fetch data from a source table and copy it into a `destination` table and then later use this `destination` table so that you can export its data to a CSV file. In order to do this, you need to first copy all columns from the `source` table to the `destination` table. This can be done with the help of the following command:

```
INSERT INTO <destination_table> SELECT * FROM <source_table>
WHERE Column=<value>;
```

Now, you can export the data of the destination table to MySQL's `datadir` folder. To check the path of this directory, type the following command:

```
SELECT @@datadir;
```

If you are using Windows, then you will get the following path in the response:

```
C:\ProgramData\MySQL\MySQL Server 8.0\Data\
```

Now, you will practice what you have learned so far and make use of the CSV storage engine to export data in the next exercise.

Exercise 13.07 – utilizing the CSV storage engine to export data

In this exercise, you will make use of the CSV storage engine to export the data of the `city` table into a CSV file. Follow the following steps to implement this:

1. Open MySQL Shell and connect to the `world` database with the help of the following command:

   ```
   USE world
   ```

2. Create a `city_export` table with the same structure as the `city` table using the following command:

   ```
   CREATE TABLE city_export LIKE city;
   ```

3. Inspect the structure of the new table with the help of the following command:

```
SHOW CREATE TABLE city_export\G
```

This produces the following output:

```
*************************** 1. row ***************************
        Table: city_export
Create Table: CREATE TABLE `city_export` (
  `ID` int NOT NULL AUTO_INCREMENT,
  `Name` char(35) NOT NULL DEFAULT '',
  `CountryCode` char(3) NOT NULL DEFAULT '',
  `District` char(20) NOT NULL DEFAULT '',
  `Population` int NOT NULL DEFAULT '0',
  PRIMARY KEY (`ID`),
  KEY `CountryCode` (`CountryCode`)
) ENGINE=InnoDB DEFAULT CHARSET=latin1
1 row in set (0.05 sec)
```

Figure 13.19 – Inspecting the structure of the new table

4. Remove auto-increments, secondary indexes, and primary keys from the new table. This is necessary, as the CSV storage engine doesn't support these. Write the following commands to implement this:

```
ALTER TABLE city_export MODIFY COLUMN 'ID' int NOT NULL;
ALTER TABLE city_export DROP KEY CountryCode;
ALTER TABLE city_export DROP PRIMARY KEY;
Change the table to use the CSV storage engine:
ALTER TABLE city_export ENGINE=CSV;
```

5. Copy all rows with CountryCode=RUS into the newly created city_export table:

```
INSERT INTO city_export SELECT * FROM city WHERE
CountryCode='RUS';
```

This produces the following output:

```
Query OK, 189 rows affected (0.14 sec)
Records: 189  Duplicates: 0  Warnings: 0
```

Figure 13.20 – Copying data from the city table to the city_export table
using INSERT INTO…SELECT…FROM

6. The result is placed in the MySQL `datadir` in a folder with the same name as the database. Write the following command to check the directory:

```
SELECT @@datadir;
```

This produces the following output:

```
+-----------------------------------------------+
| @@datadir                                     |
+-----------------------------------------------+
| C:\ProgramData\MySQL\MySQL Server 8.0\Data\   |
+-----------------------------------------------+
1 row in set (0.00 sec)
```

Figure 13.21 – Checking the directory

7. Now, write the following command to check the results of the CSV file:

```
\! type "C:\ProgramData\MySQL\MySQL Server 8.0\Data\
world\city_export.csv" | more
```

This produces the following output:

```
3580,"Moscow","RUS","Moscow (City)",8389200
3581,"St Petersburg","RUS","Pietari",4694000
3582,"Novosibirsk","RUS","Novosibirsk",1398800
3583,"Nizni Novgorod","RUS","Nizni Novgorod",1357000
3584,"Jekaterinburg","RUS","Sverdlovsk",1266300
3585,"Samara","RUS","Samara",1156100
3586,"Omsk","RUS","Omsk",1148900
3587,"Kazan","RUS","Tatarstan",1101000
3588,"Ufa","RUS","BaÜkortostan",1091200
```

Figure 13.22 – Inspecting the contents of the new CSV file

Thus, we exported the data into a CSV file with the help of the CSV storage engine.

In the next section, we will learn to use the CSV storage engine to import data.

Using the CSV storage engine to import data

After defining a CSV table, you can replace the CSV file. For that, you need to run the `FLUSH TABLE <table>` command to ensure that the server rereads this and then the data is available. Having the data in a CSV table, however, is probably not the endpoint you want to get to, as it doesn't support indexing or primary keys. So, the next step would be to use the following:

```
"INSERT INTO <new_table> SELECT * FROM <csv_table>"
```

Alternatively, you can run the following:

```
'ALTER TABLE <csv_table> ENGINE=InnoDB'
```

Either of the options can be used to convert the table to InnoDB. Once this is done, you should define a primary key and add indexes if needed. Let's see an exercise where we will make use of the CSV storage engine to import data.

> **Note**
> On Windows, always start the MySQL command-line client via the MySQL Command Line Client – Unicode entry. Non-Unicode causes text to not be displayed correctly in some cases. To verify that you are using the correct strings, run the `status` command in the MySQL client, which should return `utf8mb4` for all the four lines that show the character set.

Exercise 13.08 – utilizing the CSV storage engine to import data

In this exercise, you will be making use of the CSV storage engine to import data from the table created in the preceding exercise and display the results in MySQL Shell. Follow the following steps to implement the exercise:

1. Open MySQL Shell.

2. Connect to the `world` database:

    ```
    USE world
    ```

3. Check the contents from the `city_export` table:

    ```
    SELECT * FROM city_export WHERE District='Moskova';
    ```

This produces the following output:

```
+------+-----------------+-------------+----------+------------+
| ID   | Name            | CountryCode | District | Population |
+------+-----------------+-------------+----------+------------+
| 3673 | Podolsk         | RUS         | Moskova  |     194300 |
| 3687 | Ljubertsy       | RUS         | Moskova  |     163900 |
| 3693 | Mytištši        | RUS         | Moskova  |     155700 |
| 3694 | Kolomna         | RUS         | Moskova  |     150700 |
| 3695 | Elektrostal     | RUS         | Moskova  |     147000 |
| 3703 | Himki           | RUS         | Moskova  |     133700 |
| 3704 | Balašiha        | RUS         | Moskova  |     132900 |
| 3707 | Korolev         | RUS         | Moskova  |     132400 |
| 3708 | Serpuhov        | RUS         | Moskova  |     132000 |
| 3709 | Odintsovo       | RUS         | Moskova  |     127400 |
| 3710 | Orehovo-Zujevo  | RUS         | Moskova  |     124900 |
| 3720 | Noginsk         | RUS         | Moskova  |     117200 |
| 3726 | Sergijev Posad  | RUS         | Moskova  |     111100 |
| 3736 | Štšolkovo       | RUS         | Moskova  |     104900 |
| 3745 | Zeleznodoroznyi | RUS         | Moskova  |     100100 |
| 3756 | Zukovski        | RUS         | Moskova  |      96500 |
| 3765 | Krasnogorsk     | RUS         | Moskova  |      91000 |
| 3766 | Klin            | RUS         | Moskova  |      90000 |
+------+-----------------+-------------+----------+------------+
18 rows in set (0.00 sec)
```

Figure 13.23 – Inspecting the contents of the city_export table

The preceding command returns the rows that have Moskova as a district. Another way of getting a subset is to use LIMIT 10.

4. Modify the city_export.csv file. Change some numbers in city_export.csv or add a record; just ensure you use the exact same CSV format.

5. Check the database table again:

```
FLUSH TABLE city_export;
SELECT * FROM city_export WHERE District='Moskova';
```

This produces the following output:

```
+------+----------------+-------------+----------+------------+
| ID   | Name           | CountryCode | District | Population |
+------+----------------+-------------+----------+------------+
| 3673 | Podolsk        | RUS         | Moskova  |     194300 |
| 3687 | Ljubertsy      | RUS         | Moskova  |     163900 |
| 3693 | Mytištši       | RUS         | Moskova  |     155700 |
| 3694 | Kolomna        | RUS         | Moskova  |     150700 |
| 3695 | Elektrostal    | RUS         | Moskova  |     147000 |
| 3703 | Himki          | RUS         | Moskova  |     133700 |
| 3704 | Balašiha       | RUS         | Moskova  |     132900 |
| 3707 | Korolev        | RUS         | Moskova  |     132400 |
| 3708 | Serpuhov       | RUS         | Moskova  |     132000 |
| 3709 | Odintsovo      | RUS         | Moskova  |     127400 |
| 3710 | Orehovo-Zujevo | RUS         | Moskova  |     124900 |
| 3720 | Noginsk        | RUS         | Moskova  |     117200 |
| 3726 | Sergijev Posad | RUS         | Moskova  |     111100 |
| 3736 | Štšolkovo      | RUS         | Moskova  |     104900 |
| 3745 | Zeleznodoroznyi| RUS         | Moskova  |     100100 |
| 3756 | Zukovski       | RUS         | Moskova  |      96500 |
| 3765 | Krasnogorsk    | RUS         | Moskova  |      91000 |
| 3766 | Klin           | RUS         | Moskova  |      90000 |
+------+----------------+-------------+----------+------------+
18 rows in set (0.00 sec)
```

Figure 13.24 – The city_export table after directly modifying the city_export.csv file

This will reflect the changes you made to the file directly. This shows that creating a CSV table and then replacing the data allows you to use the CSV storage engine to import data into MySQL.

In the next section, we will solve an activity based on inserting airport records using SQL.

Searching and filtering JSON documents

To do this, we use the `worldcol` collection, which is a collection of JSON documents generated from the tables in the `world` database. Unless otherwise specified, we will use MySQL Shell in JavaScript mode.

To create this collection, we need to first run the statements in the `worldcol.js` file by writing the following query:

```
\connect —mx root@127.0.0.1:33060
\source worldcol.js
```

This outputs the following results:

```
MySQL  127.0.0.1:18018+ ssl  JS > \source worldcol.js
Default schema `world` accessible through db.
<Collection:worldcol>
Fetching table and column names from `world` for auto-completion... Press ^C to stop.
Query OK, 4079 rows affected (0.6339 sec)

Records: 4079  Duplicates: 0  Warnings: 0
```

Figure 13.25 – Output of sourcing worldcol.js

To explore, you first need to get a single document from the collection:

```
db.worldcol.find().limit(1)
```

This returns the first document of the `worldcol` collection:

```
MySQL  127.0.0.1:18018+ ssl  world  JS > db.worldcol.find().limit(1)
{
    "_id": 1,
    "name": "Kabul",
    "country": {
        "GNP": 5976,
        "code": "AFG",
        "name": "Afghanistan",
        "code2": "AF",
        "region": "Southern and Central Asia",
        "continent": "Asia",
        "local_name": "Afganistan/Afqanestan",
        "population": 22720000,
        "surface_area": 652090,
        "head_of_state": "Mohammad Omar",
        "government_form": "Islamic Emirate",
        "independence_year": 1919
    },
    "district": "Kabol",
    "language": {
        "Dari": {
            "percentage": 32.1,
            "is_official": true
        },
        "Uzbek": {
            "percentage": 8.8,
            "is_official": false
        },
        "Pashto": {
            "percentage": 52.4,
            "is_official": true
        },
        "Balochi": {
            "percentage": 0.9,
            "is_official": false
        },
        "Turkmenian": {
            "percentage": 1.9,
            "is_official": false
        }
    },
    "is_capital": true,
    "population": 1780000
}
1 document in set (0.0012 sec)
```

Figure 13.26 – The find() output with limit(1)

The `find` function accepts an argument to filter rows:

```
db.worldcol.find('name="Paris"')
```

This returns the following results:

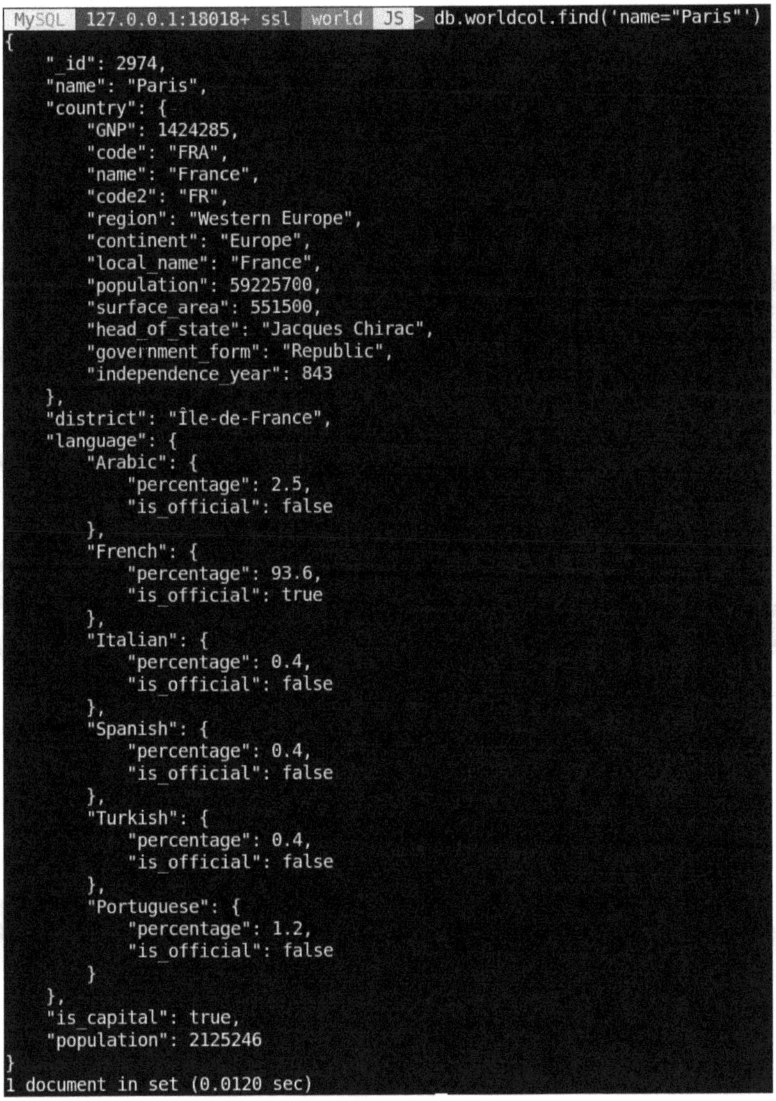

```
MySQL  127.0.0.1:18018+ ssl  world  JS > db.worldcol.find('name="Paris"')
{
    "_id": 2974,
    "name": "Paris",
    "country": {
        "GNP": 1424285,
        "code": "FRA",
        "name": "France",
        "code2": "FR",
        "region": "Western Europe",
        "continent": "Europe",
        "local_name": "France",
        "population": 59225700,
        "surface_area": 551500,
        "head_of_state": "Jacques Chirac",
        "government_form": "Republic",
        "independence_year": 843
    },
    "district": "Île-de-France",
    "language": {
        "Arabic": {
            "percentage": 2.5,
            "is_official": false
        },
        "French": {
            "percentage": 93.6,
            "is_official": true
        },
        "Italian": {
            "percentage": 0.4,
            "is_official": false
        },
        "Spanish": {
            "percentage": 0.4,
            "is_official": false
        },
        "Turkish": {
            "percentage": 0.4,
            "is_official": false
        },
        "Portuguese": {
            "percentage": 1.2,
            "is_official": false
        }
    },
    "is_capital": true,
    "population": 2125246
}
1 document in set (0.0120 sec)
```

Figure 13.27 – The find() output for name="Paris"

You might not need the whole document, so try to restrict what fields you return from the document:

```
db.worldcol.find('is_capital=true').
fields('name').
limit(5)
```

This returns five cities that are capitals of their country:

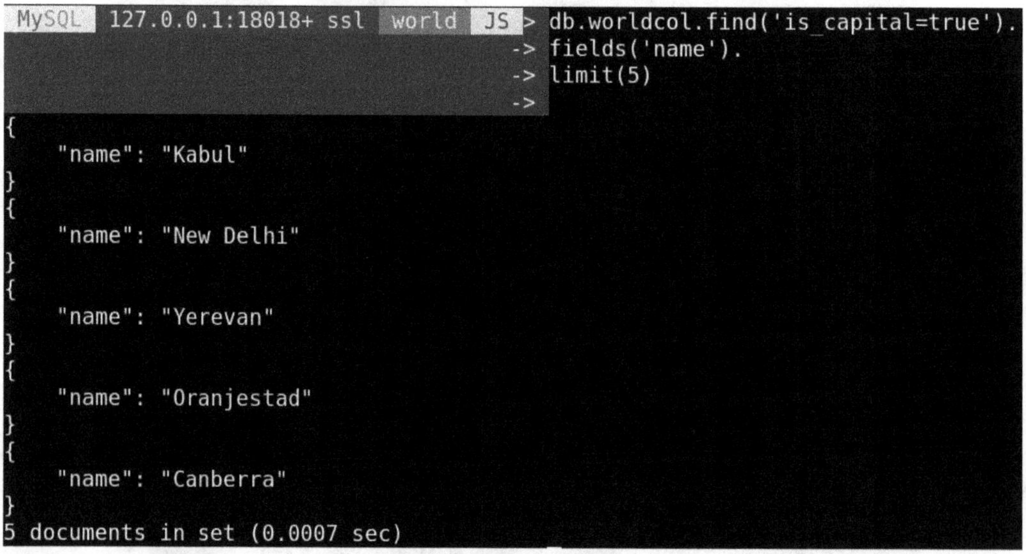

```
MySQL  127.0.0.1:18018+ ssl  world  JS > db.worldcol.find('is_capital=true').
                                     -> fields('name').
                                     -> limit(5)
                                     ->
{
    "name": "Kabul"
}
{
    "name": "New Delhi"
}
{
    "name": "Yerevan"
}
{
    "name": "Oranjestad"
}
{
    "name": "Canberra"
}
5 documents in set (0.0007 sec)
```

Figure 13.28 – The find() output for the names of capitals (limited to five results)

Now, we want to see how many cities we have in total:

```
db.worldcol.count()
```

This returns 4079 cities:

```
MySQL  127.0.0.1:18018+ ssl  world  JS > db.worldcol.count()
4079
```

Figure 13.29 – The count() output for woldcol

If you want, you can retrieve multiple fields from the document:

```
db.worldcol.find('is_capital=true').
fields('name', 'country.name').
limit(5)
```

This will produce the following output:

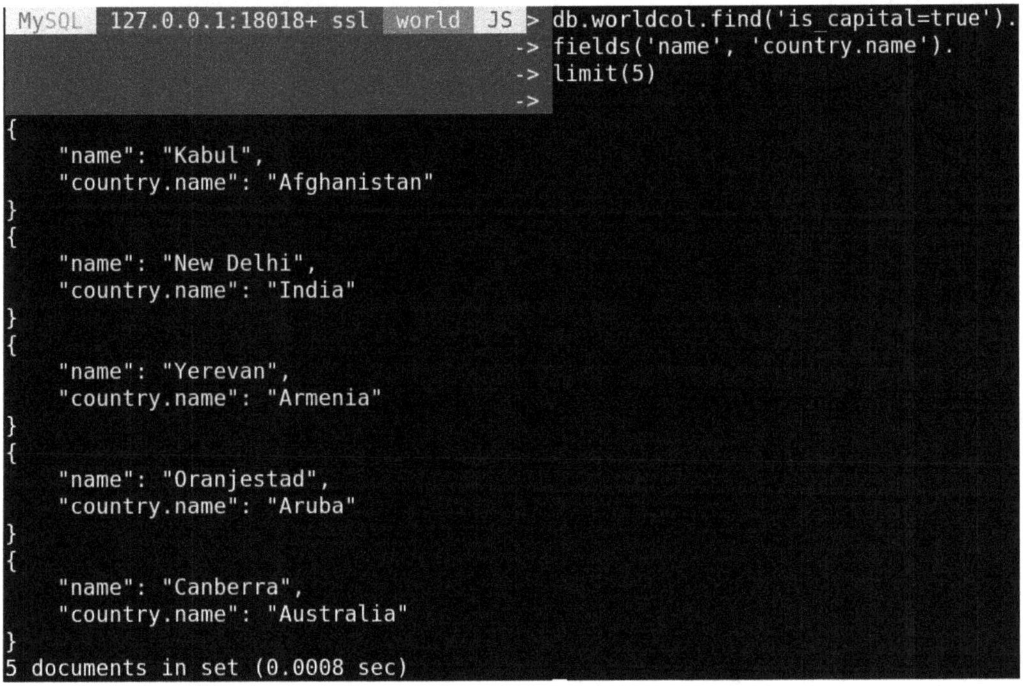

```
MySQL  127.0.0.1:18018+ ssl  world  JS > db.worldcol.find('is_capital=true').
                                     -> fields('name', 'country.name').
                                     -> limit(5)
                                     ->
{
    "name": "Kabul",
    "country.name": "Afghanistan"
}
{
    "name": "New Delhi",
    "country.name": "India"
}
{
    "name": "Yerevan",
    "country.name": "Armenia"
}
{
    "name": "Oranjestad",
    "country.name": "Aruba"
}
{
    "name": "Canberra",
    "country.name": "Australia"
}
5 documents in set (0.0008 sec)
```

Figure 13.30 – The find() output with a filter, field selection, and a limit

You can also aggregate results with `groupBy()`:

```
db.worldcol.find('language.Kazakh').
fields('country.name').
groupBy('country.name')
```

This finds the country name for cities that have `language.Kazakh`. Then, the results are grouped by `country.name`:

```
MySQL  127.0.0.1:18018+ ssl  world  JS > db.worldcol.find('language.Kazakh').
                                     -> fields('country.name').
                                     -> groupBy('country.name')
                                     ->
{
    "country.name": "Kazakstan"
}
{
    "country.name": "Kyrgyzstan"
}
{
    "country.name": "Mongolia"
}
{
    "country.name": "Turkmenistan"
}
{
    "country.name": "Uzbekistan"
}
{
    "country.name": "Russian Federation"
}
6 documents in set (0.0036 sec)
```

Figure 13.31 – Countries that use the Kazakh language

To get the top five cities in Russia by population, you can use `sort()` and `limit()` together:

```
db.worldcol.find('country.name="Russian Federation"').
fields('name', 'population').
sort('population desc').
limit(5)
```

This produces the following output:

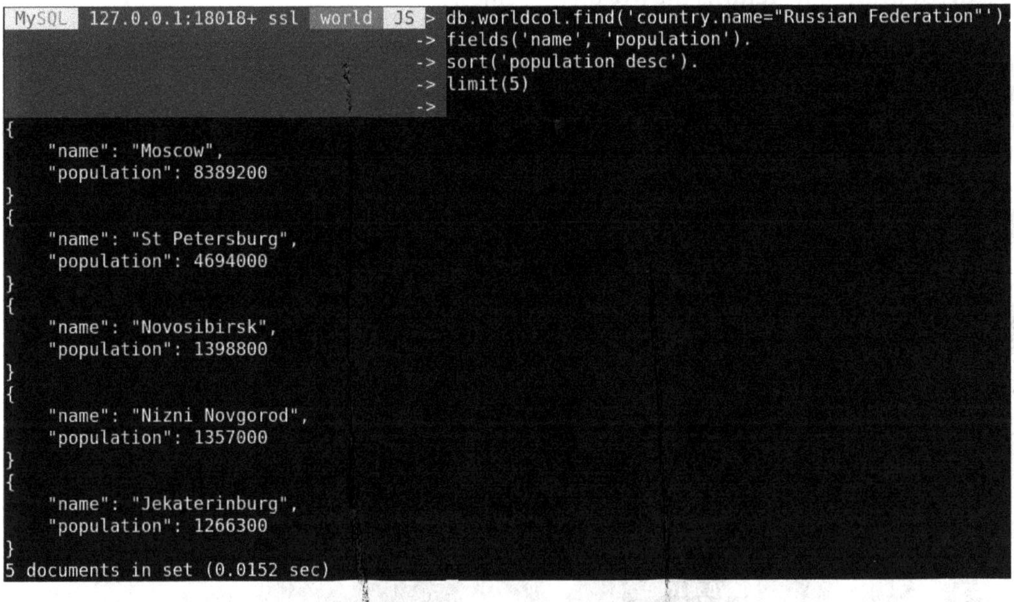

```
MySQL  127.0.0.1:18018+ ssl  world  JS > db.worldcol.find('country.name="Russian Federation"').
                                     -> fields('name', 'population').
                                     -> sort('population desc').
                                     -> limit(5)
                                     ->
{
    "name": "Moscow",
    "population": 8389200
}
{
    "name": "St Petersburg",
    "population": 4694000
}
{
    "name": "Novosibirsk",
    "population": 1398800
}
{
    "name": "Nizni Novgorod",
    "population": 1357000
}
{
    "name": "Jekaterinburg",
    "population": 1266300
}
5 documents in set (0.0152 sec)
```

Figure 13.32 – The find() output with the top five Russian cities by population

You can also do calculations inside the query:

```
db.worldcol.find('country.name="Romania"').

fields(

    'name',
    'population',
    'country.population',
    '100*population/country.population as pct_of_country'
)
.sort('population desc')
.limit(5)
```

Here, you calculate the percentage of the country's population that lives in a city in `Romania`, which only returns the five biggest cities:

```
MySQL  127.0.0.1:18018+ ssl  world  JS > db.worldcol.find('country.name="Romania"').
                                      -> fields(
                                      ->     'name',
                                      ->     'population',
                                      ->     'country.population',
                                      ->     '100*population/country.population as pct_of_country'
                                      -> )
                                      -> .sort('population desc')
                                      -> .limit(5)
                                      ->
{
    "name": "Bucuresti",
    "population": 2016131,
    "pct_of_country": 8.97833938233395,
    "country.population": 22455500
}
{
    "name": "Iasi",
    "population": 348070,
    "pct_of_country": 1.5500434192068755,
    "country.population": 22455500
}
{
    "name": "Constanta",
    "population": 342264,
    "pct_of_country": 1.5241878381688228,
    "country.population": 22455500
}
{
    "name": "Cluj-Napoca",
    "population": 332498,
    "pct_of_country": 1.4806973792612057,
    "country.population": 22455500
}
{
    "name": "Galati",
    "population": 330276,
    "pct_of_country": 1.4708022533455056,
    "country.population": 22455500
}
5 documents in set (0.0048 sec)
```

Figure 13.33 – The find() output with a calculation

In the next section, you will complete an exercise based on searching the collections and filtering the documents.

Exercise 13.09 – Searching collections and filtering documents

To expand your food palette in the Punjab region of India, you are looking to see which cities are the biggest and thus could have the biggest customer base. In this exercise, you will use the `worldcol` collection to find the biggest city in the `Punjab` region of India. You will start by connecting MySQL Shell in JavaScript mode to the `world` database, then loading the `worldcol` collection, filtering rows based on district and country name, and ordering the results by city population. Follow these steps to complete this exercise:

1. Connect MySQL Shell in JavaScript mode to the `world` database:

   ```
   \js
   \use world
   ```

   This produces the following output:

   Figure 13.34 – MySQL Shell – connecting to the world schema in JavaScript mode

2. Load the `worldcol` collection if you have not done so already:

   ```
   \source worldcol.js
   ```

   This produces the following output:

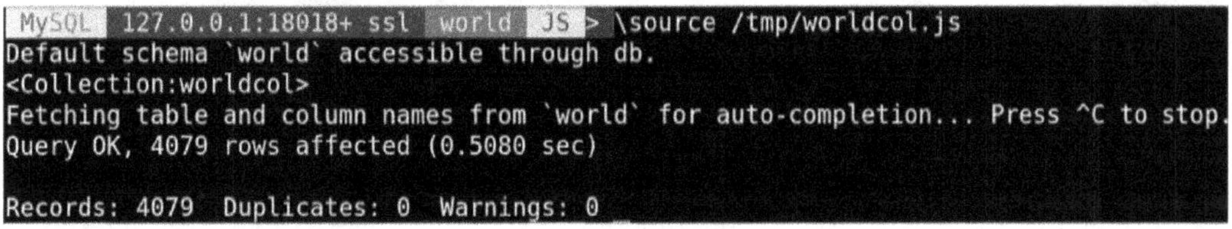

   Figure 13.35 – MySQL Shell – importing worldcol.js

3. Filter the rows based on the `district` and `country` names:

   ```
   db.worldcol.find('district="Punjab" and country.
   name="India").
   fields('name','population')
   ```

This produces the following output:

```
MySQL  127.0.0.1:18018+ ssl  world  JS  > db.worldcol.find('district="Punjab" and country.name="India"').
                                      -> fields('name','population')
                                      ->
{
    "name": "Ludhiana",
    "population": 1042740
}
{
    "name": "Amritsar",
    "population": 708835
}
{
    "name": "Jalandhar (Jullundur)",
    "population": 509510
}
{
    "name": "Patiala",
    "population": 238368
}
{
    "name": "Bhatinda (Bathinda)",
    "population": 159042
}
{
    "name": "Pathankot",
    "population": 123930
}
{
    "name": "Hoshiarpur",
    "population": 122705
}
{
    "name": "Moga",
    "population": 108304
}
{
    "name": "Abohar",
    "population": 107163
}
9 documents in set (0.0031 sec)
```

Figure 13.36 – MySQL Shell – filtering out cities in the Punjab district in India

Here, for `find()`, you filter on `district="Punjab"` and then `country.name="India"`. If you don't know what fields are available, you can run `find()`.`limit(1)` to see what the first document in the collection looks like. After filtering, you select the name and population fields by using `fields()`.

4. Order the results by `population` with the help of the following query:

```
db.worldcol.find('district="Punjab" and country.
name="India"').
fields('name','population').
sort('population desc')
```

This produces the following output:

```
MySQL  127.0.0.1:18018+ ssl  world  JS > db.worldcol.find('district="Punjab" and country.name="India"').
                                     -> fields('name','population').
                                     -> sort('population desc')
                                     ->
{
    "name": "Ludhiana",
    "population": 1042740
}
{
    "name": "Amritsar",
    "population": 708835
}
{
    "name": "Jalandhar (Jullundur)",
    "population": 509510
}
{
    "name": "Patiala",
    "population": 238368
}
{
    "name": "Bhatinda (Bathinda)",
    "population": 159042
}
{
    "name": "Pathankot",
    "population": 123930
}
{
    "name": "Hoshiarpur",
    "population": 122705
}
{
    "name": "Moga",
    "population": 108304
}
{
    "name": "Abohar",
    "population": 107163
}
9 documents in set (0.0082 sec)
```

Figure 13.37 – MySQL Shell – cities in the Punjab district of India, sorted by population

This is mostly the same as the previous step, but we add sort('population desc') to sort by population in descending order.

In this exercise, you used a condition with find() to filter the documents you wanted. Then, you filtered out the fields you were interested in with fields(). Finally, you used sort() to order the documents based on one of the fields.

In the next section, we will switch back to using SQL but use it to query JSON data. You can use collections in SQL mode. This looks like a table with an ID column and a DOC column that holds JSON data. Regular tables can have JSON columns. In both cases, we can use JSON functions and operators in SQL mode to work with this data.

Using JSON functions and operators to query JSON columns

For this, we use the MySQL client. You can also use MySQL Shell in SQL mode; just issue `\sql` after connecting with MySQL Shell.

There are many convenient functions to deal with JSON data when you are working with collections in SQL mode or tables that use JSON fields.

The first thing to do is extract and unquote fields. This is something we did in the previous chapter, so here is a quick reminder of it. The functions are `JSON_EXTRACT()` and `JSON_UNQUOTE()`. However, it is more convenient to use the operators that were created to do this – `->` to extract and `->>` to extract and unquote. You have to specify a JSON path expression to the extract function, which in its most basic form looks like `$.name`, to extract the name field.

Consider the following example:

```sql
SELECT doc->>'$.name' FROM worldcol LIMIT 5;
```

This produces the following output:

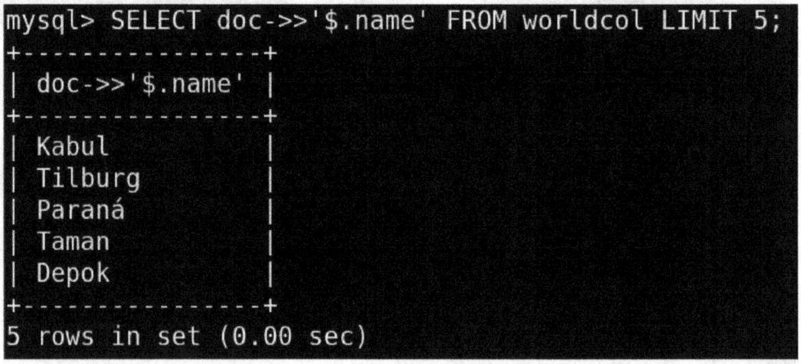

```
mysql> SELECT doc->>'$.name' FROM worldcol LIMIT 5;
+-----------------+
| doc->>'$.name'  |
+-----------------+
| Kabul           |
| Tilburg         |
| Paraná          |
| Taman           |
| Depok           |
+-----------------+
5 rows in set (0.00 sec)
```

Figure 13.38 – The SELECT output extracting the name FROM the JSON field of the worldcol collection

The same can also be written as follows:

```sql
SELECT JSON_UNQUOTE(JSON_EXTRACT(doc, '$.name')) FROM worldcol
LIMIT 5;
```

Other operations might be able to generate JSON structures from rows. This can be done with JSON_OBJECT(), a function that takes pairs – the first value is the key and the second one is the value.

Consider the following query:

```
SELECT JSON_OBJECT('name', Name, 'continent', Continent) FROM
country LIMIT 5;
```

This returns a JSON structure like the following:

```
mysql> SELECT JSON_OBJECT('name', Name, 'continent', Continent) FROM country LIMIT 5;
+----------------------------------------------------+
| JSON_OBJECT('name', Name, 'continent', Continent)  |
+----------------------------------------------------+
| {"name": "Aruba", "continent": "North America"}    |
| {"name": "Afghanistan", "continent": "Asia"}       |
| {"name": "Angola", "continent": "Africa"}          |
| {"name": "Anguilla", "continent": "North America"} |
| {"name": "Albania", "continent": "Europe"}         |
+----------------------------------------------------+
5 rows in set (0.00 sec)
```

Figure 13.39 – The SELECT output showing the JSON_OBJECT() usage

So, we used the Name and Continent columns from the table and used those as values where the keys are simply strings ("name" and "continent"). This can be very useful for converting data from tables to documents. Refer to worldcol.js for a more complete example.

We can also aggregate rows and combine results into an array. One of the functions to do this is JSON_ARRAYAGG():

```
SELECT Continent, JSON_ARRAYAGG(name) AS countries
FROM country GROUP BY continent\G
```

The result is a row for each of the seven continents and an array with the list of countries in that continent:

```
mysql> SELECT Continent, JSON_ARRAYAGG(name) AS countries
    -> FROM country GROUP BY continent\G
*************************** 1. row ***************************
Continent: Asia
countries: ["Afghanistan", "United Arab Emirates", "Armenia", "Azerbaijan", "Bangladesh", "Bahrain
", "Brunei", "Bhutan", "China", "Cyprus", "Georgia", "Hong Kong", "Indonesia", "India", "Iran", "I
raq", "Israel", "Jordan", "Japan", "Kazakstan", "Kyrgyzstan", "Cambodia", "South Korea", "Kuwait",
 "Laos", "Lebanon", "Sri Lanka", "Macao", "Maldives", "Myanmar", "Mongolia", "Malaysia", "Nepal",
"Oman", "Pakistan", "Philippines", "North Korea", "Palestine", "Qatar", "Saudi Arabia", "Singapore
", "Syria", "Thailand", "Tajikistan", "Turkmenistan", "East Timor", "Turkey", "Taiwan", "Uzbekista
n", "Vietnam", "Yemen"]
*************************** 2. row ***************************
Continent: Europe
countries: ["Albania", "Andorra", "Austria", "Belgium", "Bulgaria", "Bosnia and Herzegovina", "Bel
arus", "Switzerland", "Czech Republic", "Germany", "Denmark", "Spain", "Estonia", "Finland", "Fran
ce", "Faroe Islands", "United Kingdom", "Gibraltar", "Greece", "Croatia", "Hungary", "Ireland", "I
celand", "Italy", "Liechtenstein", "Lithuania", "Luxembourg", "Latvia", "Monaco", "Moldova", "Mace
donia", "Malta", "Netherlands", "Norway", "Poland", "Portugal", "Romania", "Russian Federation", "
Svalbard and Jan Mayen", "San Marino", "Slovakia", "Slovenia", "Sweden", "Ukraine", "Holy See (Vat
ican City State)", "Yugoslavia"]
*************************** 3. row ***************************
Continent: North America
countries: ["Aruba", "Anguilla", "Netherlands Antilles", "Antigua and Barbuda", "Bahamas", "Belize
", "Bermuda", "Barbados", "Canada", "Costa Rica", "Cuba", "Cayman Islands", "Dominica", "Dominican
 Republic", "Guadeloupe", "Grenada", "Greenland", "Guatemala", "Honduras", "Haiti", "Jamaica", "Sa
int Kitts and Nevis", "Saint Lucia", "Mexico", "Montserrat", "Martinique", "Nicaragua", "Panama",
"Puerto Rico", "El Salvador", "Saint Pierre and Miquelon", "Turks and Caicos Islands", "Trinidad a
nd Tobago", "United States", "Saint Vincent and the Grenadines", "Virgin Islands, British", "Virgi
n Islands, U.S."]
*************************** 4. row ***************************
Continent: Africa
countries: ["Angola", "Burundi", "Benin", "Burkina Faso", "Botswana", "Central African Republic",
"Côte d'Ivoire", "Cameroon", "Congo, The Democratic Republic of the", "Congo", "Comoros", "Cape Ve
rde", "Djibouti", "Algeria", "Egypt", "Eritrea", "Western Sahara", "Ethiopia", "Gabon", "Ghana", "
Guinea", "Gambia", "Guinea-Bissau", "Equatorial Guinea", "British Indian Ocean Territory", "Kenya"
, "Liberia", "Libyan Arab Jamahiriya", "Lesotho", "Morocco", "Madagascar", "Mali", "Mozambique", "
Mauritania", "Mauritius", "Malawi", "Mayotte", "Namibia", "Niger", "Nigeria", "Réunion", "Rwanda",
"Sudan", "Senegal", "Saint Helena", "Sierra Leone", "Somalia", "South Sudan", "Sao Tome and Princ
ipe", "Swaziland", "Seychelles", "Chad", "Togo", "Tunisia", "Tanzania", "Uganda", "South Africa",
"Zambia", "Zimbabwe"]
*************************** 5. row ***************************
Continent: Oceania
countries: ["American Samoa", "Australia", "Cocos (Keeling) Islands", "Cook Islands", "Christmas I
sland", "Fiji Islands", "Micronesia, Federated States of", "Guam", "Kiribati", "Marshall Islands",
 "Northern Mariana Islands", "New Caledonia", "Norfolk Island", "Niue", "Nauru", "New Zealand", "P
itcairn", "Palau", "Papua New Guinea", "French Polynesia", "Solomon Islands", "Tokelau", "Tonga",
"Tuvalu", "United States Minor Outlying Islands", "Vanuatu", "Wallis and Futuna", "Samoa"]
*************************** 6. row ***************************
Continent: Antarctica
countries: ["Antarctica", "French Southern territories", "Bouvet Island", "Heard Island and McDona
ld Islands", "South Georgia and the South Sandwich Islands"]
*************************** 7. row ***************************
Continent: South America
countries: ["Argentina", "Bolivia", "Brazil", "Chile", "Colombia", "Ecuador", "Falkland Islands",
"French Guiana", "Guyana", "Peru", "Paraguay", "Suriname", "Uruguay", "Venezuela"]
7 rows in set (0.01 sec)
```

Figure 13.40 – The SELECT output with JSON_ARRAYAGG()

> **Note**
>
> Here, we end the query with `\G`. This is used in the MySQL client and in MySQL Shell in SQL mode to display the results horizontally. Besides the difference in output, this does exactly the same as `;`, which is to send the query to the server for execution. In MySQL, you can index JSON arrays to make your queries faster. This needs a special kind of index known as a multi-valued index, which is available in MySQL 8.0.17 and higher.

The last of the functions to handle now is `JSON_PRETTY()`, which is very handy if you work with large documents. It displays JSON data in an easy-to-read format:

```
SELECT JSON_PRETTY(doc) FROM worldcol LIMIT 1\G
```

This produces the following output:

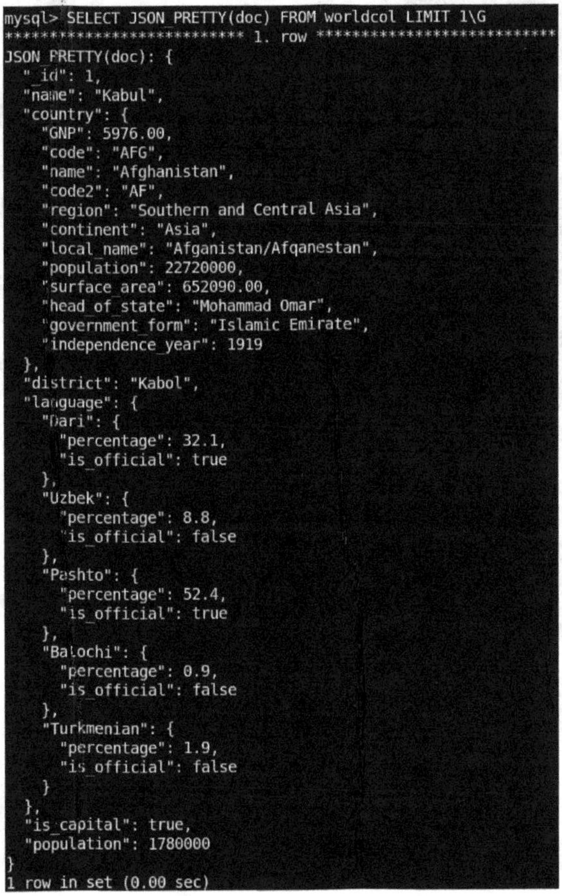

```
mysql> SELECT JSON_PRETTY(doc) FROM worldcol LIMIT 1\G
*************************** 1. row ***************************
JSON_PRETTY(doc): {
  "_id": 1,
  "name": "Kabul",
  "country": {
    "GNP": 5976.00,
    "code": "AFG",
    "name": "Afghanistan",
    "code2": "AF",
    "region": "Southern and Central Asia",
    "continent": "Asia",
    "local_name": "Afganistan/Afqanestan",
    "population": 22720000,
    "surface_area": 652090.00,
    "head_of_state": "Mohammad Omar",
    "government_form": "Islamic Emirate",
    "independence_year": 1919
  },
  "district": "Kabol",
  "language": {
    "Dari": {
      "percentage": 32.1,
      "is_official": true
    },
    "Uzbek": {
      "percentage": 8.8,
      "is_official": false
    },
    "Pashto": {
      "percentage": 52.4,
      "is_official": true
    },
    "Balochi": {
      "percentage": 0.9,
      "is_official": false
    },
    "Turkmenian": {
      "percentage": 1.9,
      "is_official": false
    }
  },
  "is_capital": true,
  "population": 1780000
}
1 row in set (0.00 sec)
```

Figure 13.41 – The SELECT output with JSON_PRETTY()

While JSON functions are mostly used in SQL mode, it is also possible to use them in JavaScript or Python mode:

```
\js
db.worldcol.find().
fields('country.name', 'json_arrayagg(name) AS cities').
groupBy('country.name').
limit(3)
```

This produces the following output:

```
MySQL  127.0.0.1:18018+ ssl  world  JS > db.worldcol.find().
                                     -> fields('country.name', 'json_arrayagg(name) AS cities').
                                     -> groupBy('country.name').
                                     -> limit(3)
                                     ->
{
    "cities": [
        "Kabul",
        "Qandahar",
        "Mazar-e-Sharif",
        "Herat"
    ],
    "country.name": "Afghanistan"
}
{
    "cities": [
        "Tirana"
    ],
    "country.name": "Albania"
}
{
    "cities": [
        "Sétif",
        "Oran",
        "Constantine",
        "Batna",
        "Ghardaïa",
        "Ech-Chleff (el-Asnam)",
        "Tiaret",
        "Béchar",
        "Tlemcen (Tilimsen)",
        "Tébessa",
        "Mostaganem",
        "Annaba",
        "Béjaïa",
        "Sidi Bel Abbès",
        "Skikda",
        "Biskra",
        "Alger",
        "Blida (el-Boulaida)"
    ],
    "country.name": "Algeria"
}
3 documents in set (0.0196 sec)
```

Figure 13.42 – The find() output with json_arrayagg()

In the next section, you will complete an exercise based on querying JSON data with SQL.

Exercise 13.10 – querying JSON data with SQL

In this exercise, you will be using the `worldcol` collection, which you created in the previous exercise. You are tasked with getting the names of the capitals of the five largest countries by surface area. Start by connecting to the `world` schema with the MySQL client. Then, build a SQL query that filters out capitals, add order and limit to sort the results, select the fields in which you are interested, and finally, run the query. Follow these steps to complete this exercise:

1. Connect to the `world` schema with the MySQL client:

    ```
    USE world;
    ```

    This produces the following output:

    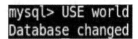

    Figure 13.43 – Connecting to the world schema

2. Build a SQL query that filters out capitals:

    ```
    FROM worldcol WHERE doc->'$.is_capital'=TRUE
    ```

    You want to use the `worldcol` table. This is a collection that doubles as a table. It has two columns, `_id` with the ID of the document and `doc` with the JSON document. You filter out the `is_capital` field from the collection and filter out rows for which the column is TRUE.

3. Add ordering and limit to sort the results:

    ```
    ORDER BY doc->'$.country.surface_area' DESC LIMIT 5
    ```

    Sort the `country.surface_area` field in descending order and limit it to five results. Note that you use `->` instead of `->>`, as you only want to extract the field. You don't want to unquote it, as that would cause the value to be a string, and sorting is different for strings.

4. Select the fields that we are interested in:

    ```
    SELECT
      doc->>'$.name' AS city_name,
      doc->>'$.country.name' AS country_name
    ```

Extract the names of the city and the country. Name the results columns to get a name that is easier to work with.

5. Run the query:

```
SELECT
    doc->>'$.name' AS city_name,
    doc->>'$.country.name' AS country_name
FROM worldcol
WHERE doc->'$.is_capital'=TRUE
ORDER BY doc->'$.country.surface_area' DESC LIMIT 5;
```

This produces the following output:

```
mysql> SELECT
    ->    doc->>'$.name' AS city_name,
    ->    doc->>'$.country.name' AS country_name
    -> FROM worldcol
    -> WHERE doc->'$.is_capital'=TRUE
    -> ORDER BY doc->'$.country.surface_area' DESC LIMIT 5;
+------------+--------------------+
| city_name  | country_name       |
+------------+--------------------+
| Moscow     | Russian Federation |
| Ottawa     | Canada             |
| Peking     | China              |
| Washington | United States      |
| Brasília   | Brazil             |
+------------+--------------------+
5 rows in set (0.01 sec)
```

Figure 13.44 – Getting the names of the capitals of the biggest countries
by surface area from the worldcol collection

Here, you used the -> and ->> operators to extract and, where needed, unquote data that is stored in a JSON column. In the next section, we will learn how to use generated columns to query and index JSON data.

Using generated columns to query and index JSON data

If you find yourself constantly extracting the same key from a JSON document in SQL mode, then it might be time to create a so-called generated column. The generated column looks like a normal column, but it has the data from whatever function you provide, usually an extract and unquote on a JSON document. The data for the generated column can either be virtual (generated on the go) or stored. The benefit of a fully virtual column is that adding or removing it is instantaneous, and it doesn't take up any storage space. With a generated column, the benefit is that it can be faster because it doesn't have to be generated every time it is used.

Take the `worldcol` collection as an example:

```
ALTER TABLE worldcol
ADD COLUMN district VARCHAR(255) AS (doc->>'$.district') NOT
NULL;
```

This extracts the district from the JSON document and places it in a generated column:

```
mysql> ALTER TABLE worldcol
    -> ADD COLUMN district VARCHAR(255) AS (doc->>'$.district') NOT NULL;
Query OK, 0 rows affected (0.04 sec)
Records: 0  Duplicates: 0  Warnings: 0
```

Figure 13.45 – The ALTER TABLE output to add a generated column

Besides now having an easy-to-query column, we have also told the database that the district can't be NULL, so if we try to add a new entry without a district, this fails. So, this can be used to place validation on the documents:

```
MySQL> INSERT INTO worldcol(doc) VALUES('{"_id": 999999 }');
 ERROR: 1048: Column 'district' cannot be null
```

One of the great features of generated columns is that they can be indexed. This is true for both virtual and stored columns.

Consider the following example:

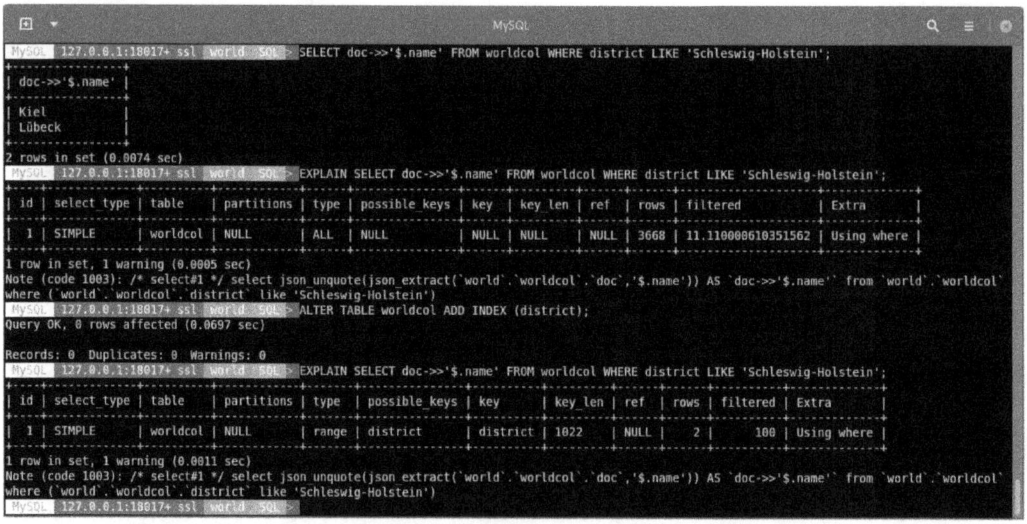

Figure 13.46 – An example of adding INDEX on a generated column

Here, you add an index on the generated `district` column that you added before:

```
ALTER TABLE worldcol ADD INDEX(district);
```

This produces the following result:

Figure 13.47 – The ALTER TABLE output to add INDEX

This is what allows MySQL to have functional indexes. This is not limited to JSON data, as you can use most functions for virtual columns.

The EXPLAIN output shows that MySQL only needs 2 rows instead of 3668 rows. This makes the query return faster. We will dive deeper into using EXPLAIN in the next chapter.

To create a generated column stored instead of a virtual one, just add the STORED keyword:

```
ALTER TABLE worldcol
ADD COLUMN name VARCHAR(255) AS (doc->>'$.name') STORED NOT
NULL;
```

This query produces the following output:

```
mysql> ALTER TABLE worldcol
    -> ADD COLUMN name VARCHAR(255) AS (doc->>'$.name') STORED NOT NULL;
Query OK, 4079 rows affected (0.35 sec)
Records: 4079  Duplicates: 0  Warnings: 0
```

Figure 13.48 – The ALTER TABLE output to add a stored column

In the next section, we will complete an activity based on the knowledge that we have gained during this chapter. In the first, you will query the `world` database; in the second, you will export some data to the CSV format.

Activity 13.01 – Exporting report data to CSV for Excel

You are working for a newspaper, and as part of an article related to the inauguration of a new king, the reporter needs a list with the heads of state of all monarchies. The requested format is CSV, as that can be loaded in Excel and later incorporated into the article. Perform the following steps to implement the activity:

1. Connect to the `world` database.

2. Select the right columns and filter out monarchies.

3. Send the result to a file in the CSV format.

After implementing these steps, the expected output looks like the following:

```
mysql> SELECT GovernmentForm FROM country
    -> WHERE GovernmentForm LIKE '%Monarchy%'
    -> GROUP BY GovernmentForm;
+------------------------------------+
| GovernmentForm                     |
+------------------------------------+
| Constitutional Monarchy            |
| Constitutional Monarchy, Federation |
| Monarchy (Emirate)                 |
| Monarchy (Sultanate)               |
| Monarchy                           |
| Constitutional Monarchy (Emirate)  |
| Parlementary Monarchy              |
+------------------------------------+
7 rows in set (0.00 sec)
```

Figure 13.49 – The SELECT output to show government forms that are monarchies

> **Note**
>
> The solution for the activity can be found in the *Appendix*.

Summary

In this chapter, you learned how to insert records into tables and documents into collections. You also imported files in the SQL, CSV, and JSON formats into the MySQL server and combined data from tables and collections. You then used the CSV storage engine to easily import and export data in the CSV format. With the CSV format, it is easy to exchange data with other applications and spreadsheets.

In the next chapter, we will continue with querying data using MySQL. This includes using some more advanced reporting capabilities such as aggregating data and using functions. We will also continue to see what MySQL can do with JSON data.

14
Manipulating User Permissions

This chapter deals with creating, modifying, and dropping user accounts in MySQL. First, we will begin with creating users, and then we will move on to setting and changing their passwords and other properties. This will be followed by granting and revoking permissions. Additionally, we will troubleshoot any connection issues that might arise when users try to connect to the database. By the end of this chapter, you will be able to use roles to grant and manage the permissions for different groups of people.

In this chapter, we will cover the following main topics:

- Introduction to user permissions
- Exploring user and accounts
- Exercise 14.01 – creating users and granting permissions
- Changing users
- Flush privileges
- Changing permissions
- Exercise 14.02 – modifying users and revoking permissions
- Using roles

- Exercise 14.03 – using roles to manage permissions
- Troubleshooting access problems
- Activity 14.01 – creating users for managing the word schema

Introduction to user permissions

In the previous chapter, we learned how to make modifications to the data stored in MySQL tables and collections. Additionally, we learned how to use the DELETE statement to delete rows and the UPDATE statement to change existing rows. For both statements, we learned how to use them together when joining multiple tables to allow for more complex changes. Additionally, we learned how to use the INSERT statement with an ON DUPLICATE KEY UPDATE clause to add new records to the database or update an existing record if it was already in the database. We learned how to use the modify() method to modify existing documents in a collection and the remove() method to remove documents from a collection.

In this chapter, we will learn how to create users to segregate and restrict access to ensure no accidental or fraudulent changes can be made to the data.

Good management of accounts and passwords is paramount to security. For example, let's suppose your company allows its customers to subscribe to a newsletter using their email addresses. These are stored in a MySQL database. The same database server also hosts an internal application for employees to register the projects they are working on. The web server and the internal application use the same account. If a bug is found in the software that is used for internal projects or if the configuration file of that application isn't guarded properly, then the internal employees can gain access not only to the database used for internal projects but also the database with the customers' email addresses. Had we used two separate accounts, the customers' email addresses would not have been accessible to someone who gained access to the database credentials of the application for internal projects. The same could happen the other way around. If there was a SQL injection found in the code running the website, then external users might have been able to gain access to the list of internal (and probably confidential) projects. Note that in a real-world situation, these applications should have their own database server.

Another example would be a developer having access to the production database for the website described earlier. The developer might need this to troubleshoot problems with articles that are published on the website and are stored in the database. For this, they don't need access to the customers table. If they are granted access to only the required tables, then the **Personally Identifiable Information (PII)** data will be more secure. If the account of this developer gets hacked or lost, then the customer data will remain safe.

The list of data breaches, which can be found at `https://en.wikipedia.org/wiki/List_of_data_breaches`, shows that this is a real problem. Each data breach can cost a company a lot of money. Managing user accounts is one of the many things you can do to reduce the chance of this happening to your company. It is recommended that you do not share accounts between users and/or applications and only grant access to what's really needed.

In the first section, we will learn a few basics regarding what user accounts are and how to connect with different user accounts. Additionally, we will learn why we should be using multiple accounts in the first place.

Exploring users and accounts

Most applications define an account as a username and a password. Then, permissions are assigned to this account. For MySQL, it is mostly the same, but there are some important differences. The first difference is that, for MySQL, an account is written as `<user>@<host>` instead of only the username. The permissions are assigned to such user and host combinations. This is important and means that `johndoe@127.0.0.1` and `johndoe@192.168.0.1` are two different accounts that can have different permissions. It also allows you to restrict access to specific hosts or IP ranges. In the next section, we will explore how to connect to MySQL with a set of credentials.

How to connect to MySQL with a set of credentials

Essentially, this is similar to what you have already been using before, but we will refresh your memory regarding this process.

To connect to the MySQL client, the code needs to be in the following format:

```
mysql -h <host> -u <user> -p <db>
```

To connect to MySQL Shell, the code needs to be in the following format:

```
mysqlsh <user>@<host>/<db>
```

As you can see here, in both cases, you are prompted to enter the password. Therefore, this is advisable over having a password on the command line as that might end up in the history of your shell.

The `<db>` part is not required, but without that, you have to use the `USE <db>` command or the `\use <db>` command to connect to the right database. So, it is more convenient to directly connect to the right schema.

Besides this, there are various other options that you could use to connect to MySQL. However, in most cases, these are not needed. We can use -p to connect to a non-standard port and –ssl-ca, --ssl-cert, and –ssl-key to specify the client certificates. Additionally, on Windows, you can use --shared-memory-base-name to connect over a shared memory connection. On Linux, you can use -S to connect over a UNIX domain socket.

> **Note**
>
> For MySQL, localhost and 127.0.0.1 are not the same. If localhost is used, then MySQL uses a UNIX domain socket instead of TCP to connect on Linux. The best practice is to use 127.0.0.1 instead of localhost to connect over TCP.

Why use multiple user accounts?

If every person and/or application that connects to the database has their own account, then you can grant different permissions to them. This reduces the chance of someone accidentally dropping or changing the data. In addition, it helps with auditing. With MySQL Enterprise Edition, or by using third-party audit plugins, you can create audit trails. However, these are not useful if everyone uses the same account. In MySQL, an account can have resource constraints in addition to permissions on schemas, tables, and columns. These resource constraints are more useful if user accounts are only used by a single application or person. And the same goes for locking and unlocking accounts, which is only useful if accounts are not shared. In the next section, we will explore how to create users.

Creating, modifying, and dropping a user

To create a user, called johndoe, who is allowed to log in from anywhere with the password of 'teigsizkudefegdec', we will write the following query:

```
CREATE USER 'johndoe'@'%' IDENTIFIED BY 'teigsizkudefegdec';
```

In this example, % is used as a wildcard. Instead of %, you can use something such as 192.168.1.%, %.example.com, 127.0.0.1, or localhost to restrict where the user can log on from.

Additionally, you can add `PASSWORD EXPIRE` to the statement to force the user to change the password once they are logged in. We can modify the details of the user by changing their password. In order to do that, we will have to use the following query:

```
ALTER USER USER() IDENTIFIED BY 'new_secure_passsword';
```

Let's say that you want to change the password of the previously created user, then in order to implement it, we will write the following query:

```
ALTER USER 'johndoe'@'%' IDENTIFIED BY 'johndoe';
```

Another thing you can do is to add `WITH MAX_USER_CONNECTIONS 10` to set a resource limit on the number of connections that are allowed for the user. This can help you to prevent the user from consuming all of the connections available to the other users from connecting to that database.

Finally, if we want to drop a user, we can use the following format:

```
DROP USER user_name;
```

Let's say we want to delete the previously created user. Then, in that case, we write the following:

```
DROP USER 'johndoe'@'%';
```

Now that we have learned how to create users, modify their details, and drop them, in the next section, we will expand that knowledge by learning how to grant permissions to them.

Granting permissions

Now that we have a user, we need to grant permissions to it.

The query to grant permission looks like this:

```
GRANT SELECT ON world.* TO 'johndoe'@'%';
```

This grants the `SELECT` permission on all the tables of the world database to the johndoe@% account.

> **Note**
>
> To grant such permissions, ensure that you have the `world` database in your local system. You can follow the instructions mentioned at `https://dev.mysql.com/doc/world-setup/en/world-setup-installation.html` to get the `world` database in your system.

It is possible to grant permissions globally (`*.*`) on a schema (`world.*`) or a specific table (`world.city`). Also, it is possible to grant access to specific columns with the following query:

```
GRANT SELECT (ID, Name) ON world.city TO 'johndoe'@'%';
```

However, this is not a very common thing to do. The most common permissions that you can grant are listed as follows:

- `SELECT`, `UPDATE`, `DELETE`, and `INSERT`: These permissions allow you to retrieve and modify data in tables.

- `CREATE`, `ALTER`, and `DROP`: These permissions allow you to create, modify, and drop tables.

- `CREATE USER`: This allows you to work with user accounts.

- `FILE`: This allows you to work with data on the filesystem.

- `PROCESS`: This allows you to manage processes, such as kill processes, and see the full process list.

- `ALL`: This allows you to grant all permissions.

Besides these permissions, there are more, less common, permissions that you can grant.

Consider that the `GRANT` statement returns the following error:

```
ERROR: 1410: You are not allowed to create a user with GRANT
```

Here, you are trying to `GRANT` permission to a user that doesn't exist. In such a scenario, you need to check the username and use the `CREATE USER` query if needed. The reason for this error message is that, in older versions, MySQL would create the user automatically if it didn't exist and you tried to grant permissions to it. The problem with this was that it often resulted in users who didn't have a password by accident, which is very insecure. In the next section, we will learn about inspecting users.

Inspecting users

We can inspect the settings for a user and the list of grants that the user has by using the following two statements:

```
SHOW CREATE USER <user>@<host>;
SHOW GRANTS FOR <user>@<host>;
```

The first query will return something similar to the following:

```
+-------------------------------------------------------------
----------------------------------------------------------+
| CREATE USER for johndoe@%

                                                           |
+-------------------------------------------------------------
-------------------------------------------------------------+
| CREATE USER 'johndoe'@'%' IDENTIFIED WITH 'caching_
INTERVAL DEFAULT PASSWORD REQUIRE CURRENT DEFAULT |
+-------------------------------------------------------------
-----------------------------------------------------------+
1 row in set (0.06 sec)
```

Figure 14.1: Inspecting the settings of the user

The second query will return something similar to the following:

```
+-----------------------------------------------------------------------+
| Grants for johndoe@%                                                  |
+-----------------------------------------------------------------------+
| GRANT USAGE ON *.* TO `johndoe`@`%`                                   |
| GRANT SELECT ON `world`.* TO `johndoe`@`%`                            |
| GRANT SELECT (`ID`, `Name`) ON `world`.`city` TO `johndoe`@`%`        |
+-----------------------------------------------------------------------+
3 rows in set (0.02 sec)
```

Figure 14.2: Inspecting the list of grants the user has

If you want to know what grants the existing user has, then you need to write the following query:

```
SHOW GRANTS;
```

Another way to get to this information is to query the information_schema tables.

The list of tables to query is as follows:

- information_schema.USER_PRIVILEGES: This holds global permissions.
- information_schema.SCHEMA_PRIVILEGES: This is used for per-schema permissions.
- information_schema.TABLE_PRIVILEGES: This is used for per-table permissions.
- information_schema.COLUMN_PRIVILEGES: This is used for per-column permissions.

Consider the following query:

```
SELECT GRANTEE FROM information_schema.USER_PRIVILEGES GROUP BY
GRANTEE;
```

The preceding query produces the following output:

```
+----------------------------------+
| GRANTEE                          |
+----------------------------------+
| 'mysql.infoschema'@'localhost'   |
| 'mysql.session'@'localhost'      |
| 'mysql.sys'@'localhost'          |
| 'root'@'localhost'               |
| 'bhavesh'@'%'                    |
| 'johndoe'@'%'                    |
+----------------------------------+
6 rows in set (0.13 sec)
```

Figure 14.3: Inspecting the USER_PRIVILEGES table

The preceding table has a row for every global permission for every user. So, for a user with 10 permissions, it will have 10 rows. However, in this query, we are only interested in the list of users that is known to the server. Therefore, we are grouping by GRANTEE.

To check the details of the SCHEMA_PRIVILEGES table, we can write the following query:

```
SELECT GRANTEE FROM information_schema.SCHEMA_PRIVILEGES GROUP
BY GRANTEE;
```

The preceding query produces the following output:

```
+------------------------------------+
| GRANTEE                            |
+------------------------------------+
| 'mysql.sys'@'localhost'            |
| 'mysql.session'@'localhost'        |
| 'johndoe'@'%'                      |
+------------------------------------+
3 rows in set (0.00 sec)
```

Figure 14.4: Inspecting the SCHEMA_PRIVILEGES table

To check the details of the TABLE_PRIVILEGES table, we can write the following query:

```
SELECT GRANTEE FROM information_schema.TABLE_PRIVILEGES GROUP
BY GRANTEE;
```

This query produces the following output:

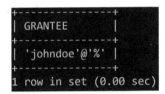

Figure 14.5: Inspecting the TABLE_PRIVILEGES table

To check the details of the COLUMN_PRIVILEGES table, we can write the following query:

```
SELECT GRANTEE FROM information_schema.COLUMN_PRIVILEGES GROUP
BY GRANTEE;
```

This query produces the following output:

```
+----------------+
| GRANTEE        |
+----------------+
| 'johndoe'@'%'  |
+----------------+
1 row in set (0.00 sec)
```

Figure 14.6: Inspecting the COLUMN_PRIVILEGES table

In the next section, you will be able to practice what you have learned so far.

Exercise 14.01 – creating users and granting permissions

You are part of a new start-up that sells electric bikes on a web page and with a mobile app. Besides you, there is the founder of the company (Patrick) and a single developer (Mike) who develops the web page and mobile app. You have been tasked with setting up accounts for the developer, the founder, and the web server.

On the database server, there are two databases: employees and ebike. The web server should be limited to 300 connections to ensure the databases are still accessible even if the website becomes overloaded.

In order to implement this exercise, first, open the MySQL client and connect to the database server. Then, create accounts for Patrick, Mike, and the web server. You will grant Patrick access to the employees and the ebike schemas, and grant Mike and the web server access to the ebike schema.

To complete this exercise, perform the following steps:

1. Open the MySQL client and connect to the database server.

    You don't need to connect to a specific database—just connecting to the server is enough.

2. Create accounts for Patrick, Mike, and the web server by writing the following queries:

    ```
    CREATE USER 'patrick'@'%' IDENTIFIED BY
    'NijTaseirpyocyea';
    CREATE USER 'mike'@'%' IDENTIFIED BY 'MyhafDixByej';
    CREATE USER 'webserver'@'%' IDENTIFIED BY 'augJigFevni'
    WITH MAX_USER_CONNECTIONS 300;
    ```

    Here, you create three users and set a randomly generated password for each of them. Additionally, for the webserver user, you set a resource limit of 300 connections.

3. Now, grant the patrick user access to the employees and the ebike schemas by writing the following queries:

    ```
    GRANT ALL ON employees.* TO 'patrick'@'%';
    GRANT ALL ON ebike.* TO 'patrick'@'%';
    ```

    Here, you grant full access to both schemas to user patrick.

4. Grant both users, mike and webserver, access to the ebike schema with the help of the following queries:

    ```
    GRANT ALL ON ebike.* TO 'mike'@'%';
    GRANT SELECT, INSERT, UPDATE, DELETE ON ebike.* TO
    'webserver'@'%';
    ```

    Here, you grant the mike user full access to the ebike schema, but you only give out specific grants to the webserver user.

In this exercise, you have mastered the skills of creating and granting users. In the upcoming section, we will focus on how to change users.

Changing users

There can be many different reasons for changing users, and there are many different things that we can change.

If a password was leaked, then the first thing you want to do is lock the account and/or change the password of the account. If you have an application account that you suspect is no longer being used, it might be smart to first lock the account before dropping it later. This allows you to simply unlock the account if it turns out that something was still relying on this account. Locking an account is also a good way to protect a shared database against a single user who is overloading the system, for example, by writing too much data or running too many heavy queries. Then, you can lock the account, ensure the application abusing the database gets fixed, and unlock the account again.

Another thing you will often need to do is periodically change passwords. For applications, you might want to create a new user with a new password but with the same permissions. Then, we might want to restart the application to use this new account and lock and drop the original account later. This allows you to change the credentials used by applications with minimal disruption.

One of the benefits of locking an account over dropping it is that the error message the clients receive is very clear:

```
ERROR 3118 (HY000): Access denied for user
'myuser'@'localhost'. Account is locked.
```

Locking and unlocking users can be done using the following queries:

```
ALTER USER <user>@<host> ACCOUNT LOCK;
ALTER USER <user>@<host> ACCOUNT UNLOCK;
```

For example, if you want to lock and unlock the 'johndoe'@'%' user, write the following queries:

```
ALTER USER 'johndoe'@'%' ACCOUNT LOCK;
ALTER USER 'johndoe'@'%' ACCOUNT UNLOCK;
```

Another thing you can do is to change the passwords. We can do this in the following way:

```
ALTER USER 'johndoe'@'%' IDENTIFIED BY 'foobar';
```

While it is possible to use SET PASSWORD..., the preferred way to do this is with ALTER USER....

Changing resource limits can be done as follows:

```
ALTER USER 'johndoe'@'%' WITH MAX_USER_CONNECTIONS 5;
```

In the next section, let's explore flushing privileges.

Flush privileges

Many tutorials and instructions to set up applications tell the users to issue FLUSH PRIVILEGES. So, what is this? And when do we need to use it?

The CREATE USER, ALTER USER, and GRANT permissions, along with many other user and permission statements, indirectly modify the system tables that are stored in the mysql schema. At startup, these tables are loaded into memory and, after every statement that modifies the users and/or permissions, these are again loaded into memory.

However, if you directly modify the tables in the mysql schema with INSERT, UPDATE, and DELETE statements, you need to force MySQL to refresh the copies of these tables it has in memory. This is where the FLUSH PRIVILEGES statement comes in. It precisely does that. Note that we do not recommend you modify these tables directly. So, as long as you stick to the supported commands to modify users, you never need to use this command.

Changing permissions

So, we have already covered granting permissions. The only other thing that we can do is to remove permissions from a user. This is done with the help of the REVOKE statement.

To remove the SELECT permission in the world.city table from the johndoe user, write the following query:

```
REVOKE SELECT ON the world.city FROM 'johndoe'@'%';
```

The REVOKE and GRANT statements look very similar but do the exact opposite of each other.

In the next section, you will solve an exercise based on what you have learned so far.

Exercise 14.02 – modifying users and revoking permissions

The ebike start-up has been very successful, and there have been a few changes. A new developer, called Sarah, was hired, and there is a new `mobileapp` schema for the mobile app to manage the bikes. Both Sarah and Mike are working on the mobile app. In addition to this, Patrick has asked you to change his password because he has forgotten what it was. You need to change the password that is used for the account used by the web server.

In order to implement this, first, connect to the database and then create an account for the new developer Sarah. Then, modify the accounts of Patrick, Mike, and the web server so that they will be able to access the new schema. Finally, you can change the password of `Patrick`.

To complete this exercise, follow the steps:

1. Connect to the database.

2. Create an account for the new developer Sarah and grant her permissions on the `ebike` and `mobileapp` schemas with the help of the following queries:

   ```
   CREATE USER 'sarah'@'%' IDENTIFIED BY 'IkbyewUgJeuj8';
   GRANT ALL ON ebike.* TO 'sarah'@'%';
   GRANT ALL ON mobileapp.* TO 'sarah'@'%';
   ```

3. Modify the accounts of Patrick, Mike, and the web server by granting them access to the `mobileapp` schema:

   ```
   GRANT ALL ON mobileapp.* TO 'mike'@'%';
   GRANT ALL ON mobileapp.* TO 'patrick'@'%';
   GRANT ALL ON mobileapp.* TO 'webserver'@'%';
   ```

4. Revoke the access of `mike` to the `ebike` schema as he no longer needs its access:

   ```
   REVOKE ALL ON ebike.* FROM 'mike'@'%';
   ```

5. Now, change the password for the `patrick` user:

   ```
   ALTER USER 'patrick'@'%' IDENTIFIED BY 'WimgeudJa';
   ```

6. Inspect all of the permissions for the `webserver` user:

   ```
   SHOW GRANTS FOR 'webserver'@'%';
   ```

This produces the following output:

```
+------------------------------------------------------------------------------
| Grants for webserver@%
+------------------------------------------------------------------------------
| GRANT USAGE ON *.* TO `webserver`@`%`
| GRANT SELECT, INSERT, UPDATE, DELETE ON `ebike`.* TO `webserver`@`%`
| GRANT ALL PRIVILEGES ON `mobileapp`.* TO `webserver`@`%`
+------------------------------------------------------------------------------
3 rows in set (0.16 sec)
```

Figure 14.7: Inspecting the currently granted permissions for the webserver user

7. Change the account for the `webserver` user. You could do this in the same way you did for Patrick, but that would likely cause some disruption between the time you changed the password on the database and the moment we reconfigured the `webserver` user to use the new password. So, create a new account, and then after reconfiguring the web server, lock the old account:

```
CREATE USER 'webserver2'@'%' IDENTIFIED BY 'dutPyicloHi'
WITH MAX_USER_CONNECTIONS 300;

GRANT SELECT, INSERT, UPDATE, DELETE ON ebike.* TO
'webserver2'@'%';

GRANT ALL PRIVILEGES ON 'mobileapp'.* TO
'webserver2'@'%';
```

8. Reconfigure the `webserver` user:

```
ALTER USER 'webserver'@'%' ACCOUNT LOCK;
```

Now, if everything is fine, you can drop the old `webserver` user. If things are not working fine, for example, the `webserver` user didn't start to use the new account, we can simply unlock the account again.

In this exercise, you changed the password for one account directly and for another account by creating a new account. Additionally, you reconfigured the application by locking the old account, thereby minimizing the downtime of the application. You also used the REVOKE statement to remove access from an account. In the next section, you will learn about how to use roles.

Using roles

Besides granting permissions to individual users, in MySQL, it is also possible to create roles and grant permissions to roles and then assign roles to users. This makes handling groups of users with similar permissions much easier.

To create a role for `webdeveloper`, we can provide the following query:

```
CREATE ROLE 'webdeveloper';
```

The next step is to assign some permissions to the role. This is done with GRANT, just like how you did for the user permissions:

```
GRANT SELECT ON mysql.user TO 'webdeveloper';
```

To assign a role to a user, we need to use GRANT as follows:

```
GRANT 'webdeveloper' TO 'johndoe'@'%';
```

An account can have no roles, a single role, or multiple roles. If a role is granted to your user, then you might need to tell MySQL which roles you want to use with the help of the following query:

```
SET ROLE 'webdeveloper';
```

Instead of having to do this every time or having to modify an application to do this after connecting to the database, you can configure a set of default roles for an account:

```
ALTER USER 'johndoe'@'%' DEFAULT ROLE 'webdeveloper';
```

Creating a user, granting it a role, and making that role the default can be done in a single statement such as the following:

```
CREATE USER 'u2'@'%' IDENTIFIED BY 'foobar' DEFAULT ROLE
'webdeveloper';
```

To see what user and role you are using, you can run the following command:

```
SELECT CURRENT_ROLE(), CURRENT_USER();
```

This will generate the following output:

```
+----------------+------------------+
| CURRENT_ROLE() | CURRENT_USER()   |
+----------------+------------------+
| NONE           | root@localhost   |
+----------------+------------------+
1 row in set (0.07 sec)
```

Figure 14.8: Inspecting the current role and user

Now that you have learned how to use roles, in the next section, you will solve an exercise based on this to hone your skills.

Exercise 14.03 – using roles to manage permissions

The company keeps growing, and there has been another set of new hires:

- **Linda**: Taking over HR responsibilities from Patrick
- **John**: Will be taking care of finance
- **Vladimir**: The mobile app developer
- **Victoria**: The designer for the website

You have been asked to start using the following roles: manager, webdeveloper, and appdeveloper.

Here, you will connect to the database, create three roles, and grant permissions to them. Following this, you will create accounts for the new hires, and then grant roles to the existing people. To implement this exercise, follow these steps:

1. Connect to the database.

2. Create three roles: manager, webdeveloper, and appdeveloper. Grant permissions to them:

    ```
    CREATE ROLE 'manager';
    GRANT ALL ON employees.* TO 'manager';
    CREATE ROLE 'webdeveloper';
    GRANT ALL ON ebike.* TO 'webdeveloper';
    CREATE ROLE 'appdeveloper';
    GRANT ALL ON mobileapp.* TO 'appdeveloper';
    ```

Each permission links a group of people to the role(s) they have in the company.

3. Now, create accounts for the new hires:

```
CREATE USER 'linda'@'%' IDENTIFIED BY 'AkFernyeisjegs'
DEFAULT ROLE manager;
CREATE USER 'john'@'%' IDENTIFIED BY 'owvurewJatkinyegod'
DEFAULT ROLE manager;
CREATE USER 'vladimir'@'%' IDENTIFIED BY 'rusvawfyoaw'
DEFAULT ROLE appdeveloper;
CREATE USER 'victoria'@'%' IDENTIFIED BY
'joigowInladdIc6' DEFAULT ROLE webdeveloper;
```

4. Grant the roles to the existing people:

```
GRANT manager, webdeveloper, appdeveloper TO
'patrick'@'%';
ALTER USER 'patrick'@'%' DEFAULT ROLE manager;
GRANT webdeveloper, appdeveloper TO 'mike'@'%';
ALTER USER 'mike'@'%' DEFAULT ROLE webdeveloper,
appdeveloper;
GRANT webdeveloper, appdeveloper TO 'sarah'@'%';
ALTER USER 'sarah'@'%' DEFAULT ROLE webdeveloper,
appdeveloper;
```

We have granted roles to accounts and set the roles they will use by default. Note that, for Patrick, only the manager role is set by default. If he wants to use the other roles, he has to switch to them.

In the next section, we will explore various issues that might arise while connecting to the database.

Troubleshooting access problems

Let's try to troubleshoot some connection issues.

We will encounter the following error if MySQL is not running or if it is running on any another machine and you have forgotten to specify the host with -h:

```
$ mysql
ERROR 2002 (HY000): Can't connect to local MySQL server through
socket '/var/lib/mysql/mysql.sock' (2)
```

The following error is similar to the one mentioned earlier as, in this case, the connection goes over TCP. This can happen if MySQL runs on a non-standard port, and you didn't specify the port with -p:

```
$ mysql -h 127.0.0.1
ERROR 2003 (HY000): Can't connect to MySQL server on
'127.0.0.1' (111)
```

If we don't supply a password, we will encounter the following error:

```
$ mysql -h 127.0.0.1
ERROR 1045 (28000): Access denied for user 'jdoe'@'localhost'
(using password: NO)
```

Here, we can reach MySQL, but we are not allowed in.

The solution is to add -p and then let the client prompt you for the password. However, we will still get an error if either the username or the password is wrong:

```
$ mysql -h 127.0.0.1 -p
Enter password:
ERROR 1045 (28000): Access denied for user 'jdoe'@'localhost'
(using password: YES)
```

In the preceding scenario, we supplied the correct username and password. However, we will get another error if a database doesn't exist or is not accessible by the user:

```
$ mysql -h 127.0.0.1 -u jdoe -p information_schemas
Enter password:
ERROR 1044 (42000): Access denied for user 'jdoe'@'%' to
database 'information_schemas'
```

Note that supplying a database name is not required. In the next section, we will perform an activity wherein we will create the users to manage our world schema.

Activity 14.01 – creating users for managing the world schema

To manage the database with cities, languages, and countries, you need to set up some accounts. The first account is for the web server user, which should be read-only. The second account is for the intranet user, which is allowed to change and create entries. The third account is for a manager, called Stewart, who is allowed to do everything. As more managers will be hired soon, this should be implemented with roles. The last account is for Sue, who is a language expert and can only change the countrylanguage table.

To complete this activity, perform the following steps:

1. Connect to the database server.
2. Create the roles.
3. Create an account for the web server user.
4. Create an account for the intranet user.
5. Create an account for Stewart.
6. Create an account for Sue.

> **Note**
> The solution for this activity can be found in the *Appendix* section.

In this activity, you have used roles to make it easier to add permissions and users later. If there are more language experts, you simply grant them access to the role and you're done. Additionally, if there are new tables to create, you don't have to grant them access to multiple accounts, just the role.

Summary

In this chapter, you learned how to create users and manage users, including locking and unlocking accounts, setting passwords, and adding resource constraints. You learned how to manage permissions by using the GRANT statement to grant specific permissions to a user and the REVOKE statement to revoke those permissions. You learned how to use roles to manage the permissions more easily for a group of people.

This allows you to control who has access to the information stored inside the database. This is a critical part of securing access to the database.

In the next chapter, you will learn how to create logical backups, which can be used to restore data after a server has crashed or after data that has been deleted by accident. Besides that, it can also be used in migrations, setting up replication, or for copying data to a development or acceptance environment.

15

Logical Backups

In this chapter, you will learn to create a backup of all data in the MySQL server, which will allow you to recover lost data, beginning with a comparative exploration of `mysqldump` and `mysqlpump` between logical and physical backups. You will make a backup copy of a single schema before learning to restore a database from a full backup or a schema from a single schema backup. We will use point-in-time restore to recover all data up to a specific point in time to minimize data loss during restoration. By the end of this chapter, you will be able to use the `mysqlbinlog` utility to inspect the contents of the `binlog` files.

This chapter covers the following concepts:

- An introduction to backups
- Understanding the basics of backups
- Logical and physical backup
- Types of restore
- Scheduling backups
- Using point-in-time recovery with `binlog` files
- Activity 15.01 – backing up and restoring a single schema
- Activity 15.02 – performing a point-in-time restore

An introduction to backups

In the previous chapter, we learned to define users in MySQL and grant permissions to restrict access to specific users and/or applications, using roles to make this task more efficient, and troubleshooted various database connection issues.

In this chapter, we will learn how to use backups to safeguard against data loss in a number of unfortunate situations, such as outages or even a software update. Besides guarding against data loss, backups also help to validate data – for example, after someone has gained unauthorized access or after a software bug has been discovered.

We will also review the basics of logical backups, before diving into `mysqldump` and `mysqlpump`, their differences, and how to create full and partial backups with both. We will then proceed with learning how to restore backups and touch upon using `binlog` files to do point-in-time restores.

Understanding the basics of backups

Backups can be used for multiple purposes. The main purpose of backups is to reduce the risk of losing data if your primary copy gets lost or damaged. Another use of backups is to seed an acceptance environment with real-life data. Depending on how you develop, you might have different setups for development, quality assurance, acceptance, and production. Restoring a backup from production to acceptance can be done to allow for performance tests and functional tests with real-life data.

This has to be done carefully, as this may or may not be allowed by regulations such as the **Health Insurance Portability and Accountability Act (HIPAA)**, the **General Data Protection Regulation (GDPR)**, and the **Payment Card Industry Data Security Standard (PCI-DSS)**. For example, if you are working with **Personally Identifiable Information (PII)**, then you may need to mask names, email addresses, and other pieces of PII with dummy values.

Also, you must not send out emails to real users from your acceptance environment. If you are dealing with data that falls under the GDPR, this can put additional constraints on what you are allowed to do with backups and restores.

So far, we have seen two uses: recovering after losing data and acceptance tests with real data. The third use is to set up replication, where you restore a backup on a second server and then configure it to replicate all the changes from your main database server. This server can then be used as a hot standby to take over if the primary server dies or to serve read-only queries from reporting systems, for example.

You want to store your backups in a safe location. This can be another disk, another server, or the cloud. In general, the greater the physical separation, the safer the data. Having multiple copies is another way of lowering the risk.

If, for example, the backup is stored on the same storage appliance as the main database, then the failure of this storage appliance will leave you with neither your main database nor the backup. If you store your backup on the storage of the same cloud provider where you are hosting your database servers, then an outage of this cloud provider might also lead to the same situation.

Unfortunately, it is not uncommon for backups to be outdated, incomplete, or completely missing when people need them. The same goes for restore procedures. Even if everything is in place, people are often not familiar with the procedures. This can cause the restore to take more time than strictly needed or cause restore failures due to human errors, which often means that the restore has to be done again or fails completely. The only way to ensure your backups work is to test restores and really use them. It is not enough to check whether the backup completed successfully. We need to ensure that we can restore the backup and that our application is able to function with it.

Here are a few risks a backup can protect us against:

- Someone accidentally drops a table or removes more rows than intended.
- There is a hardware failure of your database server and/or disks.
- Someone gains unauthorized access to your database. (Backups will only help in this situation if they are not on the same server or otherwise can't be modified from your database server.)
- There are MySQL bugs and/or OS bugs that result in data corruption.

Note that the InnoDB engine is crash-safe by default. So, if your database server suddenly loses power, it should be able to recover from that. Using hardware RAID and other redundant systems can reduce the risk, but these systems also add complexity, increasing the risk of firmware bugs, and they won't protect you against accidentally dropping a table. So, these should be used together with backups.

You should restrict access to backups in the same way you restrict access to your database server to protect against unauthorized access.

In the next section, we will investigate different types of backups to learn and understand the advantages and disadvantages of each of them.

Logical and physical backup

One of the methods to create a backup is to stop a database completely and then copy all of the files to a safe location. This is easy to do and doesn't require special tools. However, while making the backup, your database is unavailable. This type of backup is called a **physical backup**.

Another way to create a backup is to export the data for all the tables and other database objects into a file that can be imported again. This is a **logical backup** as it doesn't copy the physical files but extracts the logical objects from the database. The benefit of this is that you don't have to shut down your server while taking the backup. The drawback is that the restores generally take a lot longer than a physical backup.

There are two alternatives to taking physical backups. The first one is taking a **snapshot**. You still need to stop the database server, but the time spent waiting for the copy is generally a lot less, as is the storage space required compared to that of a full copy. This does rely on a storage system and/or an OS that has snapshot capabilities.

MySQL Enterprise Backup and **Percona XtraBackup** are both tools that are smart enough to create a copy of all data files without stopping your database and know how to make the files consistent again. They can also restore a single table if desired.

> **Note**
> In this chapter, we only cover logical backups, but it is important to know that there are other options available.

In the next section, we will learn about restores and their types.

Types of restore

There are multiple reasons why you might want to do a restore. The most obvious one is if you lost your data – for example, after accidentally deleting the wrong data or after a hardware failure. Many of the restores you do should be to test your backups, backup procedures, and restore procedures. This means you restore the data on a temporary location and then check whether the restore is working properly and if all the data you expect to be there is there. And then there are restores you do to set up a new server. The new server can then be configured to replicate from your main server, allowing you to test a new version of MySQL before upgrading. It can also be used to test an upgrade procedure for the software you are using before doing it on the actual production instance.

The simplest restore type is to just restore everything. This is a **full restore**. This is what you would use if you lost all your data.

Another option is to restore a single table or database. This is a **partial restore** and is generally useful if someone accidentally dropped a single table.

And then there is a **point-in-time restore**. This is where you do a full restore and then use `binlog` files to fast-forward to just before the point where you lost something. We will cover what `binlog` files are and what's needed for this in the *Using point-in-time recovery with binlog files* section.

Performing backups

In this section, we will look at different tools to create logical backups. There are several different methods that can be used for MySQL backups. In this section, we will look at the following tools:

- `mysqldump`
- `mysqlpump`

We will also look at various techniques designed to help schedule backups and run partial or full backups when required. To start, we will look at our first backup tool, `mysqldump`.

> **Note**
> For collections, you use the exact same tools you would use for backups of tables. There are no special tools needed to backup and restore collections. This is because collections are stored as tables in MySQL. This allows you to query collections not only with X DevAPI but also with SQL. This also means that these collections will be picked up by all the backup tools that were designed to work with tables.

Using mysqldump

This tool has been part of MySQL from the early days. Often, you run this on the same server as MySQL Server, but you can also run this over the network. You should be using the same version as the MySQL server or a newer version.

The basic use of this looks like the following:

```
mysqldump --all-databases > backup.sql
```

This creates a backup of all the databases and saves it in the `backup.sql` file.

Let's discuss some common options for `mysqldump`:

The first set of options are the same options you might use for the MySQL client. This is `-h` for `host`, `-u` to specify `user`, and `-p` to provide `password`. This might be needed to get `mysqldump` to connect to the server with the right user. If your server runs on a non-default port, you can use `-p` to specify `port number`. This is all identical to all the MySQL clients.

If you specify `--single-transaction`, then `mysqldump` won't lock the tables during the backup but, instead, use a transaction to get a consistent backup. For this to work, you need to use InnoDB or another transactional storage engine, which is the default. Another option is to use `--skip-lock-tables`, but then the tables in the backup are not guaranteed to be consistent.

Another common thing to do is to back up a single database or table. Let's say we want to back up the data of the `animals` table that is present inside the `test` database. In order to do that, we must first execute the `mysqldump.exe` file that is present in the `C:\Program Files\MySQL\MySQL Server 8.0\bin` path. This will produce the following results:

```
C:\Program Files\MySQL\MySQL Server 8.0\bin>mysqldump.exe
Usage: mysqldump [OPTIONS] database [tables]
OR       mysqldump [OPTIONS] --databases [OPTIONS] DB1 [DB2 DB3...]
OR       mysqldump [OPTIONS] --all-databases [OPTIONS]
For more options, use mysqldump --help
```

Figure 15.1 – The results of mysqldump.exe

To save the backup of the `animals` table inside the `test_animals.sql` file, we need to provide proper credentials and specify the path of the file in the following way:

```
mysqldump -u root -p test animals > "C:\Users\Desktop\test_
animals.sql"
```

> **Note**
> The path in the preceding command depends on where you want to save the backup file.

This will ask you to enter the password that you have used while installing MySQL. On entering the correct password, you will be able to create the backup of the `animals` table. If you want to backup multiple tables, you can list all tables separated by spaces.

If you are using **PowerShell on Windows**, then there is a caveat to be aware of. PowerShell will convert the output to UTF-16 if you redirect it to a file. The problem with this is that the MySQL client expects UTF-8, so this causes issues on restore.

This can be fixed in two ways. First, you can use `Get-Content -Encoding UTF8` to convert the file to UTF-8 before feeding it to the MySQL client. But the second (and best) option is to invoke `mysqldump` with the `--result-file` option, like this:

```
mysqldump --all-databases --result-file=backup.sql
```

This doesn't use output redirection. So, the output won't be converted to UTF-16.

To compress the backup to have it use less disk space, you can use various compression utilities such as `gzip`, `bzip2`, or `xz`, like this:

```
mysqldump --all-databases | gzip > backup.sql.gz
```

This heavily depends on what utilities are available on your platform. On Windows, you can use NTFS compression or use tools such as WinZip or 7-Zip to compress the files after taking the backup.

In some instances, it can be useful to only backup the structure. One example is to create a schema-only backup, which can be used on development systems where you don't have the production data, either because the size is too big for the development systems or if the regulations forbid you from doing this. To create a `schema only backup`, you can use the `--no-data` option.

Exercise 15.01 – backup using mysqldump

In this exercise, you will create a database to store the coffee preferences of your colleagues. As another department wants to do the same, you have promised to create a schema dump so that they can set up the same thing for their department. You want to create a one-time backup after you have saved all the preferences, just to be sure that you can restore to this point in case the data gets lost somehow.

You will first create the `coffeeprefs` schema and table, and then insert data into the table. Then, you will create a schema-only dump to give to the other department, create the dump of a full schema as a backup, and inspect the files you created. Follow these steps to complete this exercise:

1. Open the MySQL Client.
2. Create a new schema named `coffeeprefs` by writing the following code:

    ```
    CREATE SCHEMA coffeeprefs;
    USE coffeeprefs;
    ```

3. Create a table named `coffeeprefs` with the `name` and `preference` columns. Assign `PRIMARY KEY` to the name column:

```
CREATE TABLE coffeeprefs (
    name VARCHAR(255),
    preference VARCHAR(255),
    PRIMARY KEY(name)
);
```

4. Insert three values into the table using the following queries:

```
INSERT INTO coffeeprefs VALUES
("John", "Capuchino"),
("Sue", "Cortado"),
("Peter", "Flat White");
```

5. Open Command Prompt and run the `mysqldump.exe` file.

6. Create a schema-only dump so that you can share it with the other department. Do this by writing the following code in Command Prompt (once the `mysqldump.exe` file is executed):

```
mysqldump -u root -p --single-transaction --no-data
coffeeprefs > "C:\Users\BHAVESH\Desktop\coffeeprefs.sql"
```

> **Note**
> For these exercises, you will need to edit the file location so that it saves the dump to wherever you want.

7. Press *Enter* and enter your password as prompted. The file will be saved in the aforementioned link.

8. Create a dump of the full schema as a backup by writing the following code in Command Prompt:

```
mysqldump -u root -p --single-transaction coffeeprefs >
"C:\Users\BHAVESH\Desktop\coffeeprefs_backup.sql"
```

9. Press *Enter* and enter your password when prompted. This will save the file at the previously mentioned link.

Both files should be small and can be opened in a text editor such as Notepad. The schema-only dump (`coffeeprefs.sql`) should not have any data in it. So, if you look for John, Sue, and Peter, you shouldn't be able to find them. The backup file (`coffeeprefs_backup.sql`) should have the names in there, as it should include all the data.

In the next section, we will learn about `mysqlpump`.

Using mysqlpump

The `mysqlpump` application has been part of MySQL since version 5.7. So, it is a relatively new tool. It was created to offer a more modern and extensible alternative to `mysqldump`.

The main difference is that `mysqlpump` can create backups in parallel. This can help to reduce the time needed to take a backup.

The basic options of `mysqlpump` are identical to those of `mysqldump`. But one of the differences is that if you don't specify any options, it will create a backup of all databases. So, there is no need to use `--all-databases`.

One of the other differences is object selection. With `mysqlpump`, it is easier to select which tables to include and exclude from the backup.

With `mysqlpump`, it is also possible to use `--compress-output` to select native compression. The supported algorithms are `LZ4` and `ZLIB`. This is a very new feature, as it was added in 8.0.18.

Another difference to mention is that `mysqlpump` does progress reporting. It shows the progress in both the number of tables and rows.

One important thing to note is that `mysqlpump` won't dump the `grants` tables in the `mysql` schema by default. You need to add `--users` instead to have it write the CREATE USER statements to the backup. Let's solve an exercise in the next section to master the skills of `mysqlpump`.

Exercise 15.02 – backing up using mysqlpump

In this exercise, you will create another backup of the `coffeeprefs` schema using `mysqlpump`. This time, you are going to compress the backup. You will first create the backup of the `coffeeprefs` schema and use `zlib` compression, and then validate the created backup file. Follow these steps to complete this exercise:

1. Open Command Prompt and write the following code to create a backup of the `coffeeprefs` schema using `mysqlpump`:

```
mysqlpump -u root -p --single-transaction --set-gtid-
purged=OFF --compress-output zlib coffeeprefs --result-
file="C:\Users\BHAVESH\Desktop\Coffee\coffeeprefs.sql.gz"
```

2. Press *Enter* and provide the password as prompted. The file will be saved in the previously mentioned link.

3. Now, validate the created backup file using the following code:

```
zlib_decompress "C:\Users\BHAVESH\Desktop\Coffee\
coffeeprefs.sql.gz" ""C:\Users\BHAVESH\Desktop\Coffee\
coffeeprefs.sql"
```

You can use a text editor such as Notepad to open the resulting file and recognize the table structure and data present in the table.

In the next section, we will learn about scheduling backups.

Scheduling backups

One thing that is not included in `mysqldump` or `mysqlpump` is the scheduling of backups. These tools know how to create a backup but not when to. So, this is something where you must use the scheduling services provided by the platform you're using. This is Cron on Linux and macOS, or Task Scheduler if you are on Windows.

Besides creating backups, you probably want to automate cleaning up the oldest backups.

It might be a good idea to put the actual `mysqldump` or `mysqlpump` command into a shell script or (on Windows) in a `.bat` file. Then, you can check `returncode` of the process and send an email and/or monitoring alert if the backup fails. You can also use the same script to copy the backup to the cloud or another server.

A very basic scheduled backup on Linux can be created using `/etc/cron.d/` `mysqldump` with the following contents:

```
0 4 * * * root /user/bin/mysqldump -A > /data/backups/mysql.sql
```

This creates a backup every day at 04:00AM of all databases (`-A`) and stores this in the `/data/backups/mysql.sql` file. This overwrites the file every day. It is up to the reader to extend this with monitoring, copying the file to the cloud or another safe location.

To allow the root user of the system to have access to the database, we need to create `/root/.my.cnf` with the following content:

```
[client]
user=root
password=Biadojdogmipofilva
```

And then we have to replace `Biadojdogmipofilva` with the password you configured.

On Windows, you can use **Task Scheduler**. In the following screenshot, you can see the configuration for a very basic daily backup at 04:00:

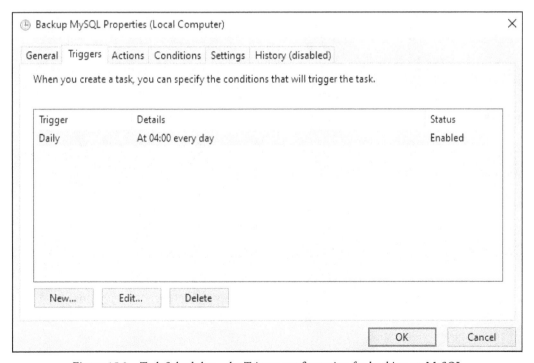

Figure 15.2 – Task Scheduler – the Trigger configuration for backing up MySQL

You can see the action that will occur on being triggered in the following screenshot:

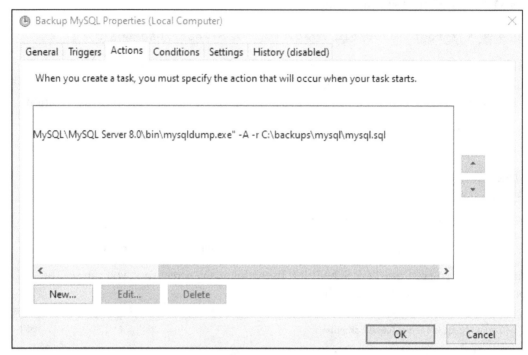

Figure 15.3 – Task Scheduler – the Action configuration for backing up MySQL

To allow the scheduled backup to work, it needs access to the `root` password of the database. On Windows, this is stored in `%APPDATA%\MySQL\.mylogin.cnf`. To create this file, you need to use the `mysql_config_editor` utility.

```
mysql_config_editor.exe set --user root -p
```

> **Note**
> On Linux and Windows, you can specify the password in the configuration of the backup schedule, but that information may be accessible to other users of the system, which wouldn't be secure.

In the next few sections, we will cover three different kinds of restores – full restores, where everything is restored; partial restores, where only a single schema is restored; and point-in-time recovery, where we restore to a specific point in time.

Full restore

If you restore a full backup on a new server, you first need to install MySQL as usual and make sure that MySQL Server is running.

Then, you can import the backup file like this:

```
mysql < backup.sql
```

If your restore contains the `mysql` schema, then you need to issue `FLUSH PRIVILEGES` to load the restored system tables for authentication into memory. Alternatively, you can add `--flush-privileges` to your `mysqldump` statement to have `mysqldump` put the command in the backup file. This is not needed with `mysqlpump` because it won't back up the system tables directly but, instead, generate `CREATE USER` statements, for which `FLUSH PRIVILEGES` is not needed.

Partial restore

If you have a backup of a single schema created with `mysqlpump`, then you can restore the backup like this:

```
mysql < backup_test.sql
```

If the backup was made with `mysqldump`, you have more options. If you have a backup of a single schema, then you need to create the schema again before doing the restore. The name of the schema doesn't have to be identical, so it is possible to restore the backup in a different schema. This might be handy if you want to use a copy of the data to work on, if you want to only restore a subset of the tables and/or rows, or if you want to compare the current content of the database with the backup. Consider the following code:

```
CREATE SCHEMA test_restore;
mysql test_restore < backup_test.sql
```

If you have a full backup and only want to restore a single database or a single table, then your best option is to do a full restore on a temporary instance of MySQL, dump only the information that you need, and then restore that on your server. The other option is to extract the right set of lines from the backup and restore that, but that's error-prone.

In the next section, we will solve an exercise based on restoring a single schema backup.

Exercise 13.03 – restore a single schema backup

You want to create a completely new version of the coffeeprefs application. To develop the new application, you want to make a copy of the database. While the production version of your database is on a central server, you might want to have the copy on a MySQL instance on your laptop to allow you to work on the new version of the application, even without network access.

In this exercise, you will create a backup of coffeeprefs schema, create a new schema named coffeeprefs_dev, restore the backup, and verify the result.

Follow these steps to complete this exercise:

1. Open Command Prompt and create a backup of the coffeeprefs schema using the following code:

    ```
    mysqldump -u root -p --single-transaction coffeeprefs >
    "C:\Users\BHAVESH\Desktop\coffeeprefs.sql"
    ```

2. Open the MySQL Client and create a new schema named coffeeprefs_dev:

    ```
    CREATE SCHEMA coffeeprefs_dev;
    ```

    This allows you to have two schemas on the same server. Even if they are not on the same server, this makes it easier to recognize whether you are working on the production database or the development one.

3. Restore the backup into a newly created schema:

    ```
    USE coffeeprefs_dev;
    SOURCE C:\Users\BHAVESH\Desktop\coffeeprefs.sql
    ```

4. Verify the results by checking the tables present in the schema:

    ```
    SHOW TABLES;
    ```

    This will produce the following result:

    ```
    +----------------------------+
    | Tables_in_coffeeprefs_dev  |
    +----------------------------+
    | coffeeprefs                |
    +----------------------------+
    1 row in set (2.93 sec)
    ```

    Figure 15.4 – Tables present in coffeeprefs_dev

In this exercise, you used backup and restore to create a copy of a schema. Besides creating a copy for development purposes, the same procedure works for other use cases, such as verifying that you can restore a backup and that it has the data you expect.

In the next section, we will learn how to use point-in-time recovery with `binlog` files.

Using point-in-time recovery with binlog files

MySQL Server is able to write all the changes made to data inside a database to a binary log file (`binlog` for short). A binary log (`binlog`) is a file written by the database server that contains all the changes made to the data, which is stored inside the database server in a specific timeframe. This is called a binary log because the changes are recorded in a binary format as opposed to a text-based format. The logs with changes can be used for multiple purposes. One of them is to stream them to a second server to keep it updated. Then, the second server can be used as a standby in case the primary server fails, or the second server can be used to offload heavy read-only queries such as reporting. But these `binlog` files can also be used to replay changes made to the database between the time of the last backup and the time of the restore point (the point just before something disastrous such as a `drop table` command happened).

For this to work, the server must be configured to write these files. In MySQL 8.0, this is done by default, and in earlier versions, you must set the `log_bin` variable to `ON` and `server_id` to `number`.

The `binlog` does take a bit of disk space and, by default, is kept for 30 days. You can set `binlog_expire_logs_seconds` to a lower value to save disk space.

There are multiple formats available for the `binlog` file: `ROW`, `STATEMENT`, and `MIXED`. The default and recommended option is `ROW` where, for every changed row, it puts a before and/or after image into `binlog`, depending on the kind of operation (for example, `INSERT`, `UPDATE`, or `DELETE`). The `STATEMENT` format instead puts the SQL statement that made the change into `binlog`. While this works and can be more efficient than the `ROW` format, it is not recommended. This is because the execution of the statement can be dependent on many factors, such as the time of the day, server settings, and so on. While these are included in `binlog`, there are still some operations that may result in different outcomes on the primary and secondary servers.

You also may want to configure MySQL to use **Global Transaction Identifiers** (**GTIDs**). This is a system where every transaction gets a globally unique ID assigned. This makes it easier to see which transactions have been processed by which server. To do this, you need to set `gtid_mode=ON` and set `enforce_gtid_consistency=ON`. This can be seen in the following sample code:

```
SET PERSIST_ONLY gtid_mode=ON;
SET PERSIST_ONLY enforce_gtid_consistency=ON;
RESTART;
```

This is one of the ways to configure these settings. You can also put these settings in the `my.cnf` configuration file and restart if you wish. The `SET PERSIST_ONLY` sets the setting in the persistent configuration but doesn't change the running configuration. `RESTART` actually restarts the server to make the setting active. In this case, the restart is not actually required, but it is the quickest way to do this.

Once this is configured, you then want to make sure you use `--master-data=2` with `mysqldump` if you are not using `GTIDs`.

If you use `mysqlpump`, you need to have GTIDs configured and use `--set-gtid-purged=ON`. If you have GTIDs configured and use `mysqldump`, you also need to use this setting.

These two settings cause `mysqldump` and `mysqlpump` to write the location of the server into the `binlog` files. This is crucial information for **Point-in-Time Recovery** (**PITR**).

Another important thing to keep in mind is where the `binlog` files are kept. If the server fails, then the `binlog` files on that server are also likely to be unavailable. So, you may want to archive them in a safe place. Also, be careful not to accidentally wipe them out when restoring a backup. It might be a good idea to make a copy before doing the restore, just in case.

To do a point-in-time restore, we need to follow these steps:

1. Find the position to which you want to restore.
2. Copy the `binlog` files to a temporary location and run `RESET MASTER;`.
3. Restore the most recent backup.
4. Apply the `binlog` files between the time of the backup and the restore point found in *Step 2*.

To find the position, you can define a date and time, or find a GTID and/or a file and a position of the time just before an incident happened.

To do this, use the following code:

```
SHOW MASTER LOGS;
SHOW BINLOG EVENTS IN '<file>';
```

This may look like this:

```
mysql> SHOW MASTER LOGS;
+---------------+-----------+-----------+
| Log_name      | File_size | Encrypted |
+---------------+-----------+-----------+
| binlog.000001 |   4107759 | No        |
| binlog.000002 |       199 | No        |
| binlog.000003 |      1006 | No        |
| binlog.000004 |       852 | No        |
| binlog.000005 |       593 | No        |
| binlog.000006 |       869 | No        |
+---------------+-----------+-----------+
6 rows in set (0.00 sec)

mysql> SHOW BINLOG EVENTS IN 'binlog.000006';
+---------------+-----+----------------+-----------+-------------+--------------------------------------------------------------------+
| Log_name      | Pos | Event_type     | Server_id | End_log_pos | Info                                                               |
+---------------+-----+----------------+-----------+-------------+--------------------------------------------------------------------+
| binlog.000006 |   4 | Format_desc    |    1      |     124     | Server ver: 8.0.18, Binlog ver: 4                                  |
| binlog.000006 | 124 | Previous_gtids |    1      |     195     | 00008017-0000-0000-0000-000000008017:1-2                           |
| binlog.000006 | 195 | Gtid           |    1      |     272     | SET @@SESSION.GTID_NEXT= '00008017-0000-0000-0000-000000008017:3'  |
| binlog.000006 | 272 | Query          |    1      |     400     | use `test`; create table foobar (id int primary key) /* xid=47 */  |
| binlog.000006 | 400 | Gtid           |    1      |     479     | SET @@SESSION.GTID_NEXT= '00008017-0000-0000-0000-000000008017:4'  |
| binlog.000006 | 479 | Query          |    1      |     554     | BEGIN                                                              |
| binlog.000006 | 554 | Table_map      |    1      |     606     | table_id: 124 (test.foobar)                                        |
| binlog.000006 | 606 | Write_rows     |    1      |     656     | table_id: 124 flags: STMT_END_F                                    |
| binlog.000006 | 656 | Xid            |    1      |     687     | COMMIT /* xid=48 */                                                |
| binlog.000006 | 687 | Gtid           |    1      |     764     | SET @@SESSION.GTID_NEXT= '00008017-0000-0000-0000-000000008017:5'  |
| binlog.000006 | 764 | Query          |    1      |     869     | drop schema test2 /* xid=49 */                                     |
+---------------+-----+----------------+-----------+-------------+--------------------------------------------------------------------+
11 rows in set (0.00 sec)
```

Figure 15.5 – The SHOW BINLOG EVENTS output

Here, in the `binlog.000006` file on position `764`, you can see `drop schema test2`. In the line just above that, you can find the GTID for this statement, which in this case is `00008017-0000-0000-0000-000000008017:5`.

GTID format

A GTID looks like `<UUID>:<number>`. Here, `UUID` is the universally unique ID assigned to the server. You can use `SELECT @@server_uuid` to see the UUID assigned to a server. If you run the command in the MySQL client, then you will get the following output:

```
+--------------------------------------+
| @@server_uuid                        |
+--------------------------------------+
| b3ba6ff4-48d1-11ea-9775-f44d304a0775 |
+--------------------------------------+
1 row in set (0.00 sec)
```

Figure 15.6 – The UUID assigned to the server

The second part is the transaction number. These two together uniquely identify a transaction. You can specify a range of transactions in this `<UUID>:<start>-<end>` format. To see which transactions a server has executed, you can use the following code:

```
SELECT @@global.gtid_executed;
```

You need to start at the position of the restored server. For this, there are the following three options:

- **Option 1**: If you are not using a GTID and you created the backup with `mysqldump` using the `--flush-logs` option, then MySQL switches to a new `binlog` file when the backup was made. So, you don't have to specify a start position, as you can start from the beginning of the first file that was created after the backup was created. At the beginning of the backup, you can find a line that looks like this:

```
CHANGE MASTER TO MASTER_LOG_FILE='binlog.000001', MASTER_
LOG_POS=155;
```

This can be used to find out which `binlog` files have already been processed by MySQL at the time of the backup.

- **Option 2**: If you didn't use `--flush-logs` when creating the backup, then you have to specify both the start and end position.

- **Option 3**: If you are using GTIDs, then you can simply look at the set of executed transactions after the restore and the last transaction you need to restore to and calculate the set that is in between. Consider the following example:

```
Executed set of the server: 00008017-0000-0000-0000-
000000008017:1-23
```
```
Accidental drop table: 00008017-0000-0000-0000-
000000008017:29
```
```
Last transaction we want to restore: 00008017-0000-0000-
0000-000000008017:28
```

Then, we need transactions `24` to `28`, which is specified as `00008017-0000-0000-0000-000000008017:24-28`.

Now, we can use the `mysqlbinlog` utility to extract the commands from the set of `binlog` files we copied away. If you have a date and time to which you want to restore, then you need to write the following:

```
mysqlbinlog --stop-datetime="2010-01-01 01:00:00" binlog.*
```

Here, you read all the `binlog` files and stop at the specified date and time. You can combine this with specifying a start position if needed.

If you are not using a GTID and have to specify start and end positions, then you need to write the following code:

```
mysqlbinlog \
 --start-position=155 \
 --stop-position=357 binlog.* > to_restore.sql
```

Here, we start in the first file at position `155` and then continue until position `357` in the last file.

If you have a GTID, then things are easier, as you have to just specify the range that you need to reapply:

```
mysqlbinlog \
 --include-gtids=00008017-0000-0000-0000-000000008017:100-200 \
 binlog.* > to_restore.sql
```

Here, we include transactions `100` to `200`.

Now, you can run the following code:

```
mysql < to_restore.sql
```

This will actually apply the changes to the server. You can also inspect the contents of the generated SQL file to make sure that it looks correct.

You should only allow applications to use the server after the full restore is done. Making a change, especially while the restore is happening, can cause the restore to fail.

When restoring data from multiple `binlog` files with `mysqlbinlog`, you have to do so as a single operation. If you try to restore one `binlog` file at a time, then transactions that span multiple files may not be applied correctly. In the next section, we will learn how to use `mysqlbinlog` to inspect `binlog` contents.

Using mysqlbinlog to inspect binlog contents

The `mysqlbinlog` utility we used to extract data from the `binlog` files can also be used to inspect changes that were made to the database. Note that the `binlog` files only contain changes to the database. So, `SELECT` statements won't appear there.

To make the data changes in the ROW format human-readable, we need to use the `--verbose` option for `mysqlbinlog`. This causes it to output this human-readable data in addition to `base64` data that can be used by MySQL to reapply the changes:

```
### INSERT INTO 'test'.'mytable'
### SET
###    @1=1
###    @2='foo'
###    @3='bar'
###    @4='baz'
```

Now, in the next activity, we will back up and restore a single schema.

Activity 15.01 – backing up and restoring a single schema

In this activity, you will create a simulated disaster in the `world` schema and recover from this disaster. For this, you will be using `mysqldump`. Perform the following steps to complete this activity:

1. Create the backup of the `world` schema.
2. Simulate the disaster. Here, delete all the rows of the `city` table.
3. Restore the backup.
4. Verify that the data has been restored.

The expected output is as follows:

Figure 15.7 – The total rows in the city table after restoring

> **Note**
> The solution for this activity can be found in the *Appendix*.

Here, we successfully restored the `world` schema after wiping out the `city` table. As there were no other changes made to the `world` schema between the time of the backup and the time of `DELETE`, we have restored all the data.

In the next activity, we will perform a point-in-time restore.

Activity 15.02 – performing a point-in-time restore

In this activity, you will use the `world` schema to perform a point-in-time restore. The server you will use for this does not have GTIDs enabled. Follow the steps here to implement this activity:

1. Reset `MASTER LOGS`.

2. Create a backup with `mysqldump`.

3. Change the population of Toulouse in the `city` table.

4. Simulate the disaster by wiping out the complete `city` table.

5. Restore the backup.

6. Reapply the changes that occurred between backup creation and the time of the disaster.

7. Validate that the data has been restored.

After implementing these steps, the expected output is as follows:

```
+------+----------+-------------+---------------+------------+
| ID   | Name     | CountryCode | District      | Population |
+------+----------+-------------+---------------+------------+
| 2977 | Toulouse | FRA         | Midi-Pyrénées |  123456789 |
+------+----------+-------------+---------------+------------+
1 row in set (0.03 sec)
```

Figure 15.8 – Inspecting the change that we made before

> **Note**
> The solution for this activity can be found in the *Appendix*.

Here, you can see that the special value we used has been restored correctly. Thus, in this activity, you created a backup, then made some additional changes, and simulated an accidental wipeout of the `city` table. Then, you restored the backup and reapplied the additional changes that you made earlier.

Summary

In this chapter, we learned why we need backups, what different options there are for creating backups, how to create backups, and how to restore them. We used `mysqldump`, `mysqlpump`, and `mysqlbinlog` in the process.

Using the tools provided by the OS, we practiced backup scheduling and performed full restores, partial restores, and point-in-time restores, in which we covered `binlog` files and GTIDs.

This helps us to keep our data safe and restore it when it is needed. It also taught us why we should test our backups. This gives us a valuable tool to copy data between servers by backing up and restoring it.

We have now covered all the fundamental concepts required to effectively use MySQL. Using the tools provided in this book, you should now have a strong foundation for working with MySQL as a database technology. Using this knowledge, you will be able to build and use a MySQL database for various purposes, such as application development and data analysis.

With this knowledge, you can now take several paths to further develop your database skills. SQL is a widely used language, which is present in many database technologies beyond MySQL. Technologies such as Microsoft SQL Server and PostgreSQL are widely used in the industry, and they are great skills to develop with further database study. Aside from learning new skills, I encourage you to try building some databases and applications to apply your skills to real-life scenarios. This will allow you to expand your horizons and become a MySQL expert!

Appendix

Solution to Activity 1.1

These are the steps to solve Activity 1.1:

1. Analyze the given table:

Hostname	Location	OperatingSystem	Layerlevel
PINKY	Ground Floor A	IOS	L2
PINKY	Ground Floor A	NXOS	L2
HERETIC	First Floor A	JUNOS	L3
HERETIC	First Floor A	NXOS	L3
HERETIC	Ground Floor B	IOS	L3

Figure 16.1 – A table of devices on the network

You can see that, currently, there does not appear to be a unique field, so this is something that needs to be added to create a 2NF table. In addition, observe that Hostname, Location, OperatingSystem, and Layerlevel are all text fields, giving you an idea of the proper data type.

2. Specify the data types for the data contained in this table:

Field Name	Type
Hostname	varchar
Location	varchar
OperatingSystem	varchar
Layerlevel	varchar

Figure 16.2 – The data types that are used for the table fields

Starting off, each column can be represented as varchar, since they are all text fields. Also, observe that the preceding table is in 1NF because every column contains a single value.

3. Create a key that uniquely identifies the table. You will have to create a composite key of Hostname and Location, since the table has common hostnames.

4. Once the composite key is made, bring the table into 2NF. The composite key creates a set of pairs that are unique in the table. Since they are unique, they are able to identify the records, giving us a 2NF table. The tables would look like the following:

OperatingSystemID	Hostname	Location	LayerLevel
1	PINKY	Ground Floor A	L2
2	HERETIC	First Floor A	L3
3	HERETIC	Ground Floor B	L3

Figure 16.3 – The 2NF of the network table

To make it easier to connect relationships between this table and other tables that contain operating systems, it is best to create an OperatingSystemID field. This gives us a simple ID field that can be used to create relationships without needing to rely on the composite key. Adding this ID gives us a quick way to create relationships in our database.

Note that the data type for OperatingSystemID is int, since it is numerical without any decimals:

OperatingSystemID	OperatingSystem
1	IOS
2	NXOS
3	JUNOS

Figure 16.4 – The 2NF of the table operating systems

5. To achieve your goal, you need to bring the tables to 3NF – that is, ensure you have no transitive functional dependencies. As a part of this process, you will also work on decomposing the table into a few smaller tables so that each table contains a single set of information. This is done to help reduce redundancies in the database. Since many entities may have a layer level or operating system, you will give them their own tables. These tables must have a primary key to preserve 3NF, so you will give them a unique ID identifier.

You are now left with the following three tables:

OperatingSystemID	Hostname	Location	LayerLevelID
1	PINKY	Ground Floor A	1
2	HERETIC	First Floor A	2
3	HERETIC	Ground Floor B	2

Figure 16.5 – The 3NF of the network devices table

This is the 3NF table for the operating systems table:

OperatingSystemID	OperatingSystem
1	IOS
2	NXOS
3	JUNOS

Figure 16.6 – The 3NF of the operating systems table

This is the 3NF table for the layer levels table:

LayerLevelID	LayerLevel
1	L2
2	L3

Figure 16.7 – The 3NF of the layer levels table

As with the other ID fields, `LayerLevelID` will also be an `int` data type, since it is numeric without decimal.

Solution to Activity 2.1

Perform the following steps to successfully execute this activity:

1. Open the EER diagram for the `autoclub` model.

2. Click on the EER diagram and press *T* to get the new table pointer. Click on the diagram to place the new table, as shown here:

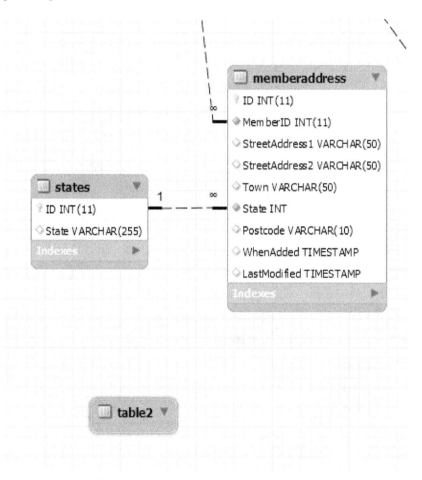

Figure 16.8 – A new table added to the EERD

3. Double-click on the new table to open the table design. It will look something like the following:

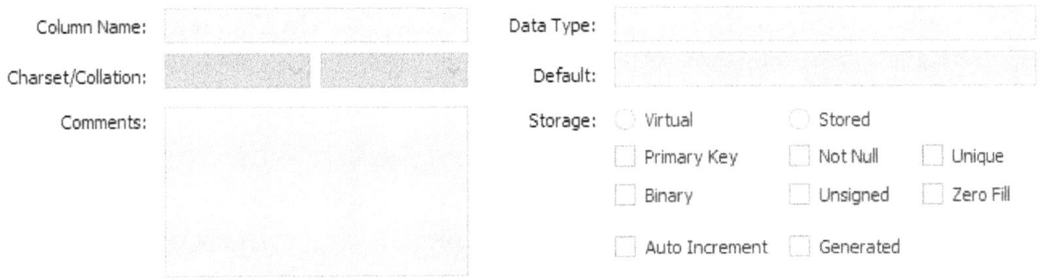

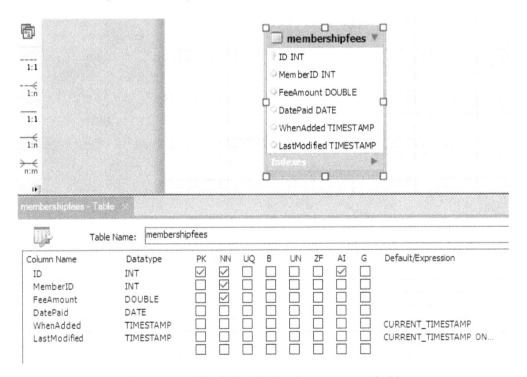

Figure 16.9 – The table design for the new table

4. Rename the table membershipfees and enter the ID, MemberID, FeeAmount, DatePaid, WhenAdded, and LastModified fields, as shown here:

Figure 16.10 – The fields added to the new renamed table

5. Click on the **Foreign Keys** tab and create a foreign key named FK_ MembershipFees_Members with 'autoclub':'members' under **Referenced Table**. In the next panel, select **MemberID** under **Column** and **ID** under **Referenced Column**. Also, set **Foreign Key Options** as RESTRICT for **On Update** and **On Delete**, as shown in the following figure:

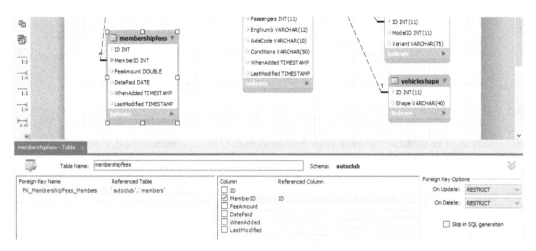

Figure 16.11 – The foreign key to members is added

6. Save the EERD using **File** and then **Save Model**.

7. Now, select **Database** and **Forward Engineer** and walk through the screen to save the changes to the model.

8. Examine the model to view the new table, **membershipfees**, as shown here:

Figure 16.12 – The new membershipfees table is in the model

9. Push the changes to the live database by selecting **Database** and **Synchronize Model**. Work through the screens of the wizard.

10. Finally, return to the **My First Connection** tab, refresh the **autoclub** database, and examine the **membershipfees** table, as shown here:

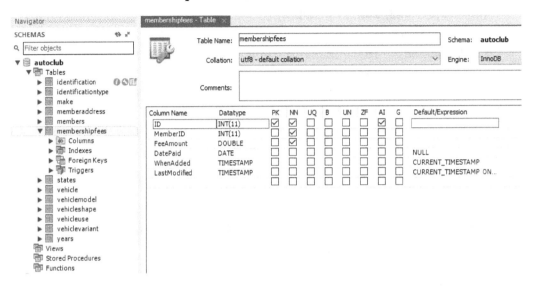

Figure 16.13 – The new membershipfees table in the production database

Solution to Activity 3.1

The solution to this activity is as follows:

1. Open a new query tab.

2. Enter the following SQL statement into the query tab:

```
-- ------------------------------------------------------
-- Table 'autoclub'.'eventmemberregistration'
-- ------------------------------------------------------
CREATE TABLE IF NOT EXISTS
'autoclub'.'eventmemberregistration' (
    'ID' INT NOT NULL AUTO_INCREMENT,
    'ClubEventID' INT NOT NULL,
    'MemberID' INT NOT NULL,
    'ExpectedGuestCount' INT NOT NULL DEFAULT 0,
    'RegistrationDate' DATE NOT NULL,
    'FeesPaid' BIT NOT NULL DEFAULT 0,
    'TotalFees' DOUBLE NOT NULL DEFAULT 0,
```

```
    'MemberAttended' BIT NOT NULL DEFAULT 0,
    'ActualGuestCount' INT NOT NULL DEFAULT 0,
    'Notes' MEDIUMTEXT NULL,
    'WhenAdded' TIMESTAMP NULL DEFAULT CURRENT_TIMESTAMP,
    'LastModified' TIMESTAMP NULL DEFAULT CURRENT_TIMESTAMP
ON UPDATE CURRENT_TIMESTAMP,
  PRIMARY KEY ('ID'),
  INDEX 'Idx_EventID' ('ClubEventID' DESC),
  INDEX 'FK_EventReg_Members_idx' ('MemberID' ASC),
  CONSTRAINT 'FK_EventReg_ClubEvents'
    FOREIGN KEY ('ClubEventID')
    REFERENCES 'autoclub'.'clubevents' ('ID')
    ON DELETE NO ACTION
    ON UPDATE NO ACTION,
  CONSTRAINT 'FK_EventReg_Members'
    FOREIGN KEY ('MemberID')
    REFERENCES 'autoclub'.'members' ('ID')
    ON DELETE NO ACTION
    ON UPDATE NO ACTION)
ENGINE = InnoDB;
```

3. Execute the SQL query by clicking the lightning bolt icon:

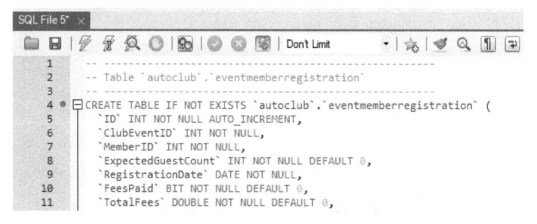

Figure 16.14 – The SQL code and the lightning bolt icon to run it

4. Refresh the **SCHEMAS** panel; you can see the new table, **eventmemberregistration**, in the table list:

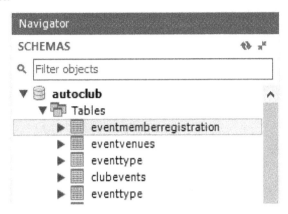

Figure 16.15 – The new table, eventmemberregistration

5. Open the **eventmemberregistration** table in the design view to examine the table design, indexes, and foreign keys:

Figure 16.16 – Right-click on the eventmemberregistration table and select Alter Table...

You will be able to see the design view as follows:

Table Name:	eventmemberregistration	Schema: **autoclub**
Collation:	Schema Default	Engine: InnoDB
Comments:		

Column Name	Datatype	PK	NN	UQ	B	UN	ZF	AI	G	Default/Expression
ID	INT(11)	☑	☑	☐	☐	☐	☐	☑	☐	
ClubEventID	INT(11)	☐	☑	☐	☐	☐	☐	☐	☐	
MemberID	INT(11)	☐	☑	☐	☐	☐	☐	☐	☐	
ExpectedGuestCount	INT(11)	☐	☑	☐	☐	☐	☐	☐	☐	'0'
RegistrationDate	DATE	☐	☑	☐	☐	☐	☐	☐	☐	
FeesPaid	BIT(1)	☐	☑	☐	☐	☐	☐	☐	☐	b'0'
TotalFees	DOUBLE	☐	☑	☐	☐	☐	☐	☐	☐	'0'
MemberAttended	BIT(1)	☐	☑	☐	☐	☐	☐	☐	☐	b'0'
ActualGuestCount	INT(11)	☐	☑	☐	☐	☐	☐	☐	☐	'0'
Notes	MEDIUMTEXT	☐	☐	☐	☐	☐	☐	☐	☐	NULL
WhenAdded	TIMESTAMP	☐	☐	☐	☐	☐	☐	☐	☐	CURRENT_TIMESTAMP
LastModified	TIMESTAMP	☐	☐	☐	☐	☐	☐	☐	☐	CURRENT_TIMESTAMP ON...
		☐	☐	☐	☐	☐	☐	☐	☐	

Figure 16.17 – The eventmemberregistration table in the design view,
with settings as defined in your SQL

The indexes can be seen here:

Table Name:	eventmemberregistration	Schema: **autoclub**
Collation:	Schema Default	Engine: InnoDB
Comments:		

Index Name	Type
PRIMARY	PRIMARY
Idx_EventID	INDEX
FK_EventReg_Members_idx	INDEX
FK_EventReg_ClubEvents	INDEX

Index Columns

Column	#	Order	Length
☐ ID		ASC	
☑ ClubEventID	1	DESC	
☐ MemberID		ASC	
☐ ExpectedGuestCou...		ASC	
☐ RegistrationDate		ASC	
☐ FeesPaid		ASC	
☐ TotalFees		ASC	
☐ MemberAttended		ASC	
☐ ActualGuestCount		ASC	
☐ Notes		ASC	
☐ WhenAdded		ASC	
☐ LastModified		ASC	

Figure 16.18 – The indexes

The foreign keys in the database can be seen here:

Table Name:	eventmemberregistration	Schema:	**autoclub**
Collation:	Schema Default	Engine:	InnoDB
Comments:			

Foreign Key Name	Referenced Table
FK_EventReg_ClubEvents	`autoclub`.`clubevents`
FK_EventReg_Members	`autoclub`.`members`

Column	Referenced Column
☐ ID	
☑ ClubEventID	ID
☐ MemberID	
☐ ExpectedGuestC...	
☐ RegistrationDate	
☐ FeesPaid	
☐ TotalFees	
☐ MemberAttended	
☐ ActualGuestCount	
☐ Notes	
☐ WhenAdded	
☐ LastModified	

Foreign Key Options
On Update: RESTRICT
On Delete: RESTRICT

☐ Skip in SQL generation

Figure 16.19 – The foreign keys

So far, you have learned how to create a table and set the indexes and foreign keys using SQL code. We still need to add a unique index to the `username` field in the `users` table and set a foreign key in the `clubevents` table to the `eventtype` table.

Solution to Activity 3.2

The solution for the activity is as follows:

1. Open a SQL tab and run the following command to fetch the ID of the new member:

   ```
   SELECT ID FROM members where 'Surname' = "Pettit"
   ```

2. Execute the SQL statement by clicking the lightning bolt icon:

Figure 16.20 – SQL to extract the user ID of a member by surname

You should get the following output:

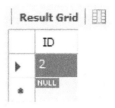

Figure 16.21 – The ID for the user Pettit

3. Download the image from GitHub and add it to the `Members/Photos` folder. Add your member ID to the name of the image. The folder should look like the following:

Figure 16.22 – The updated Photos folder

4. Create a script in a new SQL tab to update your record with the image and path:

```
UPDATE 'members'
SET
PhotoPath = "Members\\Photos\\MemberPhoto_2.jpg"
WHERE    'ID'=2;
```

5. Execute the SQL query. You should get the following result:

Figure 16.23 – The results with the second member showing the image path

6. Run the following query to fetch the full file path of the image:

```
SELECT CONCAT((SELECT 'Value' FROM 'lookups' WHERE
'Key'="ImageRepository") , 'PhotoPath') AS FullPhotoPath
FROM 'members' WHERE 'members'.'ID'=2
```

You will get the full path of the image, as shown here:

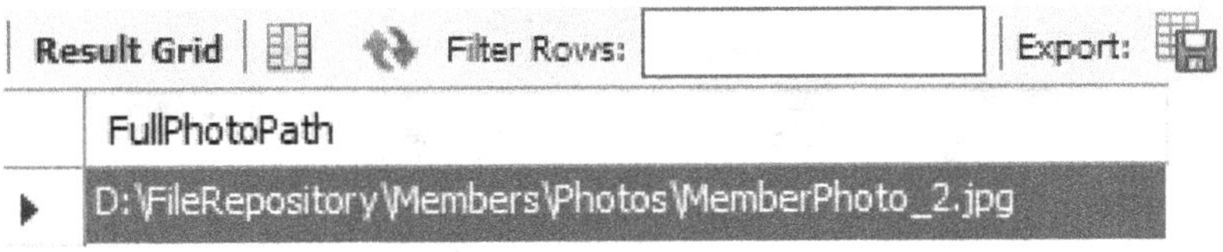

Figure 16.24 – The full file path for the image

Solution to Activity 4.1

In this activity, you will collect some information from the `world` schema to add bits of trivia to some of the articles in next month's edition of a travel magazine. Follow these steps to complete this activity:

1. Connect to the `world` schema with the MySQL client:

```
USE world;
```

This query produces the following output:

```
mysql> USE world
Database changed
```

Figure 16.25 – Connecting to the world schema

2. Write the following query to check the size of the smallest city in the database:

```
SELECT Name, Population FROM city ORDER BY Population
LIMIT 1;
```

This query produces the following output:

```
mysql> SELECT Name, Population FROM city ORDER BY Population LIMIT 1;
+-----------+------------+
| Name      | Population |
+-----------+------------+
| Adamstown |         42 |
+-----------+------------+
1 row in set (0.00 sec)
```

Figure 16.26 – The city with the smallest population

Here, you used the `city` table, sorted by `Population`, and finally, selected the `name` and `population` fields.

3. Write the following query to check the number of languages spoken in India:

```
SELECT Code FROM country WHERE name='India';
SELECT * FROM countrylanguage WHERE CountryCode='IND';
```

These queries produce the following output:

```
mysql> SELECT Code FROM country WHERE name='India';
+------+
| Code |
+------+
| IND  |
+------+
1 row in set (0.00 sec)

mysql> SELECT * FROM countrylanguage WHERE CountryCode='IND';
+-------------+-----------+-----------+------------+
| CountryCode | Language  | IsOfficial | Percentage |
+-------------+-----------+-----------+------------+
| IND         | Asami     | F         |        1.5 |
| IND         | Bengali   | F         |        8.2 |
| IND         | Gujarati  | F         |        4.8 |
| IND         | Hindi     | T         |       39.9 |
| IND         | Kannada   | F         |        3.9 |
| IND         | Malajalam | F         |        3.6 |
| IND         | Marathi   | F         |        7.4 |
| IND         | Orija     | F         |        3.3 |
| IND         | Punjabi   | F         |        2.8 |
| IND         | Tamil     | F         |        6.3 |
| IND         | Telugu    | F         |        7.8 |
| IND         | Urdu      | F         |        5.1 |
+-------------+-----------+-----------+------------+
12 rows in set (0.00 sec)
```

Figure 16.27 – The languages spoken in India

The first query is only needed to get `CountryCode` for India, so you can use that in the second query. You used * to get all fields. You could have selected only the `Language` field, but now, you can also see some other interesting information about the languages (namely, `Percentage` and whether they are official languages or not).

4. Write the following query to check the languages that are spoken in more than 20 countries:

```
SELECT Language FROM countrylanguage
GROUP BY Language HAVING COUNT(*)>20;
```

This query produces the following output:

```
mysql> SELECT Language FROM countrylanguage
    -> GROUP BY Language HAVING COUNT(*)>20;
+----------+
| Language |
+----------+
| English  |
| Spanish  |
| French   |
| Arabic   |
+----------+
4 rows in set (0.00 sec)
```

Figure 16.28 – The languages spoken in more than 20 countries

Here, you used aggregation on `Language` and filtered out rows that have more than 20 items per group.

5. Write the following query to find the five biggest cities in the `Southern and Central Asia` region:

```
SELECT doc->>'$.name'
FROM worldcol
WHERE doc->>'$.country.region' = "Southern and Central
Asia"
ORDER BY doc->'$.population' DESC
LIMIT 5;
```

This query produces the following output:

```
mysql> SELECT doc->>'$.name'
    -> FROM worldcol
    -> WHERE doc->>'$.country.region' = "Southern and Central Asia"
    -> ORDER BY doc->'$.population' DESC
    -> LIMIT 5;
+-----------------+
| doc->>'$.name'  |
+-----------------+
| Mumbai (Bombay) |
| Karachi         |
| Delhi           |
| Teheran         |
| Lahore          |
+-----------------+
5 rows in set (0.02 sec)
```

Figure 16.29 – The biggest cities in Southern and Central Asia

Here, you used the `worldcol` collection and filtered on the `country.region` field that is in the JSON column. Then, you ordered by `population` and selected the `name` column.

6. Write the following query to find cities that have a name ending with `ester`:

```
SELECT * FROM city WHERE Name LIKE '%ester';
```

This query produces the following output:

```
mysql> SELECT * FROM city WHERE Name LIKE '%ester';
+------+------------+-------------+---------------+------------+
| ID   | Name       | CountryCode | District      | Population |
+------+------------+-------------+---------------+------------+
|  462 | Manchester | GBR         | England       |     430000 |
|  467 | Leicester  | GBR         | England       |     294000 |
|  513 | Gloucester | GBR         | England       |     107000 |
|  521 | Colchester | GBR         | England       |      96063 |
|  527 | Worcester  | GBR         | England       |      95000 |
| 1848 | Gloucester | CAN         | Ontario       |     107314 |
| 3871 | Rochester  | USA         | New York      |     219773 |
| 3914 | Worcester  | USA         | Massachusetts |     172648 |
| 4011 | Manchester | USA         | New Hampshire |     107006 |
+------+------------+-------------+---------------+------------+
9 rows in set (0.01 sec)
```

Figure 16.30 – Cities with a name that ends in ester

This returns nine rows, and you get to see the information for every city. You could have used `SELECT COUNT * FROM city WHERE Name LIKE '%ester'` to only get the number instead.

Solution to Activity 5.1

In this activity, you will execute a few queries to get data that can be used by a manager for marketing and to reduce costs. Follow these steps to complete this activity:

1. Open the MySQL client and connect to the `sakila` database:

```
USE sakila;
```

2. Find the total number of films the store has with a PG rating. To do this, you need both the film table and the inventory table. For every item in the inventory table, there is a reference to a record in the film table. Then, filter on the rating, which is stored in the film table. And finally, do SELECT COUNT(*) because all you need is the total number aggregated over all rows:

```
SELECT COUNT(*)
FROM film f
JOIN inventory i ON f.film_id=i.film_id
WHERE f.rating='PG';
```

The preceding query produces the following output:

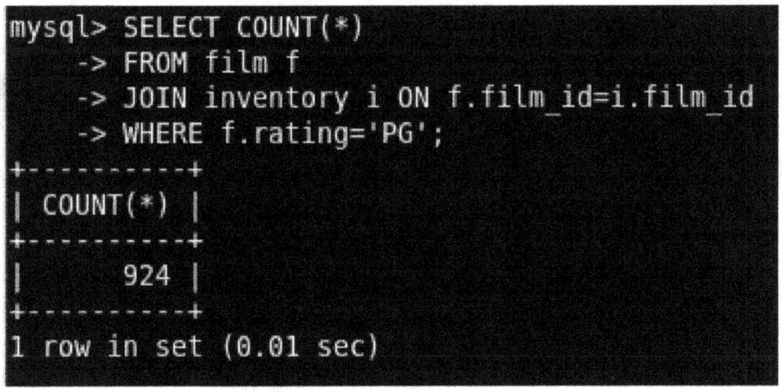

```
mysql> SELECT COUNT(*)
    -> FROM film f
    -> JOIN inventory i ON f.film_id=i.film_id
    -> WHERE f.rating='PG';
+----------+
| COUNT(*) |
+----------+
|      924 |
+----------+
1 row in set (0.01 sec)
```

Figure 16.31 – The SELECT output with the total number of PG-rated films in the inventory

3. Now, find films in which Emily Dee has performed as an actor. To do this, you need the film table and the actor table. However, there is no direct relation between them, so you need the film_actor table, which stores links between films and actors. This is because one film typically has multiple actors, and an actor can be in multiple films. Then, filter by name, which is split over the first_name and last_name columns and is stored in capitals. Finally, select the titles of the matching films:

```
SELECT f.title
FROM film f
JOIN film_actor fa ON f.film_id=fa.film_id
JOIN actor a ON a.actor_id=fa.actor_id
WHERE a.first_name='EMILY' AND a.last_name='DEE';
```

The preceding code produces the following output:

```
mysql> SELECT f.title
    -> FROM film f
    -> JOIN film_actor fa ON f.film_id=fa.film_id
    -> JOIN actor a ON a.actor_id=fa.actor_id
    -> WHERE a.first_name='EMILY' AND a.last_name='DEE';
+--------------------+
| title              |
+--------------------+
| ANONYMOUS HUMAN    |
| BASIC EASY         |
| CHAMBER ITALIAN    |
| CHRISTMAS MOONSHINE |
| DESTINY SATURDAY   |
| FUGITIVE MAGUIRE   |
| GONE TROUBLE       |
| HOLLOW JEOPARDY    |
| INVASION CYCLONE   |
| OCTOBER SUBMARINE  |
| REBEL AIRPORT      |
| SCARFACE BANG      |
| SEA VIRGIN         |
| SHREK LICENSE      |
+--------------------+
14 rows in set (0.00 sec)
```

Figure 16.32 – The SELECT output with films featuring Emily Dee

4. To find the customers who rented the most items, you need the `rental` table and the `customer` table. These are related by `customer_id`. You want to aggregate the results per customer, so use GROUP BY c.customer_id. To get the top one, use ORDER BY COUNT(*) DESC LIMIT 1:

```
SELECT c.first_name, c.last_name, COUNT(*) FROM rental r
JOIN customer c ON r.customer_id=c.customer_id
GROUP BY c.customer_id ORDER BY COUNT(*) DESC LIMIT 1;
```

The preceding code produces the following output:

```
mysql> SELECT c.first_name, c.last_name, COUNT(*) FROM rental r
    -> JOIN customer c ON r.customer_id=c.customer_id
    -> GROUP BY c.customer_id ORDER BY COUNT(*) DESC LIMIT 1;
+------------+-----------+----------+
| first_name | last_name | COUNT(*) |
+------------+-----------+----------+
| ELEANOR    | HUNT      |       46 |
+------------+-----------+----------+
1 row in set (0.02 sec)
```

Figure 16.33 – The SELECT output with the top renter

5. Now, find the film that resulted in the biggest income. The payments that make the income are stored in the `payment` table. The films are stored in the `film` table. These two tables are not directly related. So, you first have to join the `payment` table with the `rental` table and then join them to the `inventory` table. And from there, you can join it with the `film` table. Then, you have to aggregate by film by adding `GROUP BY f.film_id`. To get the top one by amount, you need to use `SUM` on the amount per group and then use `ORDER BY` with `LIMIT 1`. As output columns, select the `film` table and the sum of the amount from the `payment` table:

```
SELECT f.title, SUM(p.amount)
FROM payment p
JOIN rental r ON p.rental_id=r.rental_id
JOIN inventory i ON i.inventory_id=r.inventory_id
JOIN film f ON f.film_id=i.film_id
GROUP by f.film_id
ORDER BY SUM(p.amount) DESC LIMIT 1;
```

The preceding code produces the following output:

```
mysql> SELECT f.title, SUM(p.amount)
    -> FROM payment p
    -> JOIN rental r ON p.rental_id=r.rental_id
    -> JOIN inventory i ON i.inventory_id=r.inventory_id
    -> JOIN film f ON f.film_id=i.film_id
    -> GROUP by f.film_id
    -> ORDER BY SUM(p.amount) DESC LIMIT 1;
+-----------------+---------------+
| title           | SUM(p.amount) |
+-----------------+---------------+
| TELEGRAPH VOYAGE |        231.73 |
+-----------------+---------------+
1 row in set (0.06 sec)
```

Figure 16.34 – The SELECT output for the film generating the biggest income

6. Now, find the email address of the customer living in `Turkmenistan`. For every `customer`, store an address in the `address` table, which has a link to the `city` table, which in turn has a link to the `country` table. When you join them, you can filter out a specific country and then return the email address from the `customer` table. This would return multiple results if there were multiple customers in `Turkmenistan`, but there is only one. Write the following query to achieve this:

```
SELECT email
FROM customer cu
JOIN address a ON cu.address_id=a.address_id
```

```
JOIN city ci ON ci.city_id=a.city_id
JOIN country co ON co.country_id=ci.country_id
WHERE country='Turkmenistan';
```

The preceding code produces the following output:

```
mysql> SELECT email
    -> FROM customer cu
    -> JOIN address a ON cu.address_id=a.address_id
    -> JOIN city ci ON ci.city_id=a.city_id
    -> JOIN country co ON co.country_id=ci.country_id
    -> WHERE country='Turkmenistan'\G
*************************** 1. row ***************************
email: JEANNE.LAWSON@sakilacustomer.org
1 row in set (0.00 sec)
```

Figure 16.35 – The SELECT output with the email of the only customer in Turkmenistan

Solution to Activity 5.2

First, try to use a way of getting this data by querying the film table and using GROUP BY on the release year; however, this will only return information for years in which films have been released. In our database, all films are released in a single year. So, you want to generate a range of years and then join this with the data you have to make sure that all the years are included, even if there were no films released in that year according to our database. Follow these steps to complete this activity:

1. Open the MySQL client and connect to the sakila database:

```
USE sakila
```

This produces the following output:

```
mysql> USE sakila
Database changed
```

Figure 16.36 – The USE output

2. Inspect the result of the naive approach by writing the following query:

```
SELECT release_year, COUNT(*) FROM film
WHERE release_year BETWEEN 2005 AND 2010
GROUP BY release_year;
```

This produces the following output:

```
mysql> SELECT release_year, COUNT(*) FROM film
    -> WHERE release_year BETWEEN 2005 AND 2010
    -> GROUP BY release_year;
+--------------+----------+
| release_year | COUNT(*) |
+--------------+----------+
|         2006 |     1000 |
+--------------+----------+
1 row in set (0.00 sec)
```

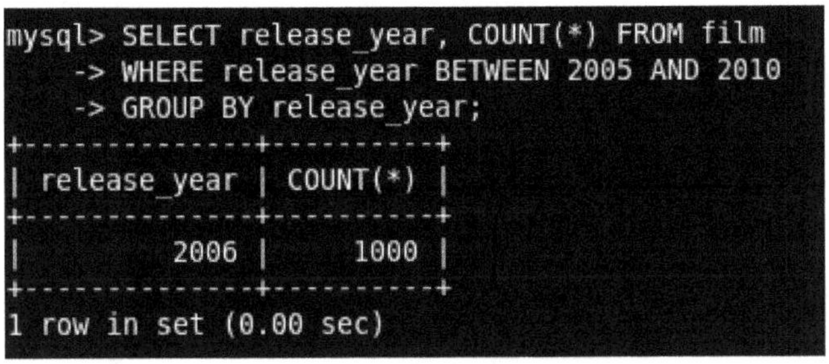

Figure 16.37 – The SELECT output for the naive approach

In the preceding screenshot, you can see matches only for the year 2006, but you want to have results for all years between 2005 and 2010.

3. Create a CTE to generate a range of years by writing the following query:

```
WITH RECURSIVE years AS (
   SELECT 2005 AS y
   UNION ALL
   SELECT y+1 FROM years WHERE y<2010
) SELECT * FROM years;
```

This produces the following output:

```
mysql> WITH RECURSIVE years AS (
    ->    SELECT 2005 AS y
    ->    UNION ALL
    ->    SELECT y+1 FROM years WHERE y<2010
    -> ) SELECT * FROM years;
+------+
| y    |
+------+
| 2005 |
| 2006 |
| 2007 |
| 2008 |
| 2009 |
| 2010 |
+------+
6 rows in set (0.00 sec)
```

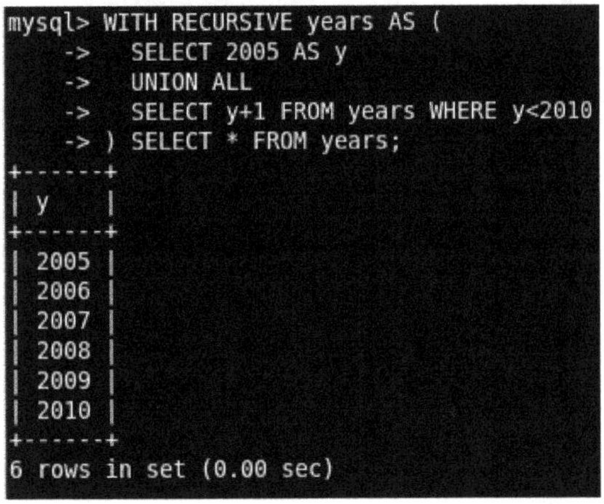

Figure 16.38 – The SELECT output for the CTE to generate the year range

Initialize the recursive CTE with 2005 as a value for y. Then, increase y by 1 until you reach the year 2010.

4. Now, join this against the list of years you have generated with the help of the following query:

```
WITH RECURSIVE years AS (
   SELECT 2005 AS y
   UNION ALL
   SELECT y+1 FROM years WHERE y<2010
)
SELECT y, COUNT(film.film_id)
FROM years
LEFT JOIN film ON years.y=film.release_year
GROUP BY years.y\G
```

This produces the following output:

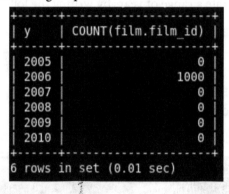

Figure 16.39 – The SELECT output with film release dates between 2005 and 2010

You now have the list you wanted. The list has all the years between 2005 and 2010 and shows how many films were released that year. There are a few things that could have gone wrong with this query. If you use JOIN instead of LEFT JOIN, then it will only match records that have entries in both tables. This would only return 2006, which is not what we wanted.

If you use COUNT(*) instead of COUNT(film. film_id), then it would show 1 for the years that should have a 0. This is because the group for that year has one record (as there is a record in the years table). It has a NULL film_id value because there is no matching film. Think of it like this:

Year	film_id
2005	NULL
2006	1
2006	2

Figure 16.40 – The years table

In the table, each color is a group, as it is the same year. The first group has one record, where `film_id` is NULL as there is no matching film. But there is a record because the year exists in the `years` table. The second group has two records, each with a year and `film_id`. Now, if you use COUNT() on `film_id`, it won't count the records where `film_id` is NULL.

Solution to Activity 6.1

The solution to this activity is as follows:

1. Right-click on the **members** table in the **Tables** list to view all records:

ID	Surname	FirstName	MiddleNames	DOB	Signature	Photo	PhotoPath
1	Bloggs	Frederick	NULL	1990-06-16	BLOB	BLOB	Members\Photos\
2	Pettit	Thomas	William	1960-10-15	NULL	NULL	Members\Photos\
5	West	Anais	Avery	1984-08-06	NULL	NULL	NULL
6	Swaniawski	Waylon	Rita	2007-10-08	NULL	NULL	NULL
7	Collins	Darby	Marielle	2019-01-11	NULL	NULL	NULL
8	Schamberger	Dexter	D'angelo	1999-10-07	NULL	NULL	NULL
9	Wintheiser	Tania	Toy	2020-01-24	NULL	NULL	NULL

Figure 16.41 – Records from the members table

2. Locate the record with the **ID** number of **7** and confirm that the record belongs to Darby Marielle Collins.

3. Examine the **DOB** column for the record and confirm that the date is NOT January 11, 1990.

4. Create a new SQL tab to create your query; your SQL will be similar to the following SQL code:

```
UPDATE vw_members_all
SET DOB = "1990-01-11"
WHERE ID=7;
```

5. After running the SQL, reload the **members** table to view the records:

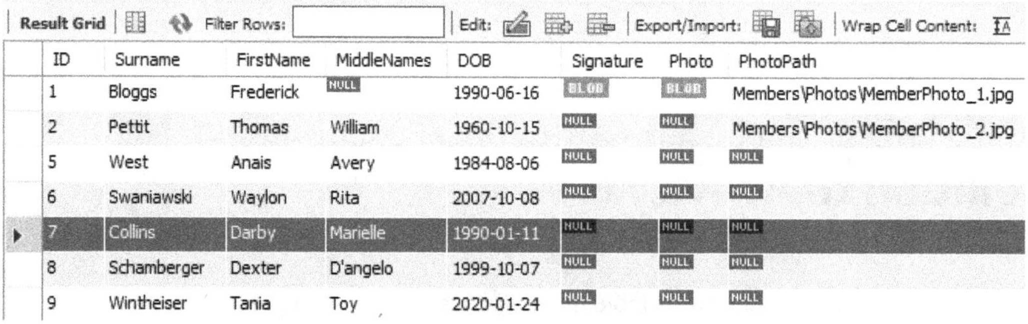

ID	Surname	FirstName	MiddleNames	DOB	Signature	Photo	PhotoPath
1	Bloggs	Frederick	NULL	1990-06-16	BLOB	BLOB	Members\Photos\MemberPhoto_1.jpg
2	Pettit	Thomas	William	1960-10-15	NULL	NULL	Members\Photos\MemberPhoto_2.jpg
5	West	Anais	Avery	1984-08-06	NULL	NULL	NULL
6	Swaniawski	Waylon	Rita	2007-10-08	NULL	NULL	NULL
7	Collins	Darby	Marielle	1990-01-11	NULL	NULL	NULL
8	Schamberger	Dexter	D'angelo	1999-10-07	NULL	NULL	NULL
9	Wintheiser	Tania	Toy	2020-01-24	NULL	NULL	NULL

Figure 16.42 – Members data reloaded after the record update

6. Examine the **DOB** column for the **ID 7** record and confirm that the date is now January 11, 1990.

You have now confirmed that you can update particular views; here, we just updated a field, but you can insert new records or delete existing records. You also confirmed that when data or records in an updatable view are modified, the modifications change the underlying tables' data or records.

Solution to Activity 7.1

The solution to this activity is as follows:

1. Using your text editor, create a Node.js script and save it as motdatabase.js. Enter the following code into the script file to create MOTdatabase:

```
var mysqlconnection = require("./mysqlconnection.js");
mysqlconnection.query("CREATE DATABASE 'MOTdatabase'",
    function (err)
        if (err) throw "Problem creating the database:- " +
            err.code;
```

```
    console.log("Database created");
    process.exit();
});
```

> **Note**
>
> The complete script can be found at https://github.com/
> PacktWorkshops/The-MySQL-Workshop/blob/master/
> Chapter07/Activity7.01/Activity_5_01_Solution_
> Create_Database.js.

This code will start by connecting to the database using the mysqlconnection.js file. Once the connection is established, the program runs a query to create the MOTdatabase database.

2. Run the motdatabase.js script in your Command Prompt. You will see the following output:

```
PS D:\MySQL Training\Nodejs> node index.js
Connected to MySQL!
Database created
```

Figure 16.43 – The result of running the code to create the database

3. Refresh the database schema in MySQL Workbench; **motdatabase** will appear in the list:

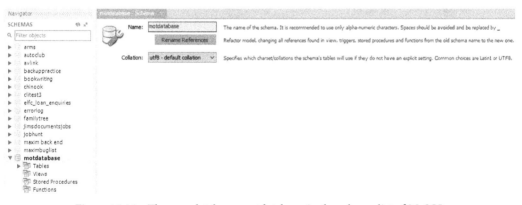

Figure 16.44 – The new database, motdatabase, in the schema list of MySQL

4. Create another Node.js file named `mottables.js` to create the two tables, `Customers` and `CustomerPurchases`, as per the requirements provided by the marketing head:

```
var mysqlconnection = require("./mysqlconnection.js");
var sql = "CREATE TABLE 'MOTdatabase'.'Customers' (
  'CustID' int(11) NOT NULL AUTO_INCREMENT,
  'CustomerName' varchar(50) NOT NULL,
  PRIMARY KEY ('CustID')
);"

mysqlconnection.query(sql, function (err) {
  if (err) throw "Problem creating the Table:- " + err.
code;
  console.log("Table created");
});

var sql = "CREATE TABLE 'MOTdatabase'.'CustomerPurchases'
( \
  'CPID' int(11) NOT NULL AUTO_INCREMENT, \
  'CustID' int(11) NOT NULL, \
  'SKU' varchar(20) NOT NULL, \
  'SaleDateTime' varchar(25) NOT NULL, \
  'Quantity' int(11) NOT NULL, \
  PRIMARY KEY ('CPID') \
);"

mysqlconnection.query(sql, function (err) {
  if (err) throw "Problem creating the Table:- " + err.
code;
  console.log("Table created");
  process.exit();
});
```

> **Note**
>
> The complete script can be found at `https://github.com/PacktWorkshops/The-MySQL-Workshop/blob/master/Chapter07/Activity7.01/Activity_5_01_Solution_Create_Tables.js`.

This code will connect to the database through the `mysqlconnection.js` file. Once the connection is created, two queries are created to run against the connection. The first query creates the `Customers` table, and the second creates the `CustomerPurchases` table. Once this is completed, both tables will be created, as specified by the marketing head.

5. Run the `mottables.js` script through the console. You should get the following output:

```
PS D:\MySQL Training\Nodejs> node index.js
Connected to MySQL!
Table created
Table created
```

Figure 16.45 – The result of running the table creation code

6. In MySQL Workbench, refresh the schema to see the two new tables, **customers** and **customerpurchases**:

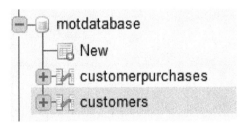

Figure 16.46 – The new tables added to motdatabase

7. After running the scripts and refreshing the schema in Workbench, locate the database and tables and select **AlterTable** for each table. You should see the following table definitions:

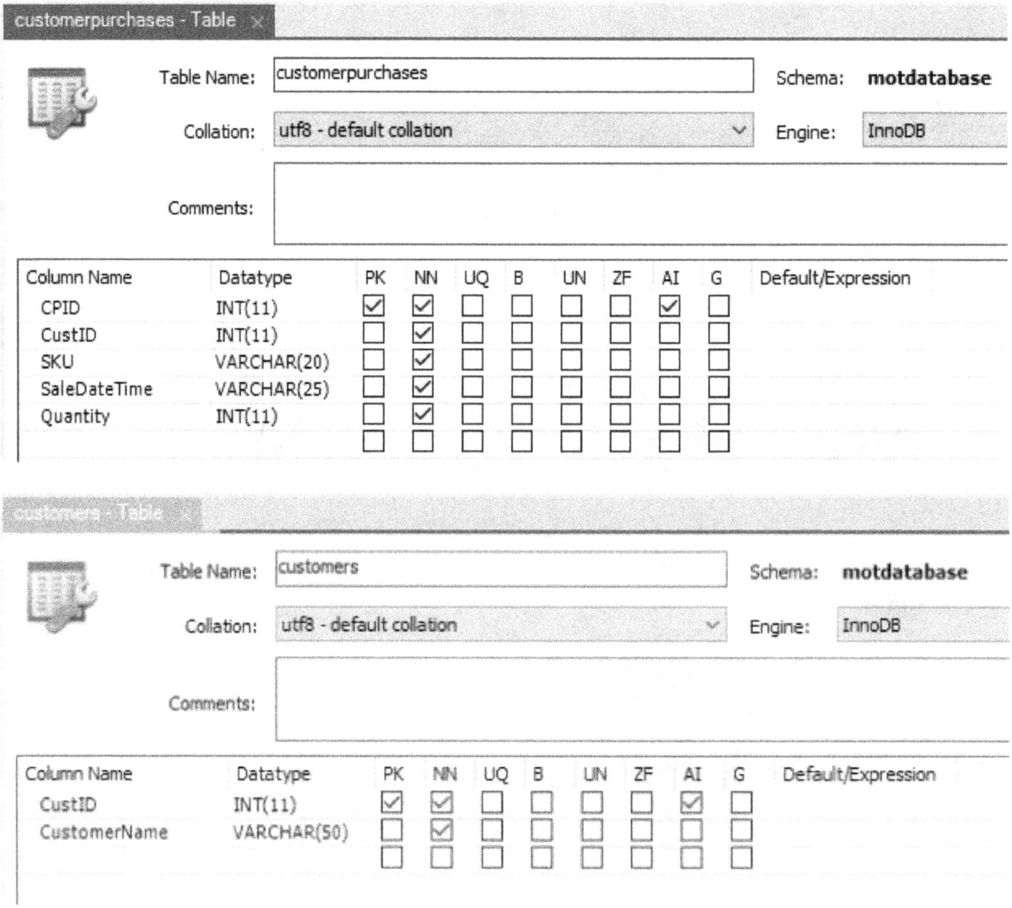

Figure 16.47 – The new tables in Workbench

The preceding output shows that the required tables have been created with the proper names, fields, and primary keys defined. With this, you can now verify that your script worked.

Solution to Activity 8.1

In this activity, you are tasked with adding new details to the existing `country` table –
that is, capital cities, countries' independence statuses, and currency types. Perform the
following steps to accomplish this:

1. Create a new script file named `Activity-MultipleUpdates.js`.

2. Add the connection module and instruct the server to use the `world_
    statistics` database. Include error handling:

    ```
    var mysqlconnection = require("./mysqlconnection.js")
    mysqlconnection.query("USE world_statistics", function
    (err, result) {
    ```

3. Deal with the error, should one occur. Each distinct task performed by the code is
    embedded in the part of the `if` statement that is executed if there is no error. This
    means that, for each task to be performed, the preceding task must be error-free. If
    an error occurs in any given task, the script will report the error and then exit, and
    none of the following tasks will be attempted:

    ```
    if (err) throw "Instructing database to use" + err.code;
    //Tell user on console
    console.log("Using World_Statistics");
    ```

4. Add the `Capital` column to the `country` table:

    ```
    var newfield = "CREATE TABLE countryalldetails(CountryID
    INT(11), ContinentID INT(11), 'Country Code' VARCHAR(5),
    'Country Name' VARCHAR(50))";
    mysqlconnection.query(newfield, function (err, result) {
    //Deal with the error should one occur
    if (err) throw "Problem creating column Capital" + err.
    code;
    //Tell user that the capital column has been created
    console.log("Column Capital created");
    ```

5. Add the `Is_Independent` column to the `country` table:

    ```
    var newfield = "ALTER TABLE countryalldetails ADD COLUMN
    Is_Independent VARCHAR(25);"
    mysqlconnection.query(newfield, function (err, result) {
    //Deal with the error should one occur
    ```

```
if (err) throw "Problem creating column Is_Independent" +
err.code;
//Tell user that the Is_Independent column has been
created
console.log("Column Is_Independent created");
```

6. Add the Currency column to the country table:

```
var newfield = "ALTER TABLE countryalldetails ADD COLUMN
Currency VARCHAR(5);"
mysqlconnection.query(newfield, function (err, result) {
//Deal with the error should one occur
if (err) throw "Problem creating column Currency" + err.
code;
//Tell user that the currency column has been created
console.log("Column Currency created");
```

7. All the columns have been created, which means that you can now insert data into the table. Each field will be updated individually using three distinct queries. Start by putting the SQL query into the updateOne variable in order to update the Capital field. Use a nested query to get the data for each record. Use LIMIT 1 to ensure that each record is updated with exactly one matching value from the temp table. The query will place the capital data from the temp table into the country table:

```
var updateOne="UPDATE countryalldetails SET Capital = "
updateOne = updateOne + "(SELECT 'Capital' FROM world_
statistics.temp WHERE 'Country Code'= 'country'.'Country
Code' LIMIT 1);"
```

8. Execute the SQL statement to update the Capital field in all records. The following code will execute the query and print the number of rows affected by the query:

```
mysqlconnection.query(updateOne, function (err, result) {
if (err) throw "Problem updating Capital" + err.code;
//Tell user that the capital is updated, and show
affectedRows
console.log("Capital is updated");
console.log("Number of rows affected : " + result.
affectedRows);
```

9. Repeat this process for the next field, `Is_Independent`. The query updates the independence status from the `temp` table, using the country code to determine which independence status to put in the `country` table:

```
var updateTwo="UPDATE countryalldetails SET Is_
Independent = "
updateTwo = updateTwo + "(SELECT 'Is_Independent'
FROM world_statistics.temp WHERE 'Country Code'=
'country'.'Country Code' LIMIT 1);"
mysqlconnection.query(updateTwo, function (err, result) {
if (err) throw "Problem updating Is_Independent" + err.
code;
//Tell user that the Is_Independent column has been
updated
console.log("Is_Independent is updated");
console.log("Number of rows affected : " + result.
affectedRows);
```

10. Repeat this process for the next field, `Currency`:

```
var updateThree="UPDATE country SET Currency = "
updateThree = updateThree + "(SELECT 'Currency'
FROM world_statistics.temp WHERE 'Country Code'=
'country'.'Country Code' LIMIT 1);"
mysqlconnection.query(updateThree, function (err, result)
{
if (err) throw "Problem updating Currency" + err.code;
//Tell user that the currency column has been updated
console.log("Currency is updated");
console.log("Number of rows affected : " + result.
affectedRows);
```

11. Now, exit the script using the following command:

```
process.exit();
```

12. Close off the brackets in reverse order:

```
        });//updateThree
      });//updateTwo
    });//updateOne
  });//Column Currency
```

```
        });//Column Is_Independent
      });//Column Capital
    });//USE world_statistics
```

13. Execute the file, `Activity-MultipleUpdates.js`, in the terminal. You should see the following output:

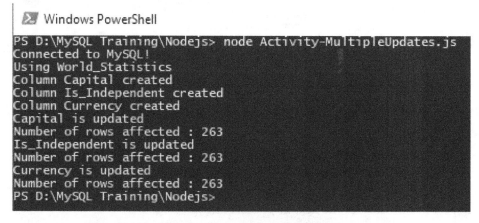

Figure 16.48 – The console messages, indicating script progress

The preceding screenshot shows that the required columns were created, and all 263 records were updated with the appropriate values.

14. In Workbench, expand the **countryalldetails** table to see the new fields:

Figure 16.49 – A schema displaying the country table with its new fields

15. Now, right-click on the **country** table and select the **Select rows** option to see the data in the table:

CountryID	ContinentID	Country Code	Country Name	Capital	Is_Independent	Currency
1	4	ABW	Aruba	Oranjestad	Part of NL	AWG
2	2	AFG	Afghanistan	Kabul	Yes	AFN
3	1	AGO	Angola	Luanda	Yes	AOA
4	3	ALB	Albania	Tirana	Yes	ALL
5	3	AND	Andorra	Andorra la Vella	Yes	EUR
6	NULL	ARB	Arab World	NULL	NULL	NULL
7	2	ARE	United Arab Emirates	Abu Dhabi	Yes	AED
8	6	ARG	Argentina	Buenos Aires	Yes	ARS

Figure 16.50 – The Select Rows view showing the new fields populated with data

This input confirms that the **Capital**, **Is_Independent**, and **Currency** fields have been populated in the **country** table.

16. Right-click on the **country** table and click **Alter Table** to verify that the structure of the table looks as pictured here:

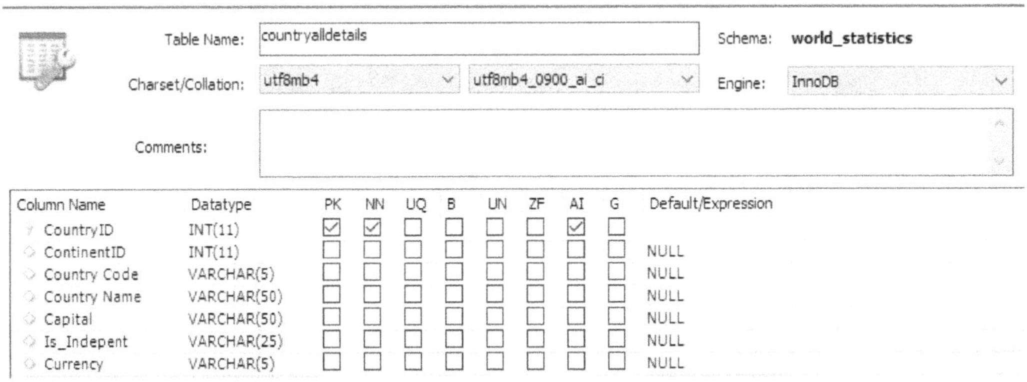

Figure 16.51 – The new fields in the Alter Table view

This output verifies that the **Capital**, **Is_Independent**, and **Currency** fields have been added with the expected data types.

Solution to Activity 8.2

The solution to this activity is as follows:

1. Using your text editor, create a script called `Activity_6_02_Solution_Populate_Tables.js`.

2. Start by connecting to the database and running a query to use `CustomerDatabase`:

```
var mysqlconnection = require("./mysqlconnection.js");
mysqlconnection.query("USE CustomerDatabase", function
(err, result) {
    if (err) throw err.code;
    console.log(result);
});
```

3. Next, create the customer records and insert them through a parameterized query:

```
var record = [['Big Company'],['Little Company'],['Old
Company'],['New Company']];

var sql = "INSERT INTO customers(CustomerName) VALUES ?;"

mysqlconnection.query(sql, [record], function (err,
result) {
    if (err) throw "Problem creating database" + err.
code;
    console.log(result);
});
```

4. Finally, create the `CustomerPurchases` records and insert them through a parameterized query:

```
record = [
            [1,'SKU001','01-JAN-2020 09:10am',3],
            [2,'SKU001','01-JAN-2020 09:10am',2],
            [3,'SKU002','02-FEB-2020 09:15am',5],
            [4,'SKU003','05-MAY-2020 12:21pm',10],
        ];
var sql = "INSERT INTO
CustomerPurchases(CustID,SKU,SalesDateTime,Quantity)
```

```
VALUES ?;"

mysqlconnection.query(sql, [record], function (err,
result) {
    if (err) throw "Problem creating database" + err.
code;
    console.log(result);
    process.exit();
});
```

After running the script, your onscreen output should be similar to the following:

```
PS D:\MySQL Training\Nodejs> node Activity_6_02_Solution_Populate_Customer.js
Connected to MySQL!
ResultSetHeader {
  fieldCount: 0,
  affectedRows: 4,
  insertId: 1,
  info: 'Records: 4  Duplicates: 0  Warnings: 0',
  serverStatus: 2,
  warningStatus: 0
}
Number of rows affected : 4
New records ID : 1
PS D:\MySQL Training\Nodejs>
```

Figure 16.52 – A console report from running the Populate Customer JS file

5. Using MySQL Workbench, right-click on the **customers** table of **MOTDatabase** and click **Select Rows**. The following data will be displayed:

Figure 16.53 – The customers table data after population

6. Next, in MySQL Workbench, right-click on the **customerpurchases** table of **MOTDatabase** and click **Select Rows**. A screen like the following should open:

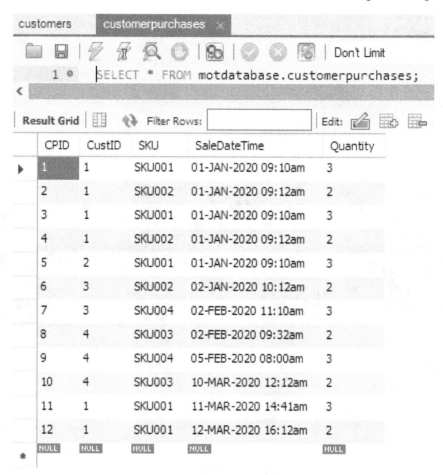

Figure 16.54 – The customerpurchases table after population

7. Finally, you need to create the ODBC connection for this database. Press the Windows *Start* button on your keyboard and type ODBC.

8. Select **ODBC Data Sources (32-bit)** and click **Yes** when prompted to allow this application to make changes. Now, the **ODBC Data Source Administrator (32-bit)** window will open.

9. Select the **System DSN** tab and click **Add...**:

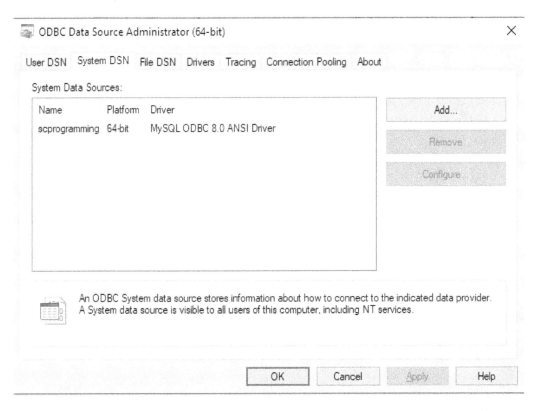

Figure 16.55 – The System DSN tab displaying the available ODBC connections

10. The driver selection window will open. Select the MySQL driver you wish to use and click **Finish**:

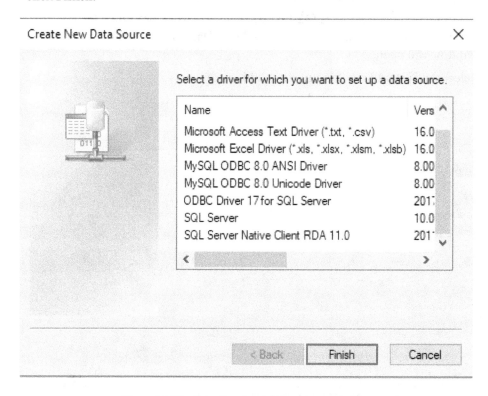

Figure 16.56 – Selecting the driver you want to use

11. For the configuration windows, enter your connection details. Your options are as follows:

- **Data Source Name**: Give it any name. In this case, `CustomerDatabase` would be fitting.

- **Description**: This is optional. It would be appropriate to add a description such as `Stores data about the customers and purchases`.

- **TCP/IP Server**: The address of the server. Since the server is on your local computer, use `localhost` or `127.0.0.1`.

- **Port**: This is already set at `3306` and can be kept as is.

- **User**: Enter the user's account name.

- **Password**: The password of the account.

- **Database**: If the IP address and port are valid, a list of databases on the server is listed. Select the database you want the connection to use and select the `CustomerDatabase` database, as shown in the following screenshot:

Figure 16.57 – The completed details; make sure to use your own

12. Test the connection by clicking **Test**. The manager attempts to connect and, if successful, displays the following result:

Figure 16.58 – A successful connection has been made

13. If you get a **Connection Successful** result, click **OK** to close the test message and click **OK** again to close the new ODBC window. If the connection failed, check your values for **TCP/IP Server**, **Port**, **User**, and **Password** and try again.

Solution to Activity 9.1

The solution to this activity is as follows:

1. Examine your MS Access table list.

2. In the navigation panel, select **Tables**. The table list will now only contain MySQL linked tables, as shown by the globe icon.

3. Open each table in turn by right-clicking and selecting **Design**; each table will be the primary key set, as indicated by the key icon next to the **Primary Key** field. You won't receive the window to select identifying fields because there was no primary key when you linked the table from MySQL.

4. Double-click on each table in turn. Each will correctly display its data. All tables will be writeable. This is indicated by the blank line and the asterisk on the last record when the table is opened to view the data:

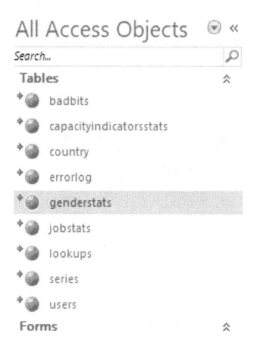

Figure 16.59 – The linked tables after all tables are linked

You did it – well done! There are a lot of things to take into account when migrating an MS Access database to MySQL, and it is not for the faint-hearted, so feel proud of what you have achieved here today.

Solution to Activity 10.1

The solution to this activity is as follows:

1. Make the necessary code changes to each SQL block.

2. For the SQL 2 block, the existing SQL statement works in Workbench, so no changes are required. Keep the code as it is:

    ```
    SQL = "SELECT Count(GenderStats.ID) AS RecCount FROM
    GenderStats;"
    ```

3. On the second line, the Set RS = statement is not required at this location, so comment it out:

    ```
    'Set RS = CurrentDb.OpenRecordset(SQL, dbOpenDynaset)
    ```

4. Call the CreatePassThrough function by passing in the SQL statement and the name for the new passthrough query, GENCount. Pass True, which indicates that the passthrough will return a value, and False because we do not want to delete the old passthrough query first; it will overwrite it:

    ```
    Call CreatePassThrough(SQL, "GENCount", True, False)
    ```

5. Place Set RS = after the query that will create it and change the recordset source to GENCount:

    ```
    Set RS = CurrentDb.OpenRecordset("GENCount",
    dbOpenDynaset)
    ```

6. Do not change the following VBA lines for the SQL 2 code block:

    ```
    RS.MoveFirst
    Me.cntGS = RS.Fields("RecCount")
    RS.Close
    ```

7. For SQL 3, don't make any changes, as the existing SQL statement works in Workbench:

    ```
    SQL = "SELECT Count(JobStats.ID) AS RecCount FROM
    JobStats;"
    ```

8. The `Set RS =` statement is not required at this location, so comment it out:

```
'Set RS = CurrentDb.OpenRecordset(SQL, dbOpenDynaset)
```

9. The call to the `CreatePassThrough` function is inserted, passing in the SQL statement and the name for the new passthrough query, `JOBCount`. `True` is passed in to indicate that the passthrough will return a value, and `False` is passed in because we do not need to delete the old passthrough query first; it will overwrite it:

```
Call CreatePassThrough(SQL, "JOBCount", True, False)
```

10. On the fourth line, place `Set RS =` after the query and change the recordset source to `JOBCount`:

```
Set RS = CurrentDb.OpenRecordset("JOBCount",
dbOpenDynaset)
```

11. Do not change the following VBA lines for the SQL 3 code block:

```
RS.MoveFirst
Me.cntGS = RS.Fields("RecCount")
RS.Close
```

12. For SQL 4, the existing `SQL` statement has spaces in the field name with surrounding square brackets and will not work in MySQL, so comment it out. You can simply change the brackets to backticks here:

```
'SQL = "SELECT Count(Country.[Country Code]) AS
RecCount FROM Country;"
```

13. The `Set RS =` statement is not required at this location, so comment it out:

```
'Set RS = CurrentDb.OpenRecordset(SQL, dbOpenDynaset)
```

14. The SQL statement is inserted with the square brackets changed to backticks:

```
SQL = "SELECT Count(Country.'Country Code') AS RecCount
FROM Country;"
```

15. The call to the `CreatePassThrough` function is inserted, passing in the SQL statement and the name for the new passthrough query, `CTRYCount`. `True` is passed in to indicate that the passthrough will return a value, and `False` is passed in because we do not need to delete the old passthrough query first; it will overwrite it:

```
Call CreatePassThrough(SQL, "CTRYCount", True, False)
```

16. After the query is created, place `Set RS = ` on the fifth line. Also, change the source of the recordset to `CTRYCount`:

```
Set RS = CurrentDb.OpenRecordset("CTRYCount",
dbOpenDynaset)
```

17. Do not change the remaining VBA lines for the SQL 3 code block:

```
RS.MoveFirst
        Me.cntCountry = RS.Fields("RecCount")
    RS.Close
```

Your results on screen and your code for the SQL 2, 3, and 4 blocks should be as follows:

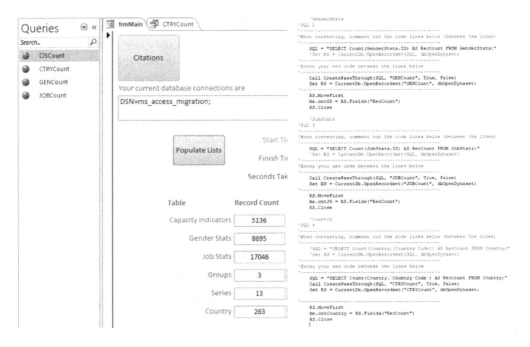

Figure 16.60 – Changes to the code and the affected onscreen controls

You should now have three new passthrough queries, and the count values should be as shown in the preceding screenshot.

We did not modify or move two of the original SQL statements. We tested them in Workbench, and they worked, so there was no need to modify them. **Country**, however, had a space in the field name, and the brackets that Access uses had to be changed to backticks, our first SQL modification. Always try to make as few changes as possible to achieve the conversion.

Solution to Activity 10.2

In this activity, you will create a function to count and assign the total groups to the `cntGroups` textbox. Follow these steps to complete this activity:

1. Copy the `fnCountSeries.sql` file and name the new file `fnCountGroups.sql`.

2. Open the file in a text editor.

3. Modify the file.

4. Instruct the server to use the `ms_access_migration` database. Always include this command so that there is no question that code will be run against the intended database. Without it, the current active database in the workbench will be used:

   ```
   USE ms_access_migration;
   ```

5. Drop the existing function named `fnCountGroups` if it already exists:

   ```
   DROP FUNCTION IF EXISTS  'fnCountGroups';
   ```

6. Set a custom delimiter so that all code between the start and end delimiters is to be treated as one set of instructions:

   ```
   DELIMITER //
   ```

7. Create the function as named and indicate that it will return a long value:

   ```
   CREATE FUNCTION 'fnCountGroups'() RETURNS long
   ```

8. Include a command to indicate that the function only reads SQL data:

   ```
   READS SQL DATA
   ```

9. Begin the function statements:

   ```
   BEGIN
   ```

10. Declare a variable to receive the results of the query:

    ```
    DECLARE TheValue Long;
    ```

11. Execute the SQL statement to count the groups and put the results in the variable:

```
SET TheValue = (SELECT Count('Group') AS RecCount FROM
(SELECT DISTINCT series.Group FROM series )   AS 'Alias');
```

12. Return the results in the variable as the output from the function:

```
RETURN(TheValue);
```

13. Signify the end of the function's statements and also the end of the custom delimiter block:

```
END//
```

14. Line 18 resets the delimiter back to the default:

```
DELIMITER ;
```

15. Save the file, load it into a Workbench query window, and run it. You should now have two functions in the schema panel for the ms_access_migration database:

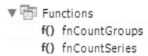

```
▼ 🔠 Functions
    f()  fnCountGroups
    f()  fnCountSeries
```

Figure 16.61 – The new function in the schema panel

16. Let's walk through the VBA code required to generate the passthrough query. Create the SQL statement to call the function for the passthrough to use. We will use SELECT and assign its returned value to a derived field name:

```
SQL = "SELECT fnCountGroups() as GroupCount"
```

17. Call the routine to create the passthrough query, pass in the name for the resulting query, and pass in the SQL statement. Set it to return values:

```
Call CreatePassThrough(SQL, "CntGroups", True, False)
```

18. Assign the passthrough query to a recordset:

```
Set RS = CurrentDb.OpenRecordset("CntGroups",
dbOpenDynaset)
```

19. Position the recordset cursor to the first (and only) record:

```
RS.MoveFirst
```

20. Assign the value of the recordset's derived field to the control receiving it:

```
Me.cntGroups = RS.Fields("GroupCount")
```

21. Close the recordset:

```
RS.Close
```

After you run by clicking **Populate Lists**, your output screen in MS Access should be as follows:

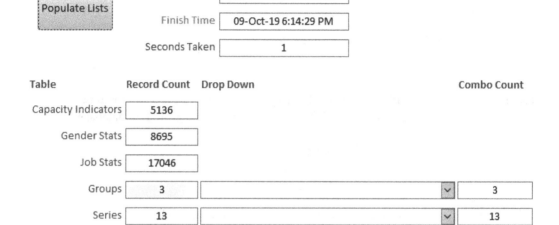

Figure 16.62 – The final output for Groups

Solution to Activity 10.3

The solution to this activity is as follows:

1. First, we need to develop SQL code to create the stored procedures. This will be done in a new SQL tab in Workbench.

2. Instruct MySQL to use the ms_access_migration database for all subsequent commands:

```
USE ms_access_migration;
```

3. Drop the existing `spGroupsList` stored procedure if it exists:

    ```
    DROP PROCEDURE IF EXISTS  spGroupsList;
    ```

4. Set the customized delimiter:

    ```
    DELIMITER //
    ```

5. Create the `spGroupsList` procedure:

    ```
    CREATE PROCEDURE spGroupsList()
    ```

6. Include a BEGIN statement to indicate where the code starts:

    ```
    BEGIN
    ```

7. The SQL statement will query the database. The records will be returned from the procedure:

    ```
    SELECT DISTINCT series.Group FROM series ORDER BY series.
    Group;
    ```

8. End the block and delimiter:

    ```
    END//
    ```

9. Reset the delimiter back to the default:

    ```
    DELIMITER ;
    ```

10. Inside `spCountryList`, instruct MySQL to use the `ms_access_migration` database for all subsequent commands:

    ```
    USE ms_access_migration;
    ```

11. Drop the existing `spCountryList` stored procedure if it exists:

    ```
    DROP PROCEDURE IF EXISTS  spCountryList;
    ```

12. Set the customized delimiter:

    ```
    DELIMITER //
    ```

13. Create the `spCountryList` procedure:

    ```
    CREATE PROCEDURE spCountryList()
    ```

14. Include a BEGIN statement to indicate where the code starts:

```
BEGIN
```

15. The SQL statement will query the database. The records will be returned from the procedure:

```
SELECT DISTINCT Country.'Country Code', Country.'Country
Name' FROM Country ORDER BY Country.'Country Name';
```

16. End the block and delimiter:

```
END//
```

17. Reset the delimiter back to the default:

```
DELIMITER ;
```

Now, we need to write the VBA code.

18. For the GroupsList combo box, create a SQL statement for the passthrough query. Note that when calling a stored procedure, we use the CALL statement, and as no parameter is required for this one, we simply use the stored procedure's name:

```
SQL = "Call spGroupsList;"
```

19. Create the passthrough query using the CreatePassThrough function, passing in SQL and the name of the new passthrough query, and indicate that it is to return results:

```
Call CreatePassThrough(SQL, "spGroupsList", True, False)
```

20. Assign the passthrough query directly to the combo box:

```
Me.cmbGroups.RowSource = "spGroupsList"
```

21. For the CountryList combo box, create a SQL statement for the passthrough query. Note that when calling a stored procedure, we use the CALL statement, and as no parameter is required, we simply use the stored procedure's name:

```
SQL = "Call spCountryList;"
```

Create the passthrough query using the CreatePassThrough function, passing in SQL and the name of the new passthrough query, and indicate that it is to return results:

```
Call CreatePassThrough(SQL, "spCountryList", True, False)
```

22. Assign the passthrough query directly to the combo box:

```
Me.cmbCountry.RowSource = "spCountryList"
```

Your lists should be populated as they were before the change:

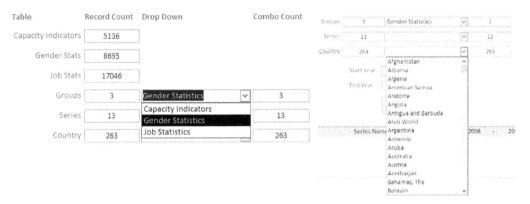

Table	Record Count	Drop Down	Combo Count
Capacity Indicators	5136		
Gender Stats	8695		
Job Stats	17046		
Groups	3	Gender Statistics	3
Series	13	Capacity Indicators / Gender Statistics / Job Statistics	13
Country	263		263

Figure 16.63 – Dropdowns displaying the lists

Solution to Activity 10.4

In this activity, we will modify the code tagged as SQL 8 to call spSeriesList_par() from a passthrough query and assign it to the cmbSeries row source. Follow these steps to implement this:

1. Locate the code tagged as SQL 8.
2. Copy and paste the original SQL line to a new line.
3. Comment out the original SQL statements.
4. Modify the new SQL line to call the spSeriesList_par stored procedure and pass in the value from cmbGroups as the parameter.
5. Save the changes.
6. Create the SQL statement. The parameter is passed in in brackets and is enclosed in single quotes. The method of constructing the parameter is identical to the VBA-based SQL:

```
SQL = "Call spSeriesList_par('" & Me.cmbGroups & "');"
```

7. Create the passthrough query with the `CreatePassThrough` function. Name the passthrough query `spSeriesFiltered`:

```
Call CreatePassThrough(SQL, "spSeriesFiltered", True,
False)
```

8. Assign the resulting passthrough query to the combo box. Because the passthrough returns a recordset, it can be assigned directly to the control:

```
Me.cmbSeries.RowSource = "spSeriesFiltered"
```

Your code for SQL 8 should now look like this:

Figure 16.64 – The old code commented out and the new code inserted for SQL 8

9. Select a group and open the series combo box. The results should change, as shown in the following screenshot. Try different group selections and ensure that the series list changes for each selection you make:

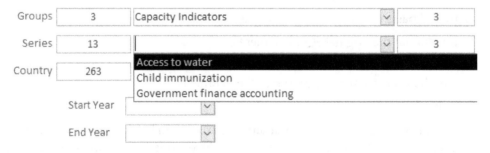

Figure 16.65 – Changing the group will change the series list values

> **Note**
>
> The **Combo Count** boxes to the right of the combo change to show the number of options in the lists. As you select different groups, you should see the **Series** counter change.

Solution to Activity 10.5

In this activity, you will create a stored procedure to determine dates, generate a passthrough query, and assign it to both date dropdowns. Follow the following steps to implement it:

1. Create a new SQL file and name it `Create Procedure spDateRange_par.sql` to generate a stored procedure.

2. Open the file named `spCountryList_par.sql` and copy and paste all of its code into the new file. You will modify this code.

3. Continuing in the new file, `spDateRange_par.sql`, there are two locations where the stored procedure's name is referenced. Change these to the new name of `spDateRange_par.sql`. They are in the **Drop Procedure** and **Create Procedure** lines.

4. Include the parameters in the **Create Procedure** line. The parameters are `Tablename` and `TheSeries`. Be sure to include `IN` and the data type declaration for both parameters.

5. Modify the SQL statement to return the `Year` field and set the series filter in the `WHERE` clause.

6. We then need to develop SQL code to create the stored procedure.

7. Include the instruction to use the `ms_access_migration` database:

```
USE ms_access_migration;
```

8. Drop the `spDateRange_par` stored procedure if it exists:

```
DROP PROCEDURE IF EXISTS  spDateRange_par;
```

9. Set the customer delimiter:

```
DELIMITER //
```

10. Create the procedure and define the `IN` parameters:

```
CREATE PROCEDURE spDateRange_par(IN TableName
VARCHAR(25),  IN
TheSeries VARCHAR(25) )
```

11. Indicate the beginning of the code:

```
BEGIN
```

12. We are building the SQL statement and including the parameter values that were passed in. Use the CONCAT method to build the string and assign it to the @t1 variable. This gives us the flexibility to use different table names:

```
SET @t1 = CONCAT(
'SELECT DISTINCT ' , TableName , '.Year ',
'FROM ', TableName , ' ',
'WHERE ' , TableName , '.'Series Code' = "' , TheSeries ,
'" ',
'ORDER BY ', TableName ,'.Year'
);
```

13. Now, we instruct the server to prepare a statement named stmt1 from the @t1 variable string. This needs to be done so that the statement can be executed:

```
PREPARE stmt1 FROM @t1;
```

14. Now, execute the stmt1 statement:

```
EXECUTE stmt1;
```

15. After the statement has been executed, we clear it from the server:

```
DEALLOCATE PREPARE stmt1;
```

16. Indicate the end of the code and customer delimiter range:

```
END//
```

17. Reset the delimiter back to the default:

```
DELIMITER ;
```

18. Save the file and then execute it. spDateRange_par.sql will appear in the schema panel after you refresh it.

The VBA code needs to be modified to use the new procedure.

19. Locate the code tagged as SQL 10.

20. Comment out all five lines building the original SQL statement.

21. Add new code to build the new SQL statement to call the stored procedure and pass in both parameters.

22. Add a new line to call the `CreatePassthroughQuery` function, passing in SQL and the `TableName` variable, and the value of the `series` dropdown.

23. Assign the `spDateRange_par` passthrough query to both `RowSource` date dropdowns. The query name must be enclosed in quotes.

    Let's work through the VBA code required to generate the passthrough query and assign it.

24. Create the SQL statement to pass to the server:

    ```
    SQL = "Call spDateRange_par('" & TableName & "','" &
    Me.cmbSeries & "');"
    ```

25. Call the `CreatePassThrough` function, passing in the SQL statement and the name of the resulting passthrough query, and indicate that it will return records:

    ```
    Call CreatePassThrough(SQL, "spDateRange_par", True,
    False)
    ```

26. Add a comment to indicate what the next few lines of code do:

    ```
    'Fill the Year dropdowns
    ```

27. Assign the `spDateRange_par` passthrough query directly to the `RowSource` `StartYear` combo boxes:

    ```
    Me.StartYear.RowSource = "spDateRange_par"
    ```

28. Assign the first element of `StartYear` as the value to the `StartYear` combo box. This will set the combo to display the earliest year:

    ```
    Me.StartYear = Me.StartYear.ItemData(0)
    ```

29. Assign the same `spDateRange_par` passthrough query directly to the `RowSource` `EndYear` combo boxes:

    ```
    Me.EndYear.RowSource = "spDateRange_par"
    ```

30. Assign the last element of `EndYear` as the value for the `EndYear` combo box. This will set the last year for the combo to display:

    ```
    Me.EndYear = Me.EndYear.ItemData(Me.EndYear.ListCount -
    1)
    ```

To test it, use the selections shown in the screenshot; your results should match:

Figure 16.66 – Both date combo boxes will change based on the series selected

Solution to Activity 11.1

One possible solution to this activity is the following:

1. The **Worksheet_Change** event subroutine should have a new `Case` statement and
    code, like this:

```
Private Sub Worksheet_Change(ByVal Target As Range)

'Test the active cell (the one that changed)
    Select Case Target.Address

        Case "$B$5"
            'The change was in the dropdown, target has the value
            Call GenreSales(Target)

            'Set the chart details Population
            Worksheets("Dashboard").ChartObjects("chrtPopulation").Activate
            With ActiveChart
                .SetSourceData Source:=Sheets("Data Sheet").Range("GenreSales"), PlotBy:=xlColumns
                .HasTitle = True
                .ChartTitle.Text = "Genre Sales - " & Target
                .SeriesCollection(1).Name = "Sales"
            End With

        Case "$P$5"

            Call ArtistTrackSales(Target)

            'Set the chart details Population
            Worksheets("Dashboard").ChartObjects("chrtArtistTrackSales").Activate
            With ActiveChart
                .SetSourceData Source:=Sheets("Data Sheet").Range("ArtistTrackSales"), PlotBy:=xlColumns
                .HasTitle = True
                .ChartTitle.Text = "Artist Track Sales - " & Target
                .SeriesCollection(1).Name = "Sales"
            End With
        Case Else
            'Nothing to work with so leave
            GoTo Leavesub
    End Select

Leavesub:
        Exit Sub


End Sub
```

Figure 16.67 – A new Case test with code

2. You will have a new function. This function is almost identical to the `GenreSales` function we created in *Exercise – load genre sales chart data* with the differences highlighted.

3. Declare the subroutine and its parameters:

```
Private Sub ArtistTrackSales(ByVal pArtist As String)
```

4. Declare the variables:

```
Dim RS As Recordset
    Dim SQL As String
    Dim MyNamedRng As Range
    Dim RS As Recordset
    Dim SQL As String
    Dim MyNamedRng As Range
```

5. Set error checking to ignore any errors. An error will occur if the range is not defined when we clear it:

```
On Error Resume Next
```

6. Clear the target area to remove the old data:

```
    Worksheets("Data Sheet").Range("ArtistTrackSales").
ClearContents
```

7. Resume normal error handling:

```
    On Error GoTo HandleError
```

8. Make the connection to MySQL and test whether it was successful:

```
    If ConnectDB_DSNless(g_Conn_DSNless) = True Then
```

9. The connection was made, so prepare the SQL statement:

```
    SQL = ""
    SQL = SQL & "SELECT  TrackName, 'Units Sold' "
    SQL = SQL & "FROM vw_Artist_Track_Sales "
    SQL = SQL & "WHERE Name = '" & pArtist & "' "
    SQL = SQL & "ORDER BY 'Units Sold' DESC"
```

10. Set the recordset variable to a new recordset:

```
    Set RS = New ADODB.Recordset
```

11. Load the recordset and pass in the SQL statement and the connection to use:

```
RS.Open SQL, g_Conn_DSNless
```

12. Test that there are records:

```
If RS.EOF And RS.BOF Then
```

13. If there is no data, exit:

```
        GoTo Leavesub
    Else
```

14. We have data, so load it in row 2 and column 12:

```
                Worksheets("Data Sheet").Cells(2,
12).CopyFromRecordset RS
```

15. Set and create a named range, covering the column with the genre name (data only). The final row for the range is the number of records + 1:

```
    Set MyNamedRng = Worksheets("Data Sheet").
Range("L2:M" & RS.RecordCount + 1)
    ActiveWorkbook.Names.Add Name:="ArtistTrackSales",
RefersTo:=MyNamedRng
```

16. Close off the If/Else blocks:

```
            End If
    Else
            'This line will be reached if there is no
data, we do nothing and drop through
        End If
```

17. Set an exit point. We close the recordset here:

```
Leavesub:
```

18. Close the recordset and exit:

```
    RS.Close
    Set RS = Nothing
    Exit Sub
```

19. Add an error-handling routine:

```
HandleError:
```

20. Display an error message and then exit:

```
MsgBox Err & " " & Error(Err)
Resume Leavesub
```

21. Close off the `Sub` block:

```
End Sub
```

22. The data in **Data Sheet** in the **L** and **M** columns is as follows:

L	M
Save The Children	2
Abraham, Martin And John	2
Seek And You Shall Find	1
Heavy Love Affair	1
You Sure Love To Ball	1
Praise	1
You've Been A Long Time Coming	1
When I Had Your Love	1

Figure 16.68 – The ArtistTrackSales output in the Data Sheet tab

23. You should have a new named range:

ArtistTrackSales

	A
1	**Name**
2	Alternative
3	Alternative & Punk

Figure 16.69 – The ArtistTrackSales named range in the range list

When the range is selected, we can see the expanded list, as follows:

L	M
Save The Children	2
Abraham, Martin And John	2
Seek And You Shall Find	1
Heavy Love Affair	1
You Sure Love To Ball	1
Praise	1
You've Been A Long Time Coming	1
When I Had Your Love	1

Figure 16.70 – ArtistTrackSales highlighted when the named range is selected

24. And finally, we have a new chart and formatting on the dashboard:

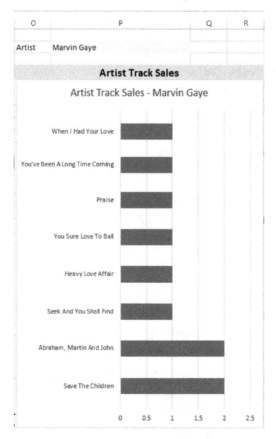

Figure 16.71 – A new chart displaying ArtistTrackSales

In this section, we learned how to create a function to read data from a MySQL database, create a new chart to consume the data, duplicate existing VBA functions to create new but similar functions, and change specific details to suit the new functions' purposes.

Solution to Activity 12.1

In this activity, we will start by creating a new database in MySQL to store our data. To achieve this, do the following:

1. Launch MySQL Workbench and connect to your local database instance:

Figure 16.72 – The connection for the local database instance

2. In your connection, create a new query defined as shown here:

```
CREATE DATABASE coffee_data
```

3. With our database created, we will move to the `CoffeeProducts.xlsx` file. Inside this file, go to the **Data** tab and click **MySQL for Excel**:

Figure 16.73 – The MySQL For Excel option in the Data tab

4. Once you click on **MySQL for Excel**, you can select the local instance connection that displays on the sidebar:

Figure 16.74 – The local instance that shows in the MySQL connection list

5. Once you have selected the local instance connection, select the coffee_data database to connect to the database we just created:

Figure 16.75 – The coffee_data database in the connection list

Now that we are connected to the database, we can move on to migrating the existing data in the **Products** sheet.

6. Select the data in the **Products** sheet:

	A	B	C	D
	ProductName	ProductPrice	ProductSize	
	Roast Coffee	1	Medium	
	Lattee	3.5	Large	
	Cappuccino	4.15	Large	
	Espresso	1.25	Small	

Figure 16.76 – The selected data in the Products sheet

7. On the **MySQL for Excel** sidebar, select the **Export Excel Data to New Table** option:

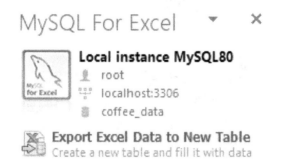

Figure 16.77 – The Export Excel Data to New Table option in MySQL for Excel

8. Set the table name as products and use the **ProductName** column as the primary key:

Figure 16.78 – The table name and primary key

9. Change the **ProductName** column to be the VarChar(50) data type to leave enough room for future product names:

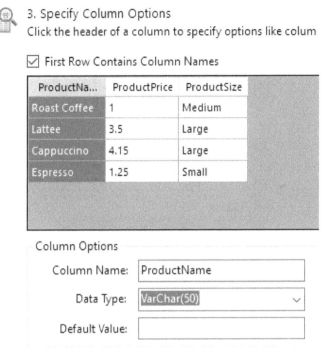

Figure 16.79 – Setting the field size to 50

10. Once this is done, click **Export data**. If everything works correctly, you should see that the operation completed successfully:

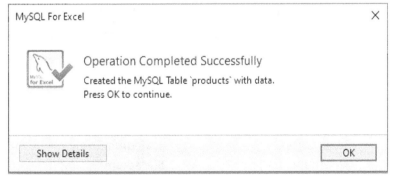

Figure 16.80 – Confirmation that the table was completed successfully

Now that the table is added, we can work on adding and updating our product data as required.

11. Go to a new sheet, select **products**, and then the **Import MySQL Data** option:

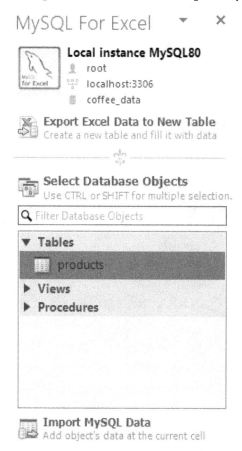

Figure 16.81 – Importing the data from MySQL Connector

12. Once the data is imported, add the Americano record to the end of the document:

ProductName ▼	ProductPrice ▼	ProductSize ▼
Cappuccino	4.15	Large
Espresso	1.25	Small
Lattee	3.5	Large
Roast Coffee	1	Medium
Americano	3.5	Medium

Figure 16.82 – The new record added to the end of the table

13. Once the data is added, click on the **Append Excel Data to Table** option in **MySQL for Excel**:

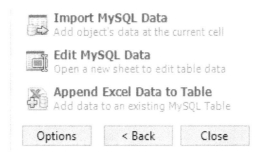

Figure 16.83 – The Append Excel Data to Table option

14. In the dialog that appears, click **Append** to add the record to the table:

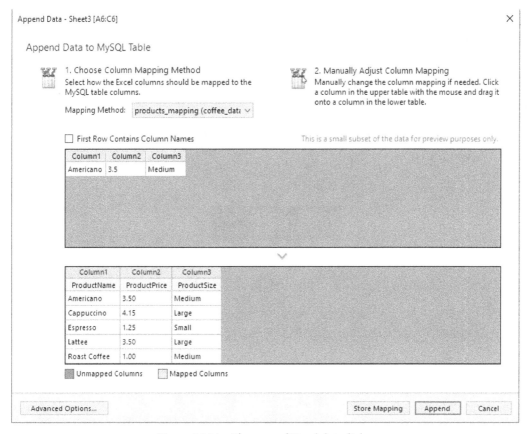

Figure 16.84 – The append Excel data dialog

15. If the operation completes successfully, you will see the dialog shown here:

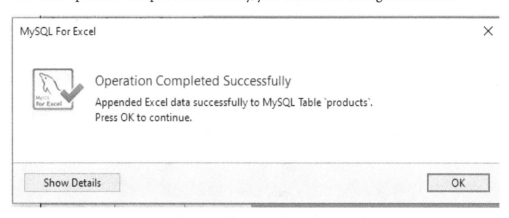

Figure 16.85 – The success notice

Solution to Activity 13.1

In this activity, you will find the list of the heads of state of all monarchies. The data will be fetched in the CSV format so that you can load it in Excel and later incorporate it into an article. Follow these steps to complete this activity:

1. Open the MySQL client and connect to the `world` database:

```
USE world
```

This query produces the following output:

```
mysql> USE world
Database changed
```

Figure 16.86 – The USE output

2. Select the columns we need and filter out the monarchies:

```
SELECT Name, HeadOfState FROM country
WHERE GovernmentForm LIKE '%Monarchy%';
```

3. Send the result to a file named `monarchy.csv`:

```
SELECT Name, HeadOfState FROM country
WHERE GovernmentForm LIKE '%Monarchy%'
INTO OUTFILE '/var/lib/mysql-files/monarchy.csv'
FIELDS TERMINATED BY ',' OPTIONALLY ENCLOSED BY '"';
```

This query produces the following output:

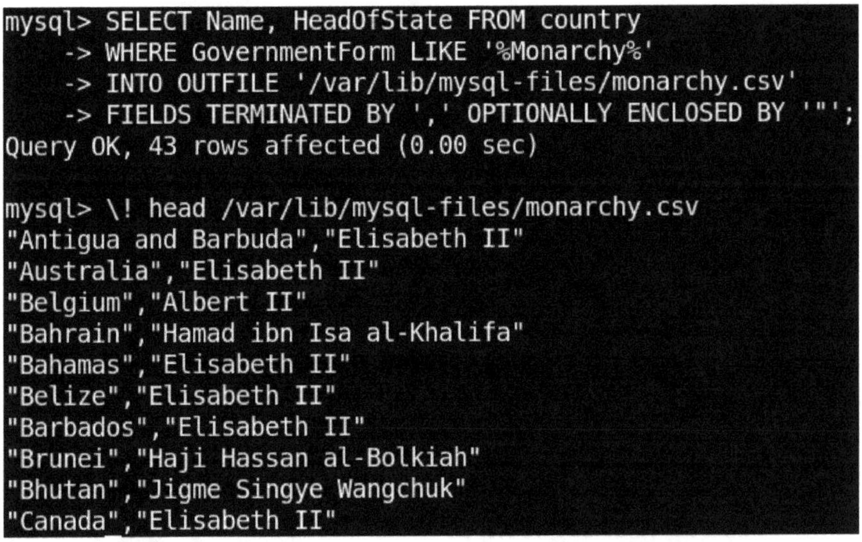

```
mysql> SELECT Name, HeadOfState FROM country
    -> WHERE GovernmentForm LIKE '%Monarchy%'
    -> INTO OUTFILE '/var/lib/mysql-files/monarchy.csv'
    -> FIELDS TERMINATED BY ',' OPTIONALLY ENCLOSED BY '"';
Query OK, 43 rows affected (0.00 sec)

mysql> \! head /var/lib/mysql-files/monarchy.csv
"Antigua and Barbuda","Elisabeth II"
"Australia","Elisabeth II"
"Belgium","Albert II"
"Bahrain","Hamad ibn Isa al-Khalifa"
"Bahamas","Elisabeth II"
"Belize","Elisabeth II"
"Barbados","Elisabeth II"
"Brunei","Haji Hassan al-Bolkiah"
"Bhutan","Jigme Singye Wangchuk"
"Canada","Elisabeth II"
```

Figure 16.87 – The SELECT...INTO OUTFILE output for the head of state query

The resulting file has the following contents:

```
"Antigua and Barbuda","Elisabeth II"
"Australia","Elisabeth II"
"Belgium","Albert II"
"Bahrain","Hamad ibn Isa al-Khalifa"
"Bahamas","Elisabeth II"
"Belize","Elisabeth II"
"Barbados","Elisabeth II"
"Brunei","Haji Hassan al-Bolkiah"
"Bhutan","Jigme Singye Wangchuk"
"Canada","Elisabeth II"
```

"Denmark","Margrethe II"

"Spain","Juan Carlos I"

"United Kingdom","Elisabeth II"

"Grenada","Elisabeth II"

"Jamaica","Elisabeth II"

"Jordan","Abdullah II"

"Japan","Akihito"

"Cambodia","Norodom Sihanouk"

"Saint Kitts and Nevis","Elisabeth II"

"Kuwait","Jabir al-Ahmad al-Jabir al-Sabah"

"Saint Lucia","Elisabeth II"

"Liechtenstein","Hans-Adam II"

"Lesotho","Letsie III"

"Luxembourg","Henri"

"Morocco","Mohammed VI"

"Monaco","Rainier III"

"Malaysia","Salahuddin Abdul Aziz Shah Alhaj"

"Netherlands","Beatrix"

"Norway","Harald V"

"Nepal","Gyanendra Bir Bikram"

"New Zealand","Elisabeth II"

"Oman","Qabus ibn Sa´id"

"Papua New Guinea","Elisabeth II"

"Qatar","Hamad ibn Khalifa al-Thani"

"Saudi Arabia","Fahd ibn Abdul-Aziz al-Sa´ud"

"Solomon Islands","Elisabeth II"

"Sweden","Carl XVI Gustaf"

"Swaziland","Mswati III"

"Thailand","Bhumibol Adulyadej"

"Tonga","Taufa'ahau Tupou IV"

"Tuvalu","Elisabeth II"

"Saint Vincent and the Grenadines","Elisabeth II"

"Samoa","Malietoa Tanumafili II"

And this is what it looks like after opening the file in a spreadsheet application:

	A	B
1	Antigua and Barbuda	Elisabeth II
2	Australia	Elisabeth II
3	Belgium	Albert II
4	Bahrain	Hamad ibn Isa al-Khalifa
5	Bahamas	Elisabeth II
6	Belize	Elisabeth II
7	Barbados	Elisabeth II
8	Brunei	Haji Hassan al-Bolkiah
9	Bhutan	Jigme Singye Wangchuk
10	Canada	Elisabeth II
11	Denmark	Margrethe II
12	Spain	Juan Carlos I
13	United Kingdom	Elisabeth II
14	Grenada	Elisabeth II
15	Jamaica	Elisabeth II
16	Jordan	Abdullah II
17	Japan	Akihito
18	Cambodia	Norodom Sihanouk
19	Saint Kitts and Nevis	Elisabeth II
20	Kuwait	Jabir al-Ahmad al-Jabir al-Sabah
21	Saint Lucia	Elisabeth II
22	Liechtenstein	Hans-Adam II
23	Lesotho	Letsie III
24	Luxembourg	Henri
25	Morocco	Mohammed VI
26	Monaco	Rainier III
27	Malaysia	Salahuddin Abdul Aziz Shah Alhaj
28	Netherlands	Beatrix
29	Norway	Harald V
30	Nepal	Gyanendra Bir Bikram
31	New Zealand	Elisabeth II
32	Oman	Qabus ibn Sa´id
33	Papua New Guinea	Elisabeth II
34	Qatar	Hamad ibn Khalifa al-Thani
35	Saudi Arabia	Fahd ibn Abdul-Aziz al-Sa´ud
36	Solomon Islands	Elisabeth II
37	Sweden	Carl XVI Gustaf
38	Swaziland	Mswati III
39	Thailand	Bhumibol Adulyadej
40	Tonga	Taufa'ahau Tupou IV
41	Tuvalu	Elisabeth II
42	Saint Vincent and the Grenadines	Elisabeth II
43	Samoa	Malietoa Tanumafili II

Figure 16.88 – monarchy.csv loaded into a spreadsheet application

While this activity seems easy, it is easy to make a mistake here. This is because an exact match on Monarchy would not match all monarchies, as there are seven government forms used in the database that are monarchies.

4. Show the different government forms with the following query:

```
SELECT GovernmentForm FROM country
WHERE GovernmentForm LIKE '%Monarchy%'
GROUP BY GovernmentForm;
```

This query produces the following output:

```
mysql> SELECT GovernmentForm FROM country
    -> WHERE GovernmentForm LIKE '%Monarchy%'
    -> GROUP BY GovernmentForm;
+------------------------------------+
| GovernmentForm                     |
+------------------------------------+
| Constitutional Monarchy            |
| Constitutional Monarchy, Federation |
| Monarchy (Emirate)                 |
| Monarchy (Sultanate)               |
| Monarchy                           |
| Constitutional Monarchy (Emirate)  |
| Parlementary Monarchy              |
+------------------------------------+
7 rows in set (0.00 sec)
```

Figure 16.89 – The SELECT output to show government forms that are monarchies

By using the INTO OUTFILE clause of the SELECT statement, you were able to export data to a CSV file that can be used in a spreadsheet application or loaded into a different application or database. You were, thereby, able to limit your export to only the columns you needed.

Solution to Activity 14.1

Perform the following steps to achieve the goal of this activity:

1. Connect to the database server. As you will be working with the `world` database, write the following:

```
USE world
```

Again, there is no need to connect to a specific schema; just connect with an account that has enough permissions to give out grants.

2. Create roles for `manager` and `language_expert` and grant permissions to `language_expert`:

```
CREATE ROLE 'manager';
GRANT ALL ON world.* TO 'manager';
CREATE ROLE 'language_expert';
GRANT ALL ON world.countrylanguage TO 'language_expert';
```

3. Create an account for `webserver` and grant permissions to the user created for `webserver`:

```
CREATE USER 'webserver'@'%' IDENTIFIED BY
'1twedByutGiawWy';
GRANT SELECT ON world.* TO 'webserver'@'%';
```

4. Create an account for `intranet` and grant permissions to the user created for `intranet`:

```
CREATE USER 'intranet'@'%' IDENTIFIED BY 'JiarjOodVavit';
GRANT INSERT, UPDATE, SELECT ON world.* TO
'intranet'@'%';
```

5. Create an account for `stewart`:

```
CREATE USER 'stewart'@'%'
IDENTIFIED BY 'UkfejmuniadBekMow4'
DEFAULT ROLE manager;
```

6. Create an account for `sue`:

```
CREATE USER 'sue'@'%'
IDENTIFIED BY 'WrawdOpAncy'
DEFAULT ROLE language_expert;
```

Solution to Activity 15.1

Perform the following steps to achieve the goal of this activity:

1. Open Command Prompt.
2. Locate and execute the `mysqldump.exe` file.

3. Create a backup of the `world` schema by writing the following code in Command Prompt:

```
mysqldump -u root -p world > "C:\Users\bhaveshb\Desktop\
world_backup.sql"
```

In the preceding code, you invoked `mysqldump` and specified the `world` schema as the only schema you want to back up. Depending on your configuration, you may have to use `-u`, `-p`, and other options to specify the credentials to connect to the database.

4. Simulate a disaster. Open the MySQL client and use the `world` database by writing the following code:

```
USE world
```

5. Now, delete all the rows from the `city` table present in the `world` database. This can be done using the following query:

```
DELETE FROM city;
```

6. To check that all the rows have been deleted from the `city` table, use the `SELECT` command:

```
SELECT * FROM city;
```

The preceding code returns the following output:

```
Empty set (0.00 sec)
```

Figure 16.90 – The empty values in the city table

As you can see from the figure here, the data inside the table has been wiped out and shows zero results.

7. Now, get back to Command Prompt and restore the `world` schema by writing the following code:

```
mysql -u root -p world < "C:\Users\bhaveshb\Desktop\
world_backup.sql"
```

This allows us to restore our data.

8. To verify whether the data has been restored correctly, switch back to the MySQL client and type the following command:

```
SELECT COUNT(*) FROM city;
```

The preceding code produces the following output:

```
+----------+
| COUNT(*) |
+----------+
|     4079 |
+----------+
1 row in set (0.20 sec)
```

Figure 16.91 – The total rows in the city table after restoring

Solution to Activity 15.2

Perform the following steps to achieve the goal of this activity:

1. Open the MySQL client and write the following command to ensure that you have a clean start position. This is not a required step, but it does make it easier to follow along:

   ```
   RESET MASTER;
   ```

   This removes all the existing `binlog` files and starts from a freshly created `binlog`.

2. Open Command Prompt.

3. Locate and execute the `mysqldump.exe` file.

4. Create a backup of the `world` schema by writing the following code in Command Prompt:

   ```
   mysqldump --master-data=2 -u root -p world > "C:\Users\
   bhaveshb\Desktop\backup_world_pitr.sql"
   ```

   Here, you specified `--master-data=2` to record the `binlog` position of the backup.

5. Switch back to the MySQL client and make some changes in the `city` table of the `world` database:

   ```
   USE world
   UPDATE city SET Population=123456789 WHERE name =
   'Toulouse';
   ```

The value here is something you can easily recognize after restoring. If you were to restore the backup without doing a point-in-time restore, then this change would be gone, as it was made after the backup.

6. Now, simulate a disaster by deleting all the records from the `city` table:

```
DELETE FROM city;
```

7. View the `MASTER LOGS` by writing the following query:

```
SHOW MASTER LOGS;
```

The preceding code produces the following output:

```
+--------------------------+-----------+-----------+
| Log_name                 | File_size | Encrypted |
+--------------------------+-----------+-----------+
| PPMUMCPU0032-bin.000001  |    134135 | No        |
+--------------------------+-----------+-----------+
1 row in set (0.00 sec)
```

Figure 16.92 – The results of the master logs

8. Find the `binlog` data to restore:

```
SHOW BINLOG EVENTS IN 'binlog.000001';
```

The preceding code produces the following output:

```
+-------------------------+-----+----------------+-----------+-------------+--------------------------------------+
| Log_name                | Pos | Event_type     | Server_id | End_log_pos | Info                                 |
+-------------------------+-----+----------------+-----------+-------------+--------------------------------------+
| PPMUMCPU0032-bin.000001 |   4 | Format_desc    |         1 |         124 | Server ver: 8.0.19, Binlog ver: 4    |
| PPMUMCPU0032-bin.000001 | 124 | Previous_gtids |         1 |         155 |                                      |
| PPMUMCPU0032-bin.000001 | 155 | Anonymous_Gtid |         1 |         234 | SET @@SESSION.GTID_NEXT= 'ANONYMOUS' |
| PPMUMCPU0032-bin.000001 | 234 | Query          |         1 |         319 | BEGIN                                |
| PPMUMCPU0032-bin.000001 | 319 | Table_map      |         1 |         383 | table_id: 331 (world.city)           |
| PPMUMCPU0032-bin.000001 | 383 | Update_rows    |         1 |         491 | table_id: 331 flags: STMT_END_F      |
| PPMUMCPU0032-bin.000001 | 491 | Xid            |         1 |         522 | COMMIT /* xid=20421 */               |
| PPMUMCPU0032-bin.000001 | 522 | Anonymous_Gtid |         1 |         602 | SET @@SESSION.GTID_NEXT= 'ANONYMOUS' |
```

Figure 16.93 – Inspecting the binlog contents to find the position to restore to

In this output, you can see that the last event before `DELETE` ended at a position of `522`.

9. Take all the changes between the start of the file and the position you found. You can do this because you initially used RESET MASTER. This is very similar to what you would have done if you had used the --flush-logs option for mysqldump. In that case, you could have started from the beginning of the file but would have had to find the right file by looking at the CHANGE MASTER TO line at the beginning of the backup file. Open Command Prompt and inspect the binlog file we created. By default, the binlog file will be stored in the C:\ProgramData\MySQL\ MySQL Server 8.0\Data path:

```
mysqlbinlog -u root -p --skip-gtids --stop-position=522
"C:\ProgramData\MySQL\MySQL Server 8.0\Data\PPMUMCPU0032-
bin.000001" > "C:\Users\bhaveshb\Desktop\restore_world_
pitr.sql"
```

10. Restore the backup_world_pitr.sql backup file by writing the following command in Command Prompt:

```
mysql -u root -p world <
"C:\Users\bhaveshb\Desktop\backup_world_pitr.sql"
```

11. Reapply the changes that occurred between backup creation and the time of the disaster by writing the following command:

```
mysql -u root -p <
"C:\Users\bhaveshb\Desktop\restore_world_pitr.sql"
```

12. Switch back to the MySQL client and verify that the data has been restored by writing the following query:

```
SELECT * FROM city WHERE name = 'Toulouse';
```

The preceding code produces the following output:

```
+------+----------+-------------+---------------+------------+
| ID   | Name     | CountryCode | District      | Population |
+------+----------+-------------+---------------+------------+
| 2977 | Toulouse | FRA         | Midi-Pyrénées |  123456789 |
+------+----------+-------------+---------------+------------+
1 row in set (0.03 sec)
```

Figure 16.94 – Inspecting the change that we made before

Index

S

X

Y

`Packt.com`

Subscribe to our online digital library for full access to over 7,000 books and videos, as well as industry leading tools to help you plan your personal development and advance your career. For more information, please visit our website.

Why subscribe?

- Spend less time learning and more time coding with practical eBooks and Videos from over 4,000 industry professionals

- Improve your learning with Skill Plans built especially for you

- Get a free eBook or video every month

- Fully searchable for easy access to vital information

- Copy and paste, print, and bookmark content

Did you know that Packt offers eBook versions of every book published, with PDF and ePub files available? You can upgrade to the eBook version at `packt.com` and as a print book customer, you are entitled to a discount on the eBook copy. Get in touch with us at `customercare@packtpub.com` for more details.

At `www.packt.com`, you can also read a collection of free technical articles, sign up for a range of free newsletters, and receive exclusive discounts and offers on Packt books and eBooks.

Other Books You May Enjoy

If you enjoyed this book, you may be interested in these other books by Packt:

Advanced MySQL 8

Eric Vanier, Birju Shah, Tejaswi Malepati

ISBN: 978-1-78883-444-5

- Explore new and exciting features of MySQL 8.0
- Analyze and optimize large MySQL queries
- Understand MySQL Server 8.0 settings
- Master the deployment of Group Replication and use it in an InnoDB cluster
- Monitor large distributed databases
- Discover different types of backups and recovery methods for your databases
- Explore tips to help your critical data reach its full potential

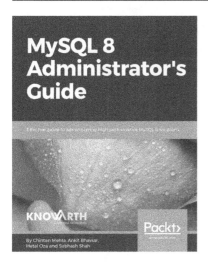

MySQL 8 Administrator's Guide

Chintan Mehta, Subhash Shah, Ankit Bhavsar, Hetal Oza

ISBN: 978-1-78839-519-9

- Understanding different MySQL 8 data types based on type of contents and storage requirements

- Best practices for optimal use of features in MySQL 8

- Explore globalization configuration and caching techniques to improve performance

- Create custom storage engine as per system requirements

- Learn various ways of index implementation for flash memory storages

- Configure and implement replication along with approaches to use replication as solution

- Understand how to make your MySQL 8 solution highly available

- Troubleshoot common issues and identify error codes while using MySQL 8

Packt is searching for authors like you

If you're interested in becoming an author for Packt, please visit `authors.packtpub.com` and apply today. We have worked with thousands of developers and tech professionals, just like you, to help them share their insight with the global tech community. You can make a general application, apply for a specific hot topic that we are recruiting an author for, or submit your own idea.

Share Your Thoughts

Now you've finished *The MySQL Workshop*, we'd love to hear your thoughts! Scan the QR code below to go straight to the Amazon review page for this book and share your feedback or leave a review on the site that you purchased it from.

https://packt.link/r/1-839-21490-2

Your review is important to us and the tech community and will help us make sure we're delivering excellent quality content.

www.ingramcontent.com/pod-product-compliance
Lightning Source LLC
Chambersburg PA
CBHW060919060326
40690CB00041B/2718